recent advances in phytochemistry

volume 38

Secondary Metabolism in Model Systems

RECENT ADVANCES IN PHYTOCHEMISTRY

Proceedings of the Phytochemical Society of North America
General Editor: John T. Romeo, *University of South Florida, Tampa, Florida*

Recent Volumes in the Series:

Volume 30 **Phytochemical Diversity and Redundancy in Ecological Interactions**
Proceedings of the Thirty-fifth Annual Meeting of the Phytochemical Society of North America, Sault Ste. Marie, Ontario, Canada, August, 1995

Volume 31 **Functionality of Food Phytochemicals**
Proceedings of the Thirty-sixth Annual Meeting of the Phytochemical Society of North America, New Orleans, Louisiana, August, 1996

Volume 32 **Phytochemical Signals and Plant-Microbe Interactions**
Proceedings of the Thirty-seventh Annual Meeting of the Phytochemical Society of North America, Noordwijkerhout, The Netherlands, April, 1997

Volume 33 **Phytochemicals in Human Health Protection, Nutrition, and Plant Defense**
Proceedings of the Thirty-eighth Annual Meeting of the Phytochemical Society of North America, Pullman, Washington, July, 1998

Volume 34 **Evolution of Metabolic Pathways**
Proceedings of the Thirty-ninth Annual Meeting of the Phytochemical Society of North America, Montreal, Quebec, Canada, July, 1999

Volume 35 **Regulation of Phytochemicals by Molecular Techniques**
Proceedings of the Fortieth Annual Meeting of the Phytochemical Society of North America, Beltsville, Maryland, June, 2000

Volume 36 **Phytochemistry in the Genomics and Post-Genomics Eras**
Proceedings of the Forty-first Annual Meeting of the Phytochemical Society of North America, Olkalohom City, Oklahoma, August, 2001

Volume 37 **Integrative Phytochemistry: From Ethnobotany to Molecular Ecology**
Proceedings of the Forty-second Annual Meeting of the Phytochemical Society of North America, Mérida, Yucatán, Mexico, July, 2002

Volume 38 **Secondary Metabolism in Model Systems**
Proceedings of the Forty-third Annual Meeting of the Phytochemical Society of North America, Peoria, Illinois, August, 2003

Cover design: "Contigs from clustering of soybean ESTs" (Chapter 9)

recent advances in phytochemistry

volume 38

Secondary Metabolism in Model Systems

Edited by

John T. Romeo
University of South Florida
Tampa, Florida, USA

2004

ELSEVIER

Amsterdam – Boston – Heidelberg – London – New York – Oxford
Paris – San Diego – San Francisco – Singapore – Sydney – Tokyo

ELSEVIER B.V.
Sara Burgerhartstraat 25
P.O. Box 211, 1000 AE Amsterdam
The Netherlands

ELSEVIER Inc.
525 B Street, Suite 1900
San Diego, CA 92101-4495
USA

ELSEVIER Ltd
The Boulevard, Langford Lane
Kidlington, Oxford OX5 1GB
UK

ELSEVIER Ltd
84 Theobalds Road
London WC1X 8RR
UK

First edition 2004

Library of Congress Cataloging in Publication Data
A catalog record is available from the Library of Congress.

British Library Cataloguing in Publication Data
A catalogue record is available from the British Library.

ISBN: 0 08 044501 2

Transferred to digital printing 2007

PREFACE

The Phytochemical Society of North America held its forty-third annual meeting in Peoria, Illinois from August 9-13, 2003. The chapters in this volume are based on the papers presented in the symposium entitled "Secondary Metabolism in Model Systems". Five mini-symposia were organized that dealt with five different model organisms: *Arabidopsis*, Maize, Legumes, Rice, and the fungus *Aspergillus*. The organizers for these mini-symposia, respectively, were: Clint Chapple, Purdue University; Erich Grotewold, Ohio State University; Mark Gijzen, Agriculture and Agri-Food, Canada; Tom Okita, Washington State University; and Susan McCormick, USDA, Peoria. They assembled an international group of speakers that concentrated their talks largely on the rapid advances in understanding of gene functions that have been catapulted onto scientific front burners as a result of the completion of recent genome projects.

The chapters on *Arabidopsis* range from using this model system for understanding volatile terpene biosynthesis, regulation, and function, to evolutionary origins of aliphatic glucosinolates, and finally to accumulation of phenylpropanoid sinapate esters. The opening chapter by Tholl *et al.* focuses on the *TPS* (terpene synthase) gene family. Although terpene biosynthesis and the sequences of the basic pathways are well-known from a number of plants, understanding regulation of biosynthesis and biological roles are likely to come from studying this model. These workers are correlating the emission of specific terpenes with the expression of *AtTPS* genes. The work is leading towards elucidating the mechanisms that regulate the process of plant-insect interactions via volatiles that operate in both the vegetative and reproductive parts of the plants. Tokuhisa *et al.*, working with glucosinolates, the largest naturally-occurring group of secondary metabolites in *Arabidopsis,* are investigating the biochemical diversity in the group, with emphasis on the aliphatic compounds derived from methionine. Glucosinolates, found in the agriculturally important Brassicaceae, have organoleptic characteristics contributing to flavor and associated health benefits. Manipulation of their levels in plants of the future is anticipated. Furthermore, links between glucosinolate biosynthesis and other plant functions, such as reduced fertility and apical dominance, are blurring the boundaries between primary and secondary metabolism as presently understood. In the chapter by Stout and Chapple, we see how the analysis of mutants of the phenylpropanoid pathway have led to numerous revisions of the pathway over the past decade. The pathway has been particularly amenable for study in *Arabidopsis* due to the accumulation of readily observable end-products coming from different branches. The new understanding, while clarifying some contradictory data of the past, has posed new questions, such as how ferulic and sinapic acid esters, components of leaves, seeds, and cell walls, are synthesized. The mutant studies have also demonstrated interactions between pathways of secondary metabolism and given insight into their evolution.

The chapters on maize address biosynthesis and evolution of two major classes of compounds – benzoxazinoids and carotenoids. Gierl *et al.* have demonstrated that gene duplications seem to be important in the evolution of secondary metabolic pathways. *TSA* (tryptophan synthase) genes from primary metabolism have been recruited for secondary pathways. Production of free indole can be used directly for signaling in tritrophic interactions with insects, or converted to a defense compound in grasses by duplicated and recruited genes for benzoxazinoids biosynthesis. The genes have been identified (*Bx*), and are expressed in a tissue-specific manner during maize development. Thus, the redundancy potential created by gene duplication does not necessarily result in functional or genetic redundancy. Benzoxazinoid biosynthesis can serve as a model for the evolution of the regulatory requirements of other secondary pathways. Wurtzel, working on biosynthesis of carotenoids (which have anti-oxidant health benefits and low levels of which in endosperms lead to vitamin A deficiency), discusses how many maize enzymes are encoded by small gene families. The pathway can be assembled on different plastid membranes. Structural and regulatory loci have been mapped by both mutant and QTL studies. Future metabolic engineering of carotenoid content and composition is dependent on our understanding of endogenous gene expression. The genetic, genomic, and germplasm resources available for maize are invaluable in this regard.

Lange and Presting have summarized the progress made to date on elucidation of specific metabolic pathways linked to key quality traits in rice. The rice genome ranks as the smallest of the major cereals and will be an important monocot model, as genes are highly conserved among cereal species. Only a few rice gene functions that are involved in metabolic pathways have been characterized in detail, which contrasts with the structurally diverse natural produces isolated form rice tissues. Like *Arabidopsis*, the capability of rice to produce secondary metabolites has either been vastly underestimated and/or gene families putatively related to secondary metabolism encode enzymes with novel functions in primary pathways. Efforts to improve aroma, texture, and starch content are discussed. The chapter by Wang *et al.* illustrates an integrative approach that uses systems biology to integrate individual components. Their work involves large-scale modeling of pathways based on genomic information and rice metabalome research. Atomic Reconstruction of Metabolism (ARM) and a Hybrid Static/Dynamic Simulation Algorithm are two of the techniques discussed. Their "e-rice" project is among the first attempts to simulate a whole plant organism.

Legume model systems have largely focused on *Medicago* (see volume 35 RAP, Dixon *et al.* and volume 36 RAP, Sumner *et al.*) and soybeans. In this volume, Maxwell *et al.* focus on engineering soybean for improved flavor and health benefits. Altering the phenylpropanoid pathway to suppress certain isoflavonoid products (those derived from liquiritigenin –glycitein and daidzein- , but not genistein) have been performed. Vector construction to suppress chalcone reductase has produced high genistein in soybean transformants. Saponin biosynthesis suppression has also been successful by suppressing β-amyrin synthase. The chapter by Strömvik *et al.* shows how mining the large soybean EST collection is enabling them to deduce knowledge about the expression of individual gene family members in regard to lectins. Additionally, by applying advanced statistical clustering analysis to global expression and microarray data, the timing of molecular events taking place during embryogenesis is becoming understood. cDNAs are differentially expressed in response to plants hormones, and such enzymes as glutathione-S-transferases, chalcone synthases and isomerases, and isoflavone synthases are affected.

The inclusion of two chapters on the economically important fungus *Aspergillus* is a natural extension of the symposium theme. The chapter from the laboratory of Keller *et a*l. reviews the contributions that *A. nidulans* has made to understanding fungal secondary metabolism. The organism produces sterigmatocystin, the precursor to aflatoxin, and penicillin. The biosynthesis has been extensively studied in this species and two gene clusters are known. A G-protein/cAMP/protein kinaseA growth pathway has been discovered that coordinates both secondary metabolism and asexual development. Lovastatin gene clusters have been moved into the species to study the regulation of its production. The contribution by Yu *et al.* discusses the aflatoxin gene cluster in *A. flavus*. This species is the most common cause of aflatoxin contamination in pre-harvest field crops and post-harvest grains. These workers are studying the molecular genetics of biosynthesis, regulation, and the factors affecting aflatoxin (derivatives of difuranocoumarins) formation. Attempts are being made to use genomics approaches to prevent contamination of grains and oil crops. Expressed Sequence Tag and microarray technologies may achieve the goal of turning aflatoxin production on and off in fungal systems as a control strategy.

Thus, the chapters presented here are a microcosm of what the recent completion, or near completion, of various genome projects are enabling biochemists to understand not only about control and regulation of secondary metabolism, and how various pathways relate to each other, but also about its relation to primary metabolism. A major paradigm shift is occurring in the way we need to view "secondary" metabolism in the future. It is also clear that model systems, such as the ones discussed in the symposium, are providing new information and insight almost faster than we can process it!

The setting of Peoria, in the heart of the grain belt, seemed indeed to be a fitting site for the chosen topic. The sunny days, the fields, lunches along the river, and a stately old hotel all made for a pleasant experience. We thank the local organizers, Mark Berhow and Susan McCormick, and the United States Department of Agriculture for making it possible. JTR, once again, thanks Darrin T. King, who because of his technical expertise makes putting this volume together a lot easier, and also the contributing authors for their cooperation and good will.

John T. Romeo
University of South Florida

CONTENTS

Chapter One

ARABIDOPSIS THALIANA, A MODEL SYSTEM FOR INVESTIGATING VOLATILE TERPENE BIOSYNTHESIS, REGULATION, AND FUNCTION

Dorothea Tholl,[1,2]* Feng Chen,[1] Jonathan Gershenzon,[2] Eran Pichersky[1]

[1]*Department of Molecular, Cellular, and Developmental Biology*
University of Michigan
Ann Arbor, MI 48109, USA

[2]*Max Planck Institute for Chemical Ecology*
Beutenberg Campus
Hans Knoell Strasse 8
D-07745, Jena, Germany

**Author for correspondence, e-mail: tholl@ice.mpg.de*

INTRODUCTION

Terpenes constitute a large and widely distributed class of natural compounds whose carbon skeleton is derived from C_5 isoprene units. (Fig. 1.1).[1-3] The biosynthesis of all terpenes follows the same general outline. First, the C_5 building blocks, isopentenyl diphosphate (IPP) and its allylic isomer dimethylallyl diphosphate (DMAPP) are each formed. In plants, this process involves two parallel

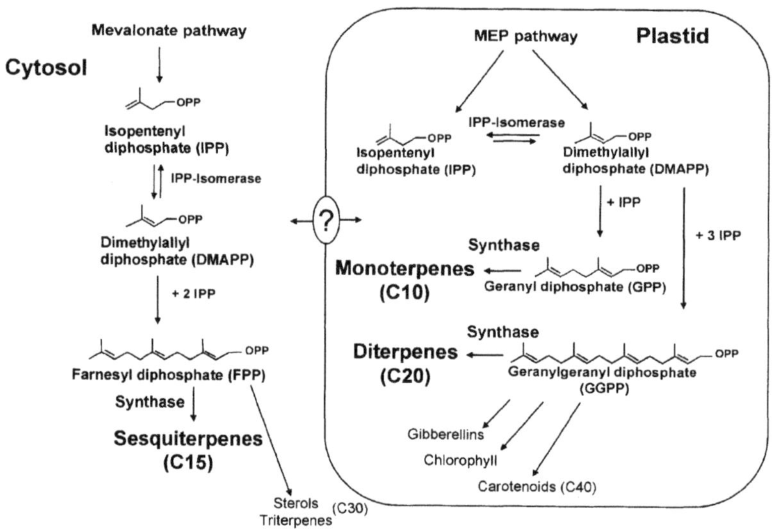

Figure 1.1: Biosynthetic pathways for the formation of terpenes in plants.

but distinct pathways, the mevalonate pathway operating in the cytosol and the methylerythritol phosphate (MEP) pathway in plastids.[3,4] Next, DMAPP is sequentially combined with varying numbers of IPP units by enzymes termed prenyltransferases to synthesize the acyclic prenyl diphosphates, geranyl diphosphate (C_{10}, GPP), farnesyl diphosphate (C_{15}, FPP), or geranylgeranyl diphosphate (C_{20}, GGPP).[5,6] These central intermediates are converted into monoterpenes (C_{10}), sesquiterpenes (C_{15}), and diterpenes (C_{20}) by a large group of enzymes called terpene synthases.[7-10] The primary products of terpene synthases may be further modified by secondary enzymatic transformations, including oxidation, reduction, and isomerization, thus producing a large number of terpene derivatives.[3] As a general rule, monoterpenes and diterpenes are synthesized in the plastids, while sesquiterpenes are synthesized in the cytosol.[11] Plant terpenes with a larger number of isoprene units, such as the C_{30}-derived triterpenes and sterols, including brassinosteroids, and C_{40} carotenoids are formed from precursors consisting of two condensed FPP (squalene) or GGPP units (phytoene) by enzymes rather unrelated to terpene synthases described above.[3,12-14] FPP and GGPP also serve as precursors of so called "meroterpenes" in which the terpene unit is attached to a non-terpene moiety such as the phytol chain in chlorophyll or the side chain of prenylated proteins.[3]

In primary metabolism, terpenes play essential roles in plant growth and development as hormones (*e.g.,* gibberellins and abscisic acid), photosynthetic pigments (phytol, carotenoids), or membrane components (sterols). However, the function of the majority of terpene secondary metabolites, which comprise mono-, sesqui-, di-, and triterpenoids, is still not well understood. Many monoterpenes, sesquiterpenes, and diterpenes are toxic to herbivores and microorganisms, and may function as direct defense compounds against such organisms.[15,16] They are often produced and stored by plants in specialized structures such as glands or resin ducts prior to any attack. Monoterpenes and sesquiterpenes, as well as a few diterpenes, volatilize readily at ambient temperature. When emitted from flowers, unmodified terpenes as well as those modified by hydroxylation, oxidation, reduction, and chain-shortening have been implicated in attracting pollinators to flowers.[16-18] Similar compounds have been found to be emitted from leaves of plants damaged by insect herbivores and are believed to serve as indirect defense compounds by attracting predators and parasitoids of such insects.[19-21] Finally, it is likely that terpenes may have additional physiological functions in plants. For example, many species emit isoprene, the smallest terpene molecule, or monoterpenes from their leaves under conditions of high light and temperature, and this emission has been proposed to mediate thermotolerance and protection against oxidative stress by quenching reactive oxygen species.[22-24]

Although terpene biosynthesis has been studied in numerous plant species and the sequences of the basic pathways are well-known, a comprehensive and detailed understanding of the regulation of biosynthesis and the biological roles of this large class of secondary metabolites will most likely come from investigating model plant species that provide extensive genetic and genomic resources. In this chapter, we describe the use of *Arabidopsis thaliana* as a model plant for terpene studies by focusing on investigations involving the large family of genes encoding terpene synthases. Employing genetic and genomic tools available for *Arabidopsis* and the latest technologies in expression and metabolite profiling has allowed us to explore the physiological and ecological significance of terpenes and basic principles of their regulation and evolution.

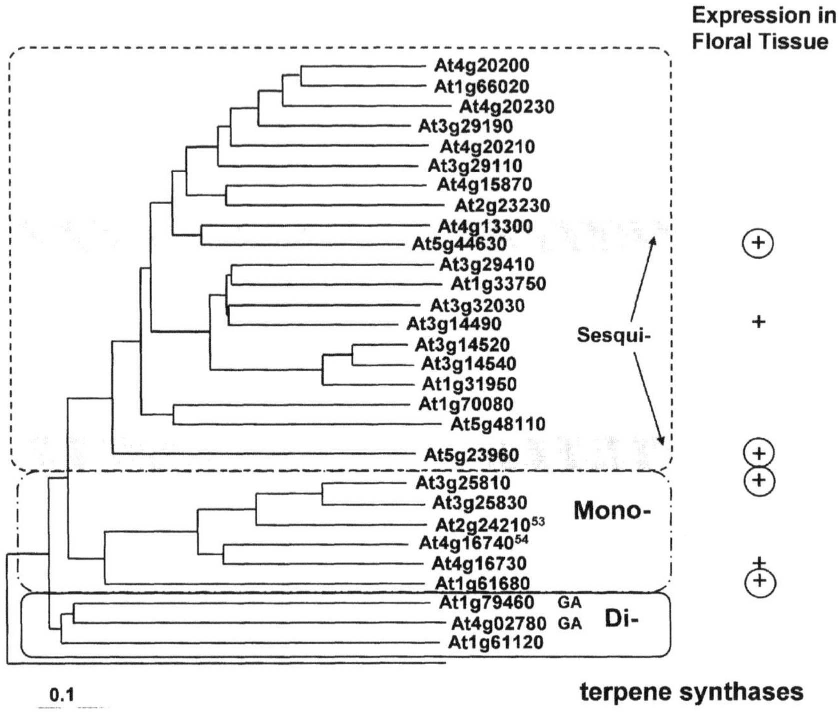

Figure 1.2: A neighbor-joining tree based on degree of sequence similarity between the members of the *Arabidopsis* terpene synthase (*TPS*) gene family. *AtTPS* genes form three major clades. Members of two clades encode proteins with high similarity to monoterpene or diterpene synthases of other angiosperms, respectively. Proteins encoded by genes of the third large clade are all likely to function as sesquiterpene synthases (highlighted in grey) or monoterpene/diterpene synthases dependent on the absence or presence of a plastidial transit peptide, respectively. Crosses indicate genes predominantly or exclusively expressed in flowers. Floral expressed genes encoding functional mono- or sesquiterpene synthases are marked with circles. GA: gibberellic acid indicating diterpene synthases involved in GA biosynthesis.

THE TERPENE SYNTHASE GENE FAMILY OF *ARABIDOPSIS THALIANA*

Previous research on certain terpene-accumulating species such as resin-producing gymnosperm trees or the herbs in the Lamiaceae family has resulted in the identification of a family of structurally related genes encoding mono-, sesqui-, and diterpene synthases.[10,25] With the completion of the sequencing of the *Arabidopsis thaliana* genome, it became possible to examine this species for the presence of terpene synthase (*TPS*) genes, even though the presence of mono-, sesqui-, or diterpenes (other than gibberellic acid (GA) derivatives) had not previously been reported. Using standard homology search methods, Aubourg *et al.*[26] showed that the *Arabidopsis* genome contains more than 30 *TPS* genes (*AtTPSs*), distributed over all five chromosomes. Our own detailed analysis (Fig. 1.2) as well as a similar analysis performed by Aubourg *et al.*[26] showed the presence of three classes. Six of the genes form one clade, and the proteins they encode are most similar to monoterpene synthases from other angiosperm species. These six genes also appear to encode proteins with a transit peptide for plastidial targeting. The genes previously determined to encode GA biosynthetic enzymes in the plastid[27,28] form a separate clade, together with a third *TPS* gene. Finally, a large clade contains all other *AtTPS* genes, some of which encode proteins with a plastid-targeting sequence (and, therefore, may be diterpene or perhaps monoterpene synthases) and some genes that encode proteins with no transit peptide (and, therefore, are probably all sesquiterpene synthases).

TERPENE BIOSYNTHESIS IN FLOWERS OF ARABIDOPSIS THALIANA

Emission of Monoterpenes and Sesquiterpenes from Flowers

We conducted a detailed analysis of the expression of all *Arabidopsis TPS* genes in the main organs of the plant (flowers, leaves, stems, roots, and siliques) using a semi-quantitative RT-PCR approach. Our results indicated that most of the *AtTPS* genes are expressed in one or more organs under normal growth conditions.[29] In particular, several *AtTPS* genes are expressed in flowers, some exclusively so (Fig. 1.2).

This observation led us to examine whether *Arabidopsis* flowers emit terpene volatiles. However, standard volatile collection and analysis techniques did not result in readily detectable levels of terpenes. We, therefore, adapted a closed loop stripping method developed initially by Donath and Boland[30] for the detection of *Arabidopsis* volatiles.[29] This method (Fig.1.3A) is based on a continuous circulation of air in the headspace of whole plants or plant parts placed in a 1-3 liter glass bell jar. Volatiles are trapped on a thin activated charcoal filter that has been fitted into a stainless steel column connected to a circulation pump. The continuous collection of volatiles for up to 12 hours in a relatively small headspace volume allows trapping of almost 100% of the emitted compounds. Alternatively, a slightly less sensitive semi-open dynamic headspace sampling system was applied (Fig.1,3B) in which purified air was pumped into a 4-liter glass jar containing the plant, and 90% of the air was actively pulled out through a charcoal filter, while the remaining air was vented through the top of the glass container.

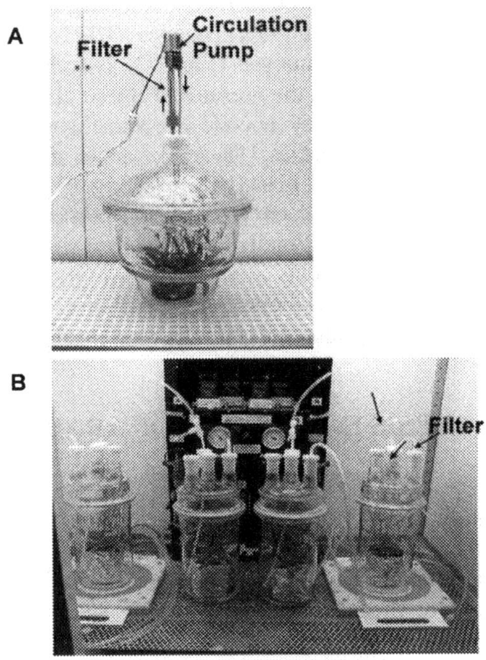

Figure 1.3: Dynamic head space sampling systems for volatile collection. A: Closed-loop stripping system according to Donath and Boland,[30] B: Semi-open collection system. The direction of the air flow is indicated by arrows.

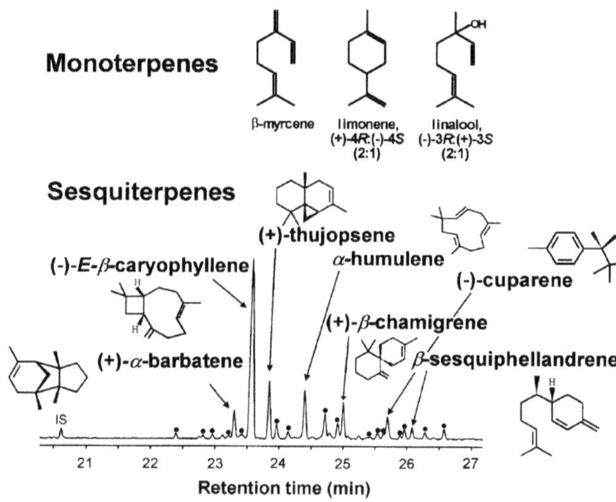

Figure 1.4: Structures and GC-MS chromatogram of monoterpene and sesquiterpene compounds emitted from inflorescences of *Arabidopsis thaliana*. Dots indicate additional sesquiterpene hydrocarbons of which 10 have been identified by comparison to authentic standards. IS: internal standard, nonyl acetate.

Using these methods in combination with gas chromatography-mass spectrometry (GC-MS), we were able to detect the emission of a number of monoterpenes as well as a large group of sesquiterpenes from whole *Arabidopsis* Columbia plants (Fig.1.4). In total, 3 monoterpenes (β-myrcene, linalool, and limonene) and over 20 sesquiterpene hydrocarbons were detected with *E*-β-caryophyllene as the predominant terpene volatile. The sesquiterpene volatiles showed a high structural diversity including acyclic, mono-, di- and tricyclic compounds. All monoterpenes and 19 sesquiterpenes were identified with certainty by mass spectra and comparison with authentic standards.

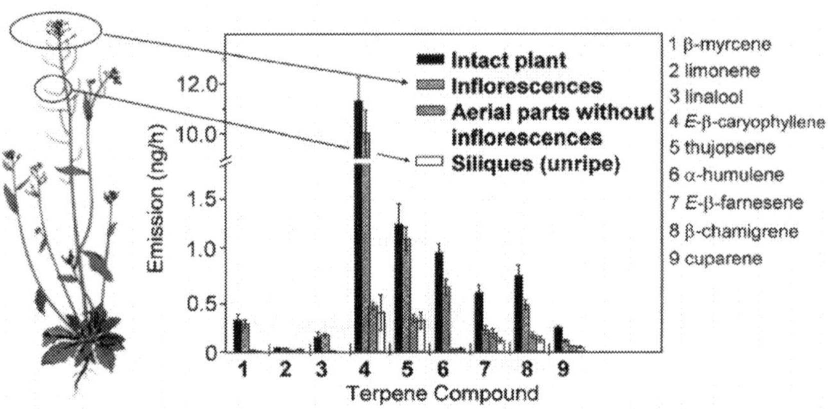

Figure 1.5: Release rates of the major terpenes from intact flowering *Arabidopsis* Col plants and parts of these plants determined by dynamic headspace sampling. Inflorescences are the main source of constitutive terpene emission.

To determine which part of the plant was responsible for the emission of each of these terpenes, we removed inflorescences or siliques and conducted head space collections of the isolated plant parts and the remaining vegetative tissue (Fig.1.5). Comparative analysis of the emitted volatiles showed that inflorescences were the main source of monoterpenes and most sesquiterpenes, together comprising more than 60% of the total amount of floral volatiles. Other volatile compounds emitted from *Arabidopsis* flowers and vegetative tissues were primarily aliphatic aldehydes and alcohols. A survey of several *A. thaliana* accessions, including ecotypes of various geographical regions, revealed distinct qualitative and quantitative differences in floral terpene emission, thus providing an extensive resource to study the mechanisms regulating natural variation and evolution of volatile terpene biosynthesis (unpublished data).

Function of Flower Specific AtTPS Genes and their Tissue Specific Expression

To determine which genes are responsible for the synthesis of the floral terpene volatiles that we had observed, we used RT-PCR to obtain full-length cDNA clones of the *AtTPS* genes shown to be expressed in flowers and predicted to encode mono- and sesquiterpene synthases. We then ligated these cDNAs into a bacterial expression vector carrying the T7 promoter and expressed them in *E. coli*. The *E. coli*–produced AtTPS proteins were tested for activity with GPP and FPP, the universal precursors of monoterpenes and sesquiterpenes, respectively (Fig.1. 1). The results indicated that the enzymes encoded by At3g25810 (*AtTPS1*) and At1g61680 (*AtTPS6*) are responsible for the synthesis of monoterpenes such as β-myrcene, β-ocimene, limonene, and linalool emitted from *Arabidopsis* flowers. The At5g23960 (AtTPS27) protein was found to catalyze the formation of the main floral sesquiterpenes *E*-β-caryophyllene and α-humulene, whereas heterologous expression of At5g44630 (*AtTPS18*) showed that the encoded enzyme is responsible for the production of most, if not all, of the other floral sesquiterpene hydrocarbons[29] (additional data unpublished). The formation of multiple enzymatic products from a single substrate is a characteristic feature of terpene synthases and can be ascribed to multiple reaction paths of the initially formed carbocationic intermediate, including differential internal electrophilic additions, hydride shifts, rearrangements, deprotonations, or addition of water.[7-10] Although several of the investigated terpene synthases are able to accept both GPP and FPP as substrates, the presence or absence of a plastidial transit peptide in mono- and sesquiterpene synthases, respectively, determine the subcellular localization of the proteins and hence the products that they make, since it is believed that GPP is available only in the plastids and FPP is available only in the cytosol.[3,11] Additionally, the *in vitro* product formation rates of these enzymes are usually higher with the compartmentally available substrate.[31,32]

To study the tissue-specific expression of floral *AtTPS* genes, we used an approach in which promoter regions of these genes were fused to the coding region of the *E. coli* β-glucuronidase (GUS) gene, and the entire construct was inserted into the *Arabidopsis* genome by *Agrobacterium*-mediated transformation.[33] The GUS reporter gene encodes an enzyme that catalyzes the formation of a blue-colored precipitating product by hydrolysis of the colorless substrate X-Gluc (5-bromo-4-chloro-3-indoyl-β-D- glucuronic acid). *In vivo* staining of transgenic plants allows for the observation of the tissue(s) in which the promoter being tested is active.[34]

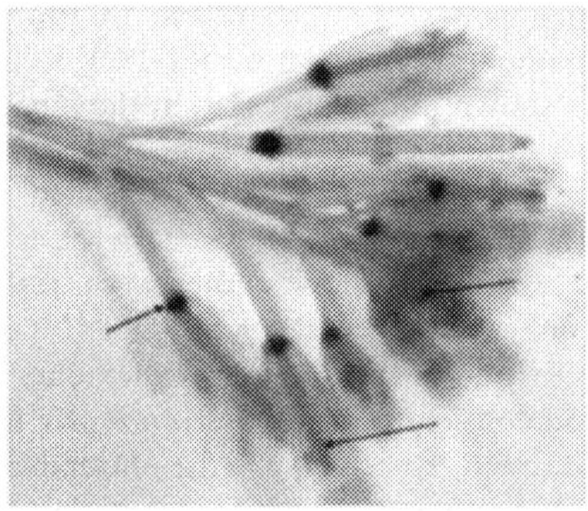

Figure 1.6: Expression patterns of the At5g44630 (*AtTPS18*)::*GUS* gene in *Arabidopsis thaliana* flowers. GUS activity was observed at the base of young and old flowers and the abscission zone of floral organs. Additional GUS staining was detected in ovaries and developing seeds. GUS staining is indicated by arrows.

Experiments with several *AtTPS* genes showed staining in various parts of the flower, verifying that these promoters are active in floral tissues. GUS activity under the control of the promoter of the monoterpene synthase gene *AtTPS1* was observed in sepals, stigma, anther filaments, and receptacles of the mature flower bud as well as the young and mature open flower.[29] In contrast, GUS expression driven by the promoter of *AtTPS18* was mainly detected at the base or receptacle of young and mature flowers and the abscission zone of siliques. Additional staining was observed in the ovules or developing seeds (Fig.1.6).

These results suggest several functions for the volatile terpenes in *Arabidopsis* flowers. The expression of terpene synthases at the stigma could be involved in protecting the moist surface area against fungal growth, since the monoterpenes produced have antimicrobial activity.[35] Similar expression patterns were found for a linalool synthase in the stigma of flowers from *Clarkia breweri.*[36] Another potential function of terpenes in this tissue may be protection against oxidative stress.[23,24] Expression of *AtTPS18* occurs at the base of the *Arabidopsis* flower, an area in which sugar producing nectaries are located.[37] The biosynthesis of several sesquiterpenes that have antimicrobial activity could, therefore, be important for defending this region against microbial infection. This might also be of significance in protecting the wound zone after abscission of the floral organs. Another obvious function of terpenes released from floral tissues is the attraction of pollinators.[18] Specifically, the observation of *AtTPS1* promoter activity in sepals, filaments and receptacles suggests such a function, since several flower tissues are involved. Interestingly, no expression of the genes investigated so far has been observed in flower petals, which have been described as the main organs of expression of non terpenoid floral scent genes in other plants like *Clarkia breweri* and *Anthirrinum majus*.[38-40] Whether or not this is due to a reduction of terpene emission as a consequence of the evolution of *A. thaliana* towards self-pollination remains to be determined.

INSECT VISITS TO A. THALIANA FLOWERS

Volatile terpenes are found in the aroma bouquet emitted from many insect-pollinated flowers.[17] A role in attracting insect pollinators was, therefore, a logical hypothesis for the emission of monoterpenes and sesquiterpenes from *A. thaliana* flowers. Although *A. thaliana*, unlike its close relative *A. lyrata*, is a self-compatible species, and, at least in the lab, it sets copious number of seeds by self-pollination, several investigators have previously reported that *A. thaliana* flowers are sometimes visited by insects like hoverflies in nature, and that a small amount of cross-pollination does occur.[41,42] These observations are consistent with findings showing that natural *A. thaliana* populations exhibit polymorphisms at tested loci and contain heterozygous individuals at frequencies that cannot be accounted for solely by mutation rates.[43,44] Cross-pollination events could be of importance in wild *Arabidopsis* populations since the progeny arising from out-crossing often have greater reproductive fitness, thereby mitigating inbreeding pressure.[45] This heterozygous advantage may have led to the retention of traits that promote out-crossing even in this mainly self-pollinating species. Indeed, the development of the *Arabidopsis* flower allows a short time window for cross pollination, when the receptive stigma protrudes from the flower petals before the anthers mature. Additionally, floral nectaries, located at the basis of the stamens, provide sugars as rewards to visiting insects.[37]

Figure 1.7: Solitary bees (Halictidae) collecting pollen from *Arabidopsis* flowers.

We examined the visitation of insects to *A. thaliana* flowers in semi-natural settings at the grounds of the botanical gardens in Halle, Germany and at Ann Arbor, Michigan, USA. While a detailed accounting of these experiments will be given elsewhere, we observed a large number and types of insects visiting the flowers. These included hover flies and other diptera, beetles, and thrips. The flowering plants of the German population were also frequently visited by solitary bees collecting and transferring flower pollen (Fig 1.7). Monitoring the frequency of these visits over the whole flowering season revealed regular daily visitation patterns that clearly corroborated the role of insects in cross pollination events in wild *Arabidopsis* populations.

It is not yet known whether the emission of terpenes from *A. thaliana* flowers is directly responsible for the attraction of these insects (as well as the efficacy of the insects in cross-pollinating the flowers). Such investigations should include GC-electroantennograms monitoring the antennal response to distinct terpene compounds of the volatile blend, and wind tunnel experiments with insect species shown to have visited the *A. thaliana* flowers. In addition, it will be useful to determine the cross-pollination rates in synthetic populations of various *Arabidopsis* ecotypes and TPS mutant lines lacking or overproducing one or several floral terpene compounds.

EMISSION OF TERPENES FROM LEAVES BY ELICITATION AND INSECT ATTACK

As described in the introduction, terpenes are often emitted from vegetative organs of plants under attack by herbivorous insects, including *Arabidopsis*.[46] The released volatiles can attract predators and parasitoids of these insects, thereby functioning as indirect defense compounds.[19-21] Terpenes have also been reported to function as antimicrobial phytoalexins accumulating in response to elicitation or pathogen attack.[2] Several groups have reported the role of phytohormones like jasmonic acid as signaling compounds in terpene induction.[47,48] However, a detailed and comprehensive picture of the process of induction is still missing. We have begun an exhaustive search to define conditions under which the emission of specific terpenes is induced in *A. thaliana*, and to correlate such emission with the induction of specific *AtTPS* genes, with the long-term goal of examining the mechanism of the regulation of this process.

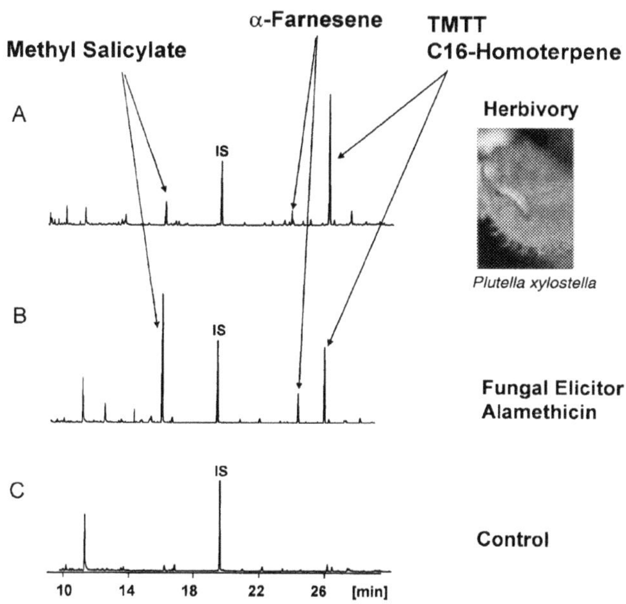

Figure 1.8: Gas chromatography of volatiles released from *A. thaliana* Col rosette leaves by feeding of *Plutella xylostella* larvae (A) or treatment with the peptaibol elicitor Alamethicin from *Trichoderma viride* (B). C: GC-chromatogram of volatiles released from leaves treated with water only. IS: internal standard, nonyl acetate.

Preliminary results indicate that under attack by caterpillars of the moth *Plutella xylostella*, rosette leaves of *Arabidopsis* Col ecotype emit at least two terpenes, α-farnesene and 4,8,12-trimethyltrideca-1,3,7,11-tetraene, a C_{16} homoterpene (Fig.1. 8A), as well as methylsalicylate.[49] A similar emission profile is observed when detached leaves are treated with alamethicin (Fig.1. 8B), a fungal peptaibol elicitor with membrane pore-forming ability.[50] We are currently investigating which genes are responsible for the synthesis of the induced compounds. This work includes screening for genes encoding cytochrome P450 enzymes that are likely to be involved in the conversion of a C_{20} isoprenoid precursor into the observed C_{16} homoterpene. Similar to floral emission, inducible volatile emission varies among *A. thaliana* ecotypes as well as between different *Arabidopsis* species. For example, we have found that *E*-β-caryophyllene, which is released only as a constitutive volatile from *A. thaliana* flowers, is inducible by insect damage of rosette leaves of some *A. lyrata* lines. Despite their close genetic relatedness, *A. thaliana* and *A. lyrata* have different life histories and breeding systems. While *A. thaliana* is a mainly a self-pollinating annual species, *A. lyrata* is a perennial species that is strictly self-incompatible.[51] The different life histories of these closely related species may have had an effect on the evolution of the roles that terpenes play in defense or attraction in these two species. We are currently investigating the regulatory mechanisms responsible for differential expression of orthologous *TPS* genes in *Arabidopsis* ecotypes and *Arabidopsis* close relatives.[52] The results should lead to exciting new insights into the evolution of functional diversity of terpene secondary metabolism in plants.

SUMMARY

Plants use volatile compounds in general, and terpenes in particular, to attract pollinators to their flowers and to ward off, directly or indirectly, harmful insect, animal, and microbial pests. We have shown that the *Arabidopsis* model system is as useful for the study of terpene biosynthesis and emission as it is for so many other areas of plant biology. The availability of the sequence of the entire *Arabidopsis* genome has allowed us to identify the complete *TPS* gene family, and to begin to correlate the emission of specific terpenes with the expression of specific *AtTPS* genes. With the modern tools available for experimentation in *Arabidopsis*, this model organism constitutes the best system to elucidate the mechanisms regulating the processes of plant-insect interaction via volatiles, which operate in both the vegetative and the reproductive parts of the plants.

ACKNOWLEDGEMENTS

We thank Wilfried Koenig for providing standards for sesquiterpene identification. This project is supported by National Science Foundation Grants MCB-9974463 and IBN-0211697 (to E.P.) and by funds from the Max Planck Society (to J.G.).

REFERENCES

1. MCGARVEY, D.J., CROTEAU, R., Terpenoid metabolism, *Plant Cell*, 1995, **7**, 1015-1026.
2. CHAPPELL, J., Biochemistry and molecular biology of the isoprenoid biosynthetic pathway in plants, *Annu. Rev. Plant Physiol. Plant Mol. Biol.*, 1995, **46**, 521-547.
3. GERSHENZON, J., KREIS, W., Biochemistry of terpenoids: Monoterpenes, sesquiterpenes, diterpenes, sterols, cardiac glycosides and steroid saponins, *in*: Biochemistry of Plant Secondary Metabolism (M. Wink, ed.), CRC Press LLC. 1999, pp. 222-299.
4. RODRIGUEZ-CONCEPTION, M., BORONAT, A., Elucidation of the methylerythritol phosphate pathway for isoprenoid biosynthesis in bacteria and plastids. A metabolic milestone achieved through genomics, *Plant Physiol.*, 2002, **130**, 1079-1089.
5. KOYAMA, T., OGURA, K., Enzymatic mechanism of chain elongation in isoprenoid biosynthesis, *in*: Comprehensive Natural Products Chemistry, Vol. 2, Isoprenoids Including Carotenoids and Steroids (D.D. Cane, ed.) Elsevier, Amsterdam. 1999, pp. 69-96.
6. KELLOGG, B.A., POULTER, C.D., Chain elongation in the isoprenoid biosynthetic pathway, *Curr. Opin. Chem. Biol.*, 1997, **1**, 570-578.
7. WISE, M., CROTEAU, R., Monoterpene biosynthesis, *in*: Comprehensive Natural Products Chemistry, Vol. 2, Isoprenoids Including Carotenoids and Steroids, (D.D. Cane, ed.) Elsevier, Amsterdam. 1999, pp. 97-153.
8. CANE, D. E., Sesquiterpene biosynthesis: Cyclization mechanisms, *in*: Comprehensive Natural Products Chemistry, Vol. 2, Isoprenoids Including Carotenoids and Steroids, (D.D. Cane, ed.) Elsevier, Amsterdam. 1999, pp. 155-200.
9. MACMILLAN, J., BEALE, M. H., Diterpene biosynthesis, *in* Comprehensive Natural Products Chemistry, Vol. 2, Isoprenoids Including Carotenoids and Steroids, (D.D. Cane, ed.) Elsevier, Amsterdam. 1999, pp. 217-243.
10. DAVIS, E.M., CROTEAU, R., Cyclization enzymes in the biosynthesis of monoterpenes, sesquiterpenes, and diterpenes, *Top. Curr. Chem.*, 2000, **209**, 53-95.
11. LICHTENTHALER, H. K., The 1-deoxy-D-xylulose-5-phosphate pathway of isoprenoid biosynthesis in plants, *Annu. Rev. Plant Physiol. Plant Mol. Biol.*, 1999, **50**, 47-65.
12. OSBOURN, A.E., HARALAMPIDIS, K., Triterpenoid saponin biosynthesis in plants, *in*: Recent Advances in Phytochemistry, Phytochemistry in the Genomics and Post-Genomics Eras (J.T. Romeo and R.A. Dixon, eds.), Pergamon Press, New York. 2002, pp. 81-93.

13. FUJIOKA, S., YOKOTA, T., Biosynthesis and metabolism of brassinosteroids, *Annu. Rev. Plant Biol.*, 2003, **54**, 137-164.
14. CUNNINGHAM, F.X. JR., GANTT, E., Genes and enzymes of carotenoid biosynthesis in plants, *Annu. Rev. Plant Physiol. Plant Mol. Biol.*, 1998, **49**, 557-583.
15. LANGENHEIM, J.H., Higher plant terpenoids: A phytocentric overview of their ecological roles, *J. Chem. Ecol.*, 1994, **20**, 1223-1280.
16. PICHERSKY, E., GERSHENZON, J., The formation and function of plant volatiles: perfumes for pollinator attraction and defense, *Curr. Opin. Plant Biol.*, 2002, **5**, 237-243.
17. KNUDSEN, J.T., TOLLSTEN, L., BERGSTROM, G., Floral scents - a checklist of volatile compounds isolated by head-space techniques, *Phytochemistry*, 1993, **33**, 253-280.
18. DUDAREVA, N., PICHERSKY, E., Biochemical and molecular genetic aspects of floral scents, *Plant Physiol.*, 2000, **122**, 627-633.
19. PARE, P.W., TUMLINSON, J.H., Plant volatiles as a defense against insect herbivores, *Plant Physiol.*, 1999, **121**, 325-331.
20. DICKE, M., VAN LOON, J.J.A., Multitrophic effects of herbivore-induced plant volatiles in an evolutionary context, *Entomol. Exp. Appl.*, 2000, **97**, 237-249.
21. KESSLER, A., BALDWIN, I.T., Defensive function of herbivore-induced plant volatile emissions in nature, *Science*, 2001, **291**, 2141-2144.
22. SHARKEY, T.D., YEH, S., Isoprene emission from plants, *Annu. Rev. Plant Physiol. Plant Mol. Biol.*, 2001, **52**, 407-436.
23. CALOGIROU, A., LARSEN, B. R., KOTZIAS, D., Gas-phase terpene oxidation products: A review, *Atmos. Environ.*, 1999, **33**, 1423-1439.
24. LORETO, F., VELIKOVA, V., Isoprene produced by leaves protects the photosynthetic apparatus against ozone damage, quenches ozone products, and reduces lipid peroxidation of cellular membranes, *Plant Physiol.*, 2001, **127**, 1781-1787.
25. BOHLMANN, J., MEYER-GAUEN, G., CROTEAU, R., Plant terpenoid synthases: Molecular biology and phylogenetic analysis, *Proc. Natl. Acad. Sci. USA*, 1998, **95**, 4126-4133.
26. AUBOURG, S., LECHARNY, A., BOHLMANN, J., Genomic analysis of the terpenoid synthase (*AtTPS*) gene family of *Arabidopsis thaliana*, *Mol. Genet. Genomics*, 2002, **267**, 730-745.
27. SUN, T. P., KAMIYA, Y., The Arabidopsis GA1 locus encodes the cyclase *ent*-kaurene synthetase A of gibberellin biosynthesis, *Plant Cell*, 1994, **6**, 1509-1518.
28. YAMAGUCHI, S., SUN, T. P., KAWAIDE, H., KAMIYA, Y., The GA2 locus of *Arabidopsis thaliana* encodes *ent*-kaurene synthase of gibberellin biosynthesis, *Plant Physiol.*, 1998, **116**, 1271-1278.
29. CHEN, F., THOLL, D., D'AURIA, J.C., FAROOQ, A., PICHERSKY, E., GERSHENZON, J., Biosynthesis and emission of terpenoid volatiles from Arabidopsis flowers, *Plant Cell*, 2003, **15**, 481-494.
30. DONATH, J., BOLAND, W., Biosynthesis of acyclic homoterpenes - enzyme selectivity and absolute configuration of the nerolidol precursor, *Phytochemistry*, 1995, **39**, 785-790.

31. SCHNEE, C., KOLLNER, T.G., GERSHENZON, J., DEGENHARDT, J., The maize gene terpene synthase 1 encodes a sesquiterpene synthase catalyzing the formation of (*E*)-beta-farnesene, (*E*)-nerolidol, and (*E,E*)-farnesol after herbivore damage, *Plant Physiol.*, 2002, **130**, 2049-2060.
32. CROCK, J., WILDUNG, M., CROTEAU, R., Isolation and bacterial expression of a sesquiterpene synthase cDNA from peppermint (*Mentha* x *piperita*, L.) that produces the aphid alarm pheromone (*E*)-β-farnesene, *Proc. Natl. Acad. Sci. USA*, 1997, **94**, 12833-12838.
33. BECHTOLD, N., ELLIS, J., PELLETIER, G., *In planta* Agrobacterium mediated gene-transfer by infiltration of adult *Arabidopsis thaliana* plants, *C. R. Acad. Sci. Paris Life Sci.*, 1993, **316,** 1194-1199.
34. JEFFERSON, R. A., KAVANAGH, T. A., BEVAN, M. W., Gus fusions - beta-glucuronidase as a sensitive and versatile gene fusion marker in higher-plants, *EMBO J.*, 1987, **6**, 3901-3907.
35. DEANS, S. G., WATERMAN, P. G., Biological activity of volatile oils, *in*: Volatile Oil Crops: Their Biology, Biochemistry and Production (R.K.M. Hay and P.G. Waterman, eds.), Longman Scientific and Technical, Essex, England. 1993, pp. 97-111.
36. DUDAREVA, N., CSEKE, L., BLANC, V. M., PICHERSKY, E., Evolution of floral scent in *Clarkia*: Novel patterns of *S*-linalool synthase gene expression in the *C. breweri* flower, *Plant Cell*, 1996, **8**, 1137-1148.
37. DAVIS, A. R., PYLATUIK, J. D., PARADIS, J. C., LOW, N. H., Nectar-carbohydrate production and composition vary in relation to nectary anatomy and location within individual flowers of several species of Brassicaceae, *Planta*, 1998, **205**, 305-318.
38. WANG, J., DUDAREVA, N., BHAKTA, S., RAGUSO, R. A., **PICHERSKY E.,** Floral scent production in *Clarkia breweri* (Onagraceae) II. Localization and developmental modulation of the enzyme S-adenosyl-L-methionine:(Iso)eugenol O-methyltransferase and phenylpropanoid emission, *Plant Physiol.*, 1997, **114,** 213-221.
39. DUDAREVA, N., D'AURIA, J. C., NAM, K. H., RAGUSO, R. A., PICHERSKY E., Acetyl-CoA:benzylalcohol acetyltransferase: An enzyme involved in floral scent production in *Clarkia breweri*, *Plant J.*, 1998, **14**, 297-304.
40. DUDAREVA, N., MURFITT, L. M., MANN, C. J., GORENSTEIN, N., KOLOSOVA, N., KISH, C. M., BONHAM, C., WOOD, K., Developmental regulation of methyl benzoate biosynthesis and emission in snapdragon flowers, *Plant Cell*, 2000, **12**, 949-961.
41. JONES, M. E., Population genetics of *Arabiodopsis thaliana.* 1. Breeding system, *Heredity*, 1971, **27**, 39-50.
42. SNAPE, J. W., LAWRENCE, M. J., Breeding system of *Arabidopsis thaliana*, *Heredity*, 1971, **27**, 299-301.
43. LORIDON, K., COURNOYER, B., GOUBELY, C., DEPEIGES, A., PICARD, G., Length polymorphism and allele structure of trinucleotide microsatellites in natural accessions of *Arabidopsis thaliana*, *Theor. Appl. Genet.*, 1998, **97**, 591-604.
44. ABBOTT, R. J., GOMES, M. F., Population genetic structure and outcrossing rate of *Arabidopsis thaliana* (L) Heynh., *Heredity*, 1989, **62**, 411-418.

45. AGREN, J., SCHEMSKE, D. W., Outcrossing rate and inbreeding depression in 2 annual monoecious herbs, *Begonia hirsuta* and *B semiovata*, *Evolution*, 1993, **47**, 125-135.
46. VAN POECKE, R. M. P., POSTHUMUS, M. A., DICKE, M., Herbivore-induced volatile production by *Arabidopsis thaliana* leads to attraction of the parasitoid *Cotesia rubecula*: Chemical, behavioral, and gene-expression analysis, *J. Chem. Ecol.*, 2001, **27**, 1911-1928.
47. KOCH, T., KRUMM, T., JUNG, V., ENGELBERTH, J., BOLAND, W., Differential induction of plant volatile biosynthesis in the lima bean by early and late intermediates of the octadecanoid-signaling pathway, *Plant Physiol.*, 1999, **121**, 153-162.
48. MARTIN, D., THOLL, D., GERSHENZON, J., BOHLMANN, J., Methyl jasmonate induces traumatic resin ducts, terpenoid resin biosynthesis, and terpenoid accumulation in developing xylem of Norway spruce stems, *Plant Physiol.*, 2002, **129**, 1003-1018.
49. CHEN, F., D'AURIA, J.C., THOLL, D., ROSS, J.R., GERSHENZON, J., NOEL, J.P., PICHERSKY, E., An *Arabidopsis thaliana* gene for methylsalicylate biosynthesis, identified by a biochemical genomics approach, has a role in defense, *Plant J.*, 2003, **36**, 577-588.
50. ENGELBERTH, J., KOCH, T., SCHULER, G., BACHMANN, N., RECHTENBACH, J., BOLAND, W., Ion channel-forming alamethicin is a potent elicitor of volatile biosynthesis and tendril coiling. Cross talk between jasmonate and salicylate signaling in lima bean, *Plant Physiol.*, 2001, **125**, 369-377.
51. SCHIERUP, M.H., MABLE, B.K., AWADALLA, P., CHARLESWORTH, D., Identification and characterization of a polymorphic receptor kinase gene linked to the self-incompatibility locus of *Arabidopsis lyrata*, *Genetics*, 2001, **158**, 387-399.
52. MITCHELL-OLDS, T., *Arabidopsis thaliana* and its wild relatives: A model system for ecology and evolution, *Trends Ecol. Evol.*, 2001, **16**, 693-700.
53. BOHLMANN, J., MARTIN, D., OLDHAM, N. J., GERSHENZON, J., Terpenoid secondary metabolism in *Arabidopsis thaliana*: cDNA cloning, characterization, and functional expression of a myrcene/(E)-β-ocimene synthase, *Arch. Biochem. Biophys.*, 2000, **375**, 262-269.
54. FALDT, J., ARIMURA, G, I., GERSHENZON, J., TAKABAYASHI, J., BOHLMANN, J., Functional identification of *AtTPS03* as (*E*)-beta-ocimene synthase: A new monoterpene synthase catalyzing jasmonate- and wound-induced volatile formation in *Arabidopsis thaliana*, *Planta*, 2003, **216**, 745-751.

Chapter Two

THE BIOCHEMICAL AND MOLECULAR ORIGINS OF ALIPHATIC GLUCOSINOLATE DIVERSITY IN *ARABIDOPSIS THALIANA*

Jim Tokuhisa,* Jan-Willem de Kraker, Susanne Textor, and Jonathan Gershenzon

Max Planck Institute for Chemical Ecology
Winzwerlaer Str. 10
07745 Jena, Germany

**Author for correspondence: tokuhisa@ice.mpg.de*

INTRODUCTION

Glucosinolates are a diverse class of secondary metabolites found principally in plants of the order Brassicales (formerly Capparales).[1] Many agriculturally important plants are found in this order, and glucosinolates contribute both positively as well as negatively to human uses of these plants.[2] As a consequence, efforts to understand and manipulate glucosinolate composition have attracted many researchers. For example, nearly 40 years ago, Canadian researchers developed Canola, a rapeseed type with low glucosinolate levels in the seed that reduced the adverse goitrogenic potential of the oil and residual seed meal, making these available for food production and animal feed production, respectively.[3] More recently, the benefits of glucosinolates have been recognized in studies of cover crops for use as green manures or soil fumigants.[4] The organoleptic characteristics of some glucosinolates contribute to the flavors associated with brassicaceous vegetables, including cabbage, kale, broccoli, and radish and make them the principals in condiments such as mustard, horseradish, and wasabi.[5] These crop species have often been bred for modified glucosinolate levels. With a broader understanding of biosynthesis, more sophisticated manipulations of plant glucosinolate composition can be anticipated. For example, individual glucosinolates have been implicated as precursors of effective cancer prevention agents that act by inducing the synthesis of a set of enzymes in humans that can detoxify potential carcinogens.[6] Thus, the health benefits of eating brassicaceous vegetables could be enhanced by altering glucosinolate quantity and composition.

Although glucosinolates are not widespread in the plant kingdom, most species within the Brassicales contain them, and over 130 different structures have been reported.[1,7] These structures include a wide range of different functional groups and chain lengths, despite the fact that glucosinolates are derived from a limited number of amino acids. This review describes some of the biochemical and molecular bases of this structural diversity. The ecological factors contributing to diversity are not discussed here, although the variety of glucosinolates present undoubtedly reflects selective pressures for their roles in defense against herbivores and pathogens.[8] Since glucosinolate hydrolysis products are thought to be primarily responsible for the biological activity of this compound class,[9] the structural types of the parent glucosinolate found are likely to have been selected for their ability to form specific hydrolysis products.

Glucosinolate Structure

The 130-plus glucosinolates have several common structural features (Fig. 2.1), including an oxime group derived from the α-carbon and the amino group of the parent amino acid. A glucose moiety is attached to the oxime carbon by a β-thio-linkage, and the hydroxyl function, which has a *Z*-configuration relative to the thioglucose residue, is esterified with a sulfate group. The various classes of glucosinolates are distinguished by variable R groups attached to the oxime carbon that are derived from the side chain of the particular amino acid precursor.

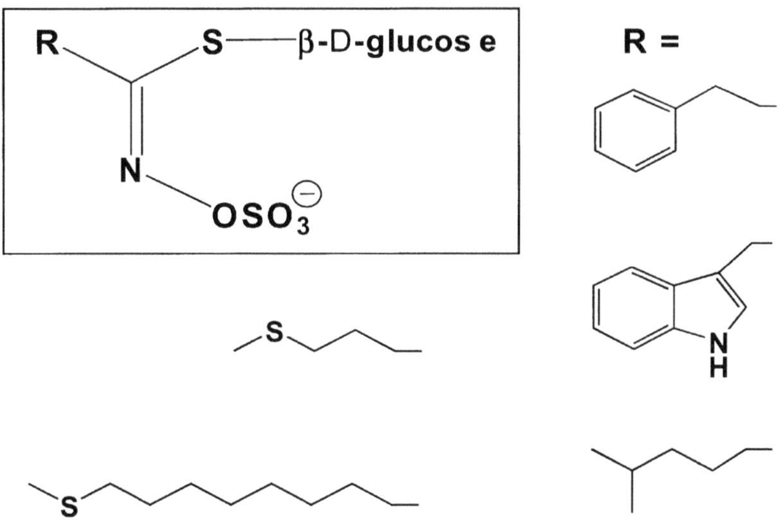

Figure 2.1: General Glucosinolate Structure (inset) and examples of R groups.

Glucosinolates are divided into three classes based on the general chemical properties of the amino acid precursors. Aliphatic glucosinolates variously contain a straight carbon chain derived from methionine or a branched chain from isoleucine, leucine or valine. Indole glucosinolates are formed from tryptophan, and the aromatic glucosinolates are derived from phenylalanine or tyrosine.[10]

Glucosinolate Biosynthesis

In the last 40 years, a variety of classical approaches, including precursor feeding experiments, enzymological investigations, and genetic studies have been employed to elucidate the general pathway of glucosinolate biosynthesis.[2] Recently, these have been supplemented with studies on glucosinolate biosynthetic genes. To the enormous good fortune of glucosinolate researchers, *Arabidopsis thaliana*, the first model system for molecular genetics in higher plants, produces over 35 different glucosinolates.[7,11] Thus, the molecular genetic tools available from the *Arabidopsis* community have been exploited to substantiate and clarify previous work and to extend our understanding of glucosinolate biosynthesis and its role in plant biology.[12-14]

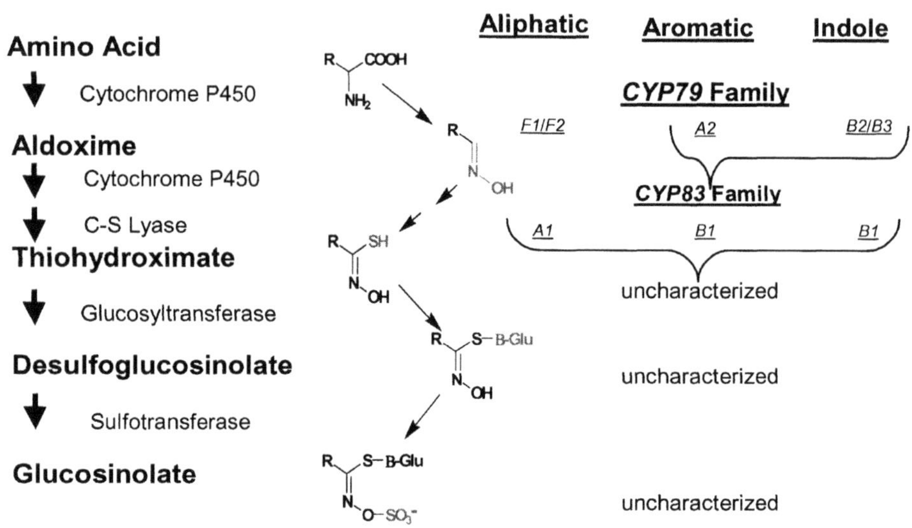

Figure 2.2: Steps of the Core Biosynthetic Pathway. The lighter shaded structural domains indicate the changes at each enzymatic step. The genes of *A. thaliana* characterized for particular steps of the pathway are listed on the right arranged by their predominant activities for each glucosinolate class.

Taken together, the classic and molecular genetic approaches have led to the following general understanding of glucosinolate biosynthesis.[15] Amino acids are converted to glucosinolates in a core pathway involving five enzymatic steps (Fig. 2.2). The initial step involves the oxidation of the amino function to an aldoxime, catalyzed by cytochrome P450 mixed function oxygenases specific to each class of amino acids. The second step is another cytochrome P450-catalyzed oxidation with broader substrate specificities. The aldoxime is converted to a reactive *aci*-nitro intermediate that acquires a thiol group through the conjugation of the α-carbon with the thiol group of cysteine followed by C-S lyase-mediated cleavage to release a thiohydroximic acid and alanine. Finally, a glucose residue is conjugated via a β-linkage to the thiol group by uridine diphosphate thiohydroximate glucosyltransferase, and a sulfate group is esterified to the free hydroxyl group of the oxime by the activity of a phosphoadenosine phosphosulfate desulfoglucosinolate sulfotransferase.

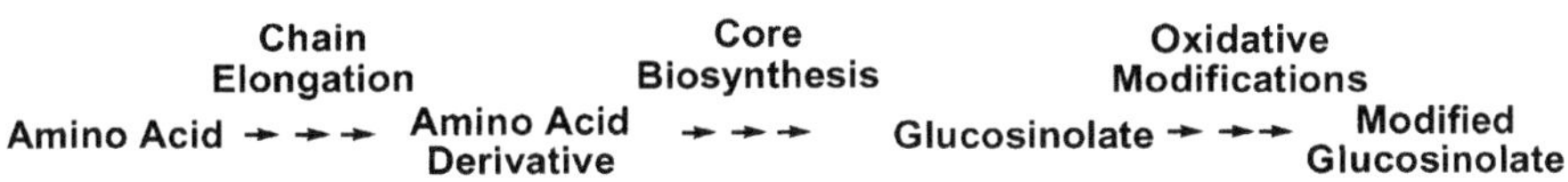

Figure 2.3: Major Stages of the Glucosinolate Biosynthetic Pathway.

In this review, we emphasize the biosynthesis of the 60+ glucosinolates derived from methionine, which includes the majority of glucosinolates in most of the economically-important glucosinolate-containing species. We highlight recent results that have identified biochemical and genetic features of glucosinolate biosynthesis that are associated with glucosinolate diversity and natural variation. The core pathway for glucosinolate biosynthesis from methionine is augmented by two sets of reactions that generate the skeletal diversity of end products (Fig. 2.3). One set of reactions modifies the amino acid precursor by extending the carbon chains, thereby increasing the number of amino acid substrates available to the core pathway. Another set of reactions modifies the product of the core pathway by oxidative processes in the side chain. As is frequently recognized for enzymes of secondary metabolism, the activities catalyzing these reactions are encoded by genes recruited from primary metabolism through gene duplication with subsequent functional divergence.[16] We describe results indicating further gene duplications and functional divergences that contribute to glucosinolate diversity. These duplications and their arrangement in the genome are likely to be responsible for the high amount of natural variation in glucosinolate content observed among the different accessions of *A. thaliana*.

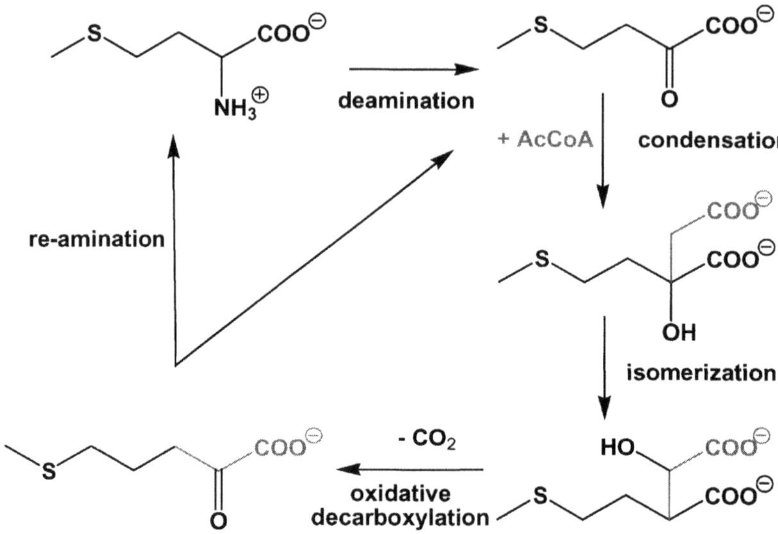

Figure 2.4: Methionine Chain Elongation Pathway.

MODIFICATION OF THE AMINO ACID PRECURSORS

The amino acid precursors for glucosinolate biosynthesis are subject to chain elongation. In the case of methionine, this results in the incorporation of 1-9 additional methylene groups in the carbon skeleton. As early as 1962,[17] Chisholm and coworkers provided evidence by using *in vivo* feeding studies showing that radiolabeled acetate was incorporated into methionine-derived glucosinolates as additional methylene groups. These and other results allowed a pathway for chain elongation to be proposed, which was confirmed by more recent *in vivo* studies with stable isotope-labeled precursors (Fig. 2.4).[18-20] Initially, methionine is deaminated to generate a 2-oxo acid derivative. This is followed by a three step cycle of methylene incorporation: 1) Condensation of acetyl-CoA to the carbonyl carbon atom of the 2-oxo acid derivative to generate a dicarboxylic acid, 2) Isomerization of the resulting hydroxyl group from C2 to C3, 3) Oxidative decarboxylation regenerating a 2-oxo acid with an additional methylene group. The product can be re-aminated to an amino acid and channeled to glucosinolate biosynthesis, or undergo another condensation with acetyl-CoA followed by another isomerization and oxidative decarboxylation. The pathway is similar to the single methylene incorporation that occurs in the leucine biosynthetic pathway catalyzed by isopropylmalate synthase (IPMS). However, the methionine chain-elongation machinery can catalyze additional cycles of methylene incorporation to produce not only homomethionine but also, di-, tri-, tetra-, up to nona-homomethionine.

The biochemical characterization of methionine chain elongation has been challenging. The initial deamination reaction in *Brassica carinata* was shown to be catalyzed by a methionine-glyoxylate transaminase.[21] However, the steps of the elongation cycle have proven more elusive. The first step, the condensation of acetyl-CoA with the 2-oxo acid, considered to be the critical and committed step of the cycle, was not detectable in initial studies although the proposed product of the reaction, 2-(2'-methylthio)ethylmalate, was isolated.[22] Only recently has an acetyl-CoA condensation activity been demonstrated in crude extracts of *Eruca sativa* and *A. thaliana*.[23,24] The remaining two steps of the chain elongation cycle have not been characterized, but are presumed to be homologous with the parallel reactions in leucine biosynthesis.

Formation of Chain Elongated Analogs of Methionine

Mutant analysis, genetic mapping, and the biochemical characterization of heterologously expressed genes have provided alternative and successful approaches to the investigation of the methionine chain elongation cycle. Haughn and coworkers carried out a screen for mutants of *A. thaliana* with altered glucosinolate profiles.[13] From 1200 progeny (M_2) of an ethylmethane sulfonate-mutagenized population, six lines were shown to have altered glucosinolate profiles that were stably inherited. For the *gsm1* mutant, the altered profile and the products formed by the administration of radiolabeled putative-precursors indicated a mutation in the chain elongation pathway. Although further characterizations were not done, the mutants were made available publicly through the *Arabidopsis* Biological Resource Center.

Differences in the total content and profile of glucosinolates among varieties and cultivars of the amphidiploid *Brassica napus* were exploited to identify loci associated with glucosinolate biosynthesis.[25,26] The segregation pattern of glucosinolate chain length in the F_2 progeny of crosses between synthetic and cultivated *B. napus* lines identified three to four loci that were determinants of propyl-, butyl-, or pentylglucosinolate chain length (where glucosinolate chain length refers to the number of methylene groups in the R group). This genetic approach was extended to *A. thaliana*,[14,27] which has also been shown to have extensive variation in glucosinolate content and profile among the various accessions.[28,29] These studies used recombinant inbred lines (RIL) of a cross between the Columbia and Landsberg *erecta* (L*er*) accessions to map the variation of the chain length of the predominant glucosinolate, either propyl- or butylglucosinolates. This trait mapped to the upper arm of chromosome (Chr) V designated *ELONG*.

Four *A. thaliana* genes were identified that could encode the enzyme catalyzing the initial step of the elongation cycle based on sequence similarity to genes that encode *IPMS*, the enzyme catalyzing the condensation reaction for the three-step methylene incorporation in leucine biosynthesis.[30] Two of these genes are on Chr I (At1g74040, At1g18500) and share about 90% identity with each other and have approximately 60% identity to microbial *IPMS* sequences. The other two (At5g23010 and At5g23020) display lower identity to the microbial *IPMS* genes but they share 85% identity and are identical in intron/exon structure.[30] Based on their proximity to the *ELONG* region of Chr V, these latter two genes were regarded as strong candidates for encoding the initial condensation step of methionine chain elongation, and were thus subjected to further study.

Three different approaches addressed the function of At5g23010.[30] First, fine-scale mapping within the *ELONG* region identified At5g23010 as the locus for variation in the predominance of propyl- and butylglucosinolates. Second, the reduced levels of butyl glucosinolates observed in two allelic mutant lines (*gsm1-1, gsm1-2)*[13] were shown to be caused by base substitution mutations in the At5g23010 locus. Third, initial biochemical characterizations of the enzyme activity generated by heterologous expression of this gene in *E. coli* indicated the ability to condense the 2-oxo-acid derivative of methionine with acetyl-CoA to produce 2-(2'-methylthio)ethylmalate. Similar biochemical characterization of the mutated protein from the *gsm1-1* mutant did not detect any activity.[23] Thus, At5g23010 was designated *methylthioalkylmalate synthase 1* (*MAM1*) based on the activity of the encoded enzyme. Subsequently, a more detailed characterization of the MAM1 protein showed that the enzyme also accepts the 2-oxo acid derivative of homomethionine as a substrate for the condensation reaction, but does not accept derivatives of longer chain methionine analogs nor the substrate used by IPMS in leucine biosynthesis.[23] Kinetic analyses with the two accepted 2-oxo-acid substrates indicated a 4.5-fold lower K_m for the homomethionine derivative compared to the methionine derivative. Coupled with the lack of any measurable activity with the next larger substrate, dihomomethionine, these data are consistent with the greater levels of butyl glucosinolates, compared to propyl or pentyl glucosinolates, in the Columbia accession.

The MAM1 enzyme does not account for all of the chain elongation evident from the aliphatic glucosinolate profile of the Columbia accession. The *gsm1-1* mutant line, which has a mutated *MAM1* that does not function *in vitro*,[23] showed a 4- to 6-fold increase in propyl glucosinolates and a slight increase in the longer-chained heptyl- and octylglucosinolates relative to wild-type plants.[30] These results indicated the presence of at least two additional methionine chain-elongating activities. Preliminary results from other mutant lines and biochemical characterizations indicate that At5g23020, designated *MAM-L* for *MAM-like*, has a significant role in methionine chain elongation (de Kraker, Textor, Tokuhisa, and Gershenzon, unpublished results). Thus, the range of chain-elongated, methionine-derived glucosinolates observed in the Brassicaceae is probably due to at least two enzymes with methylthioalkylmalate synthase activities that have different velocities for substrates of different chain length (Fig. 2.5).

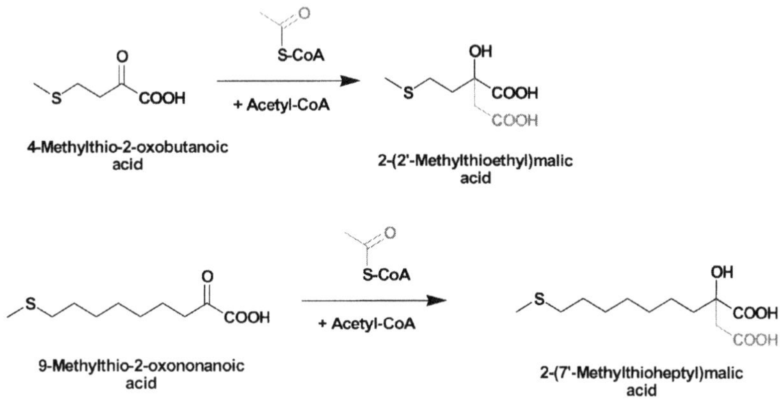

Figure 2.5: Condensation Reactions of the Chain Elongation Pathway for the Shortest and Longest 2-Oxo Acid Derivatives of Methionine in *Arabidopsis*.

Molecular Basis for Natural Variation in Chain Length

Extensive natural variation has been observed in the composition of chain-elongated glucosinolates in *A. thaliana*,[29] with the various ecotypes having either propyl- or butylglucosinolates as their predominant class. Underlying this simple biochemical variation is a complex polymorphism in the organization of the *ELONG* region. Analysis of this region in different *A. thaliana* accessions shows seven major classes of insertion/deletion (indel) arrangements observed among 25 accessions.[31] An archetypal gene arrangement is present in the Sorbo accession, consisting of three genes, of which two, designated *MAM1* and *MAM2*, have 95% identity, and the third, *MAML*, is more distantly related with approximately 85% identity to the other two genes. The other indel classes reflect partial or complete deletion of either *MAM1* or *MAM2* sometimes accompanied by the duplication of the remaining locus. Sequence comparisons among different accessions reveal further polymorphism between and within *MAM1* and *MAM2* due to extensive intra- and interlocus gene conversions.

Among all these different arrangements, the presence of a full-length copy of the Sorbo-like *MAM1* gene is consistently associated with the accumulation of butyl glucosinolates.

To address whether this polymorphism is a result of natural selection or neutral change, the sequence variations in the *MAM2* gene from different accessions were compared with variations in the surrounding genes.[31] The variation within the coding region of *MAM2* rejects a neutral evolutionary model, whereas the changes in the surrounding genes were consistent with neutrality. One potential selective force that could maintain variation of the *MAM2* locus was identified by a quantitative trait locus analysis for glucosinolate content and resistance to insect herbivory. Increased propylglucosinolate content associated with the Landsberg *MAM2* allele was correlated with reduced herbivore damage by the generalist herbivore *Spodoptera exigua*.[31] The determination of other selective forces involved in the natural variation of MAM enzymes will require further work on the functional significance of different glucosinolate profiles.

SUBSTRATE SPECIFICITIES IN THE CORE PATHWAY OF GLUCOSINOLATE BIOSYNTHESIS

The first two steps of the core glucosinolate biosynthetic pathway are catalyzed by cytochrome P450 enzymes belonging to the *CYP79* and *CYP83* families, respectively, and result in the sequential N-oxidation of the amino group and the formation of a cysteine conjugate. The cytochrome P450 superfamily of *A. thaliana* contains approximately 275 characterized or putative genes in 45 families and 70 subfamilies (NSF 2010: Functional Genomics of *Arabidopsis* P450s; http://arabidopsis-p450.biotec.uiuc.edu/abstract.shtml). In plants as well as animals, these enzymes are associated with xenobiotic detoxification as well as biosynthesis, and catalyze a wide variety of oxidations including hydroxylations, epoxidations, and heteroatom oxidations.[32] It has been suggested that P450 enzymes devoted to biosynthesis have narrow substrate specificities whereas those involved with detoxification of xenobiotics have broad substrate specificities.[33] Indeed, the first characterized cytochrome P450 enzymes of glucosinolate biosynthesis, CYP79A2, CYP79B2, and CYP79B3 have narrow substrate specificities.[34] Biochemical characterizations of CYP79F1, CYP79F2, CYP83A1, and CYP83B1 indicate that narrow specificities may be the exception rather than the rule for the cytochromes P450 of glucosinolate biosynthesis. The remaining enzymes of the pathway appear to have broad substrate specificities for all classes of glucosinolate precursors, but this remains to be rigorously tested.

Cytochromes P450

The P450 family designated *CYP79* includes at least five genes involved with the conversion of amino acids into their corresponding aldoximes in *A. thaliana.*[35] Three genes participate in aromatic (*CYP79A2*) and indole (*CYP79B2* and *B3*) glucosinolate biosynthesis. The remaining two genes, *CYP79F1* (At1g16410) and *CYP79F2* (At1g16400), are tandemly arrayed gene duplications on Chr I and have roles in aliphatic glucosinolate biosynthesis. Halkier and coworkers have shown that *CYP79F1*, heterologously expressed and purified from *E. coli*, accepts as substrates all chain-elongated methionine derivatives, from homomethionine to hexahomomethionine, whereas *CYP79F2*, similarly expressed and isolated from *Saccharomyces cerevisiae*, accepts only the longer penta- and hexahomomethionines.[36]

The second step in glucosinolate formation generates an unstable *aci*-nitro intermediate that becomes conjugated with the thiol group of cysteine via the α-carbon atom. This reaction is catalyzed by two enzymes encoded by the *CYP83* family. The *CYP83B1* gene (At4g31500) has a primary role in the metabolism of the aldoxime derivative of tryptophan whereas *CYP83A1* (At4g13770) appears to have a broad specificity for aldoximes, including those derived from chain-elongated methionine derivatives. Initial studies with heterologously expressed *CYP83A1* indicated a broad catalytic ability to metabolize the aldoxime derivatives of tryptophan, tyrosine and phenylalanine.[37] Further investigations of CYP83A1 with aliphatic aldoxime substrates indicated that they are the principal substrates for *CYP83A1.*[38] These results are supported by the glucosinolate profile of the *ref2 A. thaliana* lines that contain mutations in *CYP83A1* and were isolated in a screen for mutants of phenylpropanoid metabolism.[39] In these mutants, the leaf and seed glucosinolate profiles showed significantly lower levels of all aliphatic glucosinolates. This profile is consistent with *CYP83A1* encoding a catalytic activity for methionine-derived aldoximes and having a limited effect on tryptophan-derived aldoximes. The residual level of aliphatic glucosinolates in the *ref2* mutants indicated a cryptic metabolic activity perhaps due to *CYP83B1*, which has 63% identity to *CYP83A1* at the amino acid level.[39]

Further Steps of the Core Pathway

The remaining steps of glucosinolate biosynthesis involve enzymes that are thought to accommodate nearly all glucosinolate precursors regardless of their R groups.[40] Broad specificities of these enzymes are indicated by the ability of brassicaceous plants to metabolize a variety of xenobiotic aldoximes to the corresponding artificial glucosinolates.[41] C-S lyase activities isolated from *B. napus* hydrolyze cysteine conjugates that are precursors of benzyl- and 2-phenylethylglucosinolates but are unable to hydrolyze the precursor for the unnatural phenylglucosinolate. The ability to cleave the benzyl-cysteine conjugate is surprising as benzyl glucosinolates have not been detected in *B. napus.*[42] Enzyme activities for glycosylation, uridine diphosphate thiohydroximate glucosyltransferase, and sulfation, a 3'-phosphadenosine 5'-phosphosulfate:desulfoglucosinolate sulfotransferase, have been characterized in several crucifers and partially purified.[40] While the corresponding genes in *A. thaliana* have not been characterized, it is likely that such studies will be undertaken in the near future, providing additional information on enzyme specificity in the pathway. In summary, both the initial and later enzymes of the core glucosinolate pathway have broad specificities for substrates derived from a variety of amino acids.

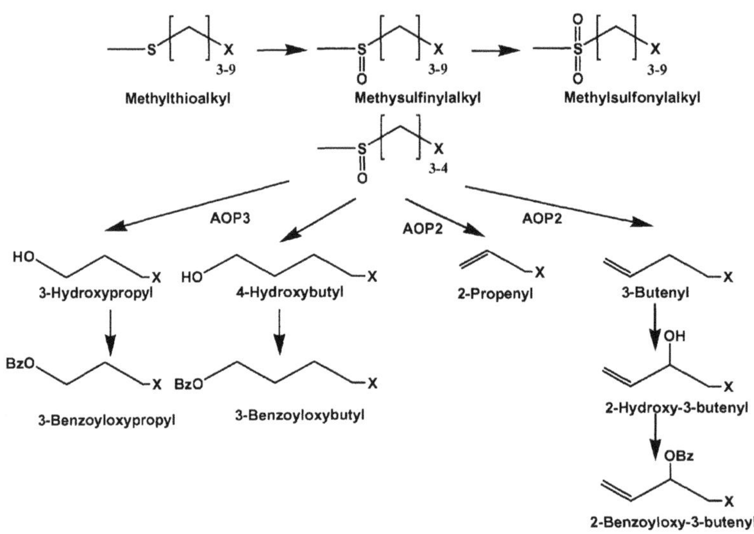

Figure 2.6: R Group Structures and Enzymes Involved in the Formation of Common Modified Glucosinolates of *Arabidopsis*.

FURTHER OXIDATIVE MODIFICATIONS

The formation of aliphatic glucosinolates does not end with the sulfation step. The various substituents of the glucosinolate molecule, especially the R group, can be modified further as illustrated in Figure 2.6. Based on the different glucosinolate profiles in various tissues of *A. thaliana*, these modifications occur in organ- and developmental-specific patterns.[43,44] The R group of the chain-elongated methionine-derived glucosinolates has a terminal methylthio group whose sulfur atom can be sequentially oxidized to a methylsulfinyl and then a methylsulfonyl group. In *A. thaliana*, methylsulfinylalkyl glucosinolates are common, while methylsulfonylalkyl glucosinolates have not been detected. The enzymology of this sulfur oxidation is currently unknown, and the process could even occur spontaneously in an appropriate redox environment.

In the seeds of the Columbia accession, there is a high proportion of methylthioalkyl glucosinolates with respect to methylsulfinylalkyl glucosinolates,[43,44] in contrast to the situation in the vegetative parts. This suggests that of the glucosinolates imported into the seeds from the rest of the plant,[44] the methylsulfinylalkylglucosinolates would need to be reduced *in situ*. In fact, there is precedence for the reduction of sulfinyl groups arising from the oxidation of the thiol groups of methionine residues in proteins by a specific methionine reductase.[45] Since the ratio of the reduced to oxidized forms in the seed is similar for all methionine-derived glucosinolates,[43,44] the reduction process must have low substrate specificity.

2-Oxoglutarate-Dependent Dioxygenases

Another major set of modifications involves the cleavage of the terminal methylsulfinyl group and its replacement either by a terminal hydroxyl group or a terminal double bond on the remaining side chain. These reactions are catalyzed by 2-oxoglutarate-dependent dioxygenases and in *A. thaliana* are confined to the short-chained (propyl- and/or butyl-) glucosinolates. Mithen and coworkers investigated the genetics of these modifications and their variation in *B. napus*.[26] In a series of communications,[28,46,47] they mapped the locus responsible for these oxidations and proposed a biochemical pathway. More recently they mapped the activity forming the terminal hydroxyl group using an RIL population from a cross of the Columbia and L*er* accessions of *A. thaliana* segregating for methylsulfinyl and hydroxylglucosinolates. This region, a 54 kBp region on the upper arm of Chr IV designated *ALK-OHP*, contains three genes encoding members of the 2-oxoglutarate-dependent dioxygenase family. The activity forming a terminal double bond (to produce alkenyl glucosinolates) was mapped to the homologous region in *B. oleracea*.[48] This coincidence of glucosinolate modification traits at the *ALK-OHP* region was identified as well by a mapping analysis that used the same RIL population plus another one derived from a cross of the Cape Verde Island (CVI) and L*er* accessions that segregates for alkenyl- and hydroxylglucosinolates.[49] In this latter study, the candidate genes, designated *AOP1*, *AOP2*, and *AOP3*, were cloned and heterologously expressed in *E. coli*.

The genomic organization of the *ALK-OHP* locus is reminiscent of the *ELONG* locus.[49] It consists of a pseudogene and three transcribed genes (*AOP1, AOP2, and AOP3*) that have approximately 75% nucleotide identity and identical intron/exon structure. Transcript levels of these genes were measured in rosette leaves of the Columbia, CVI and the L*er* accessions, the parental lines of the two RIL populations. The *AOP1* gene was transcribed in all three accessions, but the biochemical activity of the encoded protein remains to be determined. In contrast, *AOP2* was transcribed in both Columbia and CVI, but the transcript sequences indicated that only the CVI transcript could produce a complete protein. The *AOP3* transcript was present only in L*er*.[49]

The gene family coding for 2-oxoglutarate-dependent dioxygenases encompasses about 100 members in *Arabidopsis*.[50] The described activities of this group include a variety of oxidations in gibberellin, flavonoid, and alkaloid biosyntheses. In contrast to the other enzymes of glucosinolate biosynthesis, the AOP proteins appear to have narrower substrate specificities. For example, *AOP3* transcripts, isolated from accessions with hydroxylated glucosinolates, have been expressed heterologously in *E. coli*. These catalyze the cleavage of the methylsulfinyl group and the hydroxylation of the new terminal carbon atom. 3-Methylsulfinylpropylglucosinolate is accommodated as a substrate, but the butyl and longer chain-elongated homologs are not.[49] The glucosinolate profile of the *A. thaliana* tissues containing *AOP3* transcripts is consistent with this single functionality; only 3-hydroxypropylglucosinolate is detected even though 4-methylsulfinylbutylglucosinolate, a possible substrate for the formation of 4-hydroxybutylglucosinolate, is present. The presence of 4-hydroxybutyl-glucosinolate in the seeds of *A. thaliana* is considered to be the product of a different enzyme activity in these tissues.[49] The alkenyl-forming reaction has a slightly broader substrate specificity; since heterologous expression of *AOP2* results in enzyme activity forming 2-propenyl- and 3-butenylglucosinolates from 3-methylsulfinyl- and 4-methylsulfinyl precursors, respectively.

Other Modifications

The biochemical and genetic information about other glucosinolate modifications is limited. The hydroxylation resulting in the formation of 2-hydroxy-3-butenylglucosinolate is prominent in 10 of approximately 40 *A. thaliana* accessions analyzed,[29] and has been approximately mapped in this species.[46] The esterification of hydroxyalkyl glucosinolates by benzoic acid occurs in all *A. thaliana* accessions analyzed to date. Benzoyloxy-glucosinolates are found principally in the seeds, where they can represent 20% or more of the total glucosinolate content,[29] but they also persist in very young seedlings.[43] The incorporation of isotopically-labeled precursors in developing seeds shows that the benzoate function is derived from phenylalanine via benzoic acid, while the aliphatic portion is exclusively from chain-elongated methionine derivatives, usually 3-hydroxypropyl- and 4-hydroxybutylglucosinolates.[51]

SUMMARY AND FUTURE DIRECTIONS

Aliphatic glucosinolates derived from methionine are the major class of glucosinolates in *A. thaliana* and many other species of the Brassicaceae. As we have shown in the present survey, the structural diversity of this group can be attributed to three significant features of the biosynthetic pathway. The first feature is the evolution of an iterative cycle of methylene additions to methionine resulting in glucosinolates with side chains possessing anywhere from 1-9 additional methylene groups. The second feature is the recruitment of oxidizing enzymes to glucosinolate biosynthesis from two large enzyme families, the cytochrome P450 mixed function oxygenases and the 2-oxoglutarate-dependent dioxygenases. Representatives of these families are capable of catalyzing a large variety of oxidative processes on a diversity of substrates. The third feature is the broad specificity of the various enzymes of the core biosynthetic pathway. The last three steps of this sequence appear to be catalyzed by individual enzymes that can each accommodate all aliphatic, aromatic, and indole glucosinolate precursors. Even the first two steps, catalyzed by members of the cytochrome P450 superfamily, have broad specificity for the R group. The two CYP79F enzymes of the first step together use all six methionine derivatives of different chain lengths as substrates. For the second step, one enzyme, CYP83A1, appears to accommodate the metabolism of all aliphatic aldoximes while CYP83B1 is responsible for indole and aromatic aldoximes.

These same features appear to be responsible for creating diversity in other secondary metabolic pathways. For example, in both polyketide and terpene formation, repetitive addition of either C_2 or C_5 carbon subunits leads to the formation of a variety of carbon skeletons. In addition, in nearly all groups of secondary metabolites, including alkaloids, phenylpropanoids, and terpenes, the initially-formed products are subjected to a wide variety of oxidative modifications. Thus, despite the seemingly large and chaotic assemblage of secondary metabolites found in plants, their formation may be governed by a few common principles.

Further research on aliphatic glucosinolate biosynthesis will provide even more information on the molecular and biochemical bases of diversity in this class of compounds. For example, it is now clear that some glucosinolate-containing species possess more than two genes of the *MAM* family.[31] Knowledge of the expression profiles of these genes and the catalytic abilities of their encoded proteins should broaden our picture on how glucosinolate chain length is controlled. Furthermore, while several important enzymes of aliphatic glucosinolate biosynthesis remain uncharacterized, there are also gaps in our knowledge of the pathways leading to other classes of glucosinolates. In addition, as knowledge of genes of other metabolic pathways accumulates, the evolutionary origins of individual glucosinolate pathway genes should become more apparent, allowing the links between glucosinolate and primary metabolism to be explored in more depth. As discussed in this chapter, the core enzymes of indole glucosinolate biosynthesis, CYP79B2 and CYP79B3, also participate in auxin biosynthesis.[52] The *ref2* mutant with a lesion in *CYP83A1,* a gene of the core pathway, has a pleiotropic phenotype exhibiting reduced levels of aliphatic glucosinolates and sinapate esters derived from the phenylpropanoid pathway.[39] Another example demonstrating links between glucosinolate biosynthesis and other plant functions involves the CYP79F1 gene for which mutant and transgenic plant lines with reduced transcript levels show not only reduced levels of aliphatic glucosinolates, but also reduced fertility and reduced apical dominance.[53-55] Indeed, as our general knowledge of plant metabolism improves, the boundaries between primary and secondary metabolism are becoming more and more blurred. The entire concept of secondary metabolism as presently understood is likely to undergo profound changes in light of future molecular and functional studies on glucosinolates and other plant metabolites.

ACKNOWLEDGEMENTS

The research was supported by the Deutsche Forschungsgemeinschaft (grant FOR383) and the Max Planck Gesellschaft.

REFERENCES

1. FAHEY, J.W., ZALCMANN, A.T., TALALAY, P., The chemical diversity and distribution of glucosinolates and isothiocyanates among plants., *Phytochemistry*, 2001, **56**, 5-51.
2. MITHEN, R.F., DEKKER, M., VERKERK, R., RABOT, S., The nutritional significance, biosynthesis and bioavailability of glucosinolates in human foods., *J Sci. Food Agric.*, 2000, **80**, 967-984.
3. BUSCH, L., GUNTER, V., MENTELE, T., TACHIKAWA, M., TANAKA, K., Socializing nature: technoscience and the transformation of rapeseed into canola., *Crop Sci.*, 1994, **34** , 607-614.
4. GARDINER, J.B., MORRA, M.J., EBERLEIN, C.V., BROWN, P.D., BOREK, V., Allelochemicals released in soil following incorporation of rapeseed (*Brassica napus*) green manures., *J. Agric. Food Chem.*, 1999, **47**, 3837-3842.
5. YU, E.Y., PICKERING, I.J., GEORGE, G.N., PRINCE, R.C., *In situ* observation of the generation of isothiocyanates from sinigrin in horseradish and wasabi., *Biochim. Biophys. Acta.*, 2001, **1527**, 156-160.
6. FAHEY, J.W., ZHANG, Y., TALALAY, P., Broccoli sprouts: An exceptionally rich source of inducers of enzymes that protect against chemical carcinogens., *Proc. Natl. Acad. Sci. USA*, 1997, **94**, 10367-10372.

7. REICHELT, M., BROWN, P.D., SCHNEIDER, B., OLDHAM, N.J., STAUBER, E., TOKUHISA, J., KLIEBENSTEIN, D.J., MITCHELL-OLDS, T., GERSHENZON, J., Benzoic acid glucosinolate esters and other glucosinolates from *Arabidopsis thaliana*., *Phytochemistry*, 2002, **59**, 663-671.
8. WITTSTOCK, U., KLIEBENSTEIN, D.J., LAMBRIX, V., REICHELT, M., GERSHENZON, J., Glucosinolate hydrolysis and its impact on generalist and specialist insect herbivores, *in*: Phytochemistry As Integrative Biology: From Ethnobotany to Molecular Ecology (J.T. Romeo, ed.) Elsevier Science, Amsterdam. 2003, pp. 101-125
9. CHEW, F.S., Biological effects of glucosinolates, *in*: Biologically Active Natural Products for Potential Use in Agriculture (H.G. Cutler, ed.), American Chemical Society, Washington, D.C. 1988, pp. 155-181
10. LARSEN, P.O., Glucosinolates, *in*: Secondary Plant Products (E.E. Conn, ed.), Academic Press, New York. 1981, pp. 501-525
11. HOGGE, L.R., REED, D.W., UNDERHILL, E.W., HAUGHN, G.W., HPLC separation of glucosinolates from leaves and seeds of *Arabidopsis thaliana* and their identification using thermospray liquid chromatography-mass spectrometry., *J. Chromatog. Sci.*, 1988, **26**, 551-556.
12. BAK, S., OLSEN, C.E., PETERSEN, B.L., MOLLER, B.L., HALKIER, B.A., Metabolic engineering of ρ-hydroxybenzylglucosinolate in *Arabidopsis* by expression of the cyanogenic CYP79A1 from *Sorghum bicolor*,170., *Plant J.*, 1999, **20**, 663-671.
13. HAUGHN, G.W., DAVIN, L., GIBLIN, M., UNDERHILL, E.W., Biochemical genetics of plant secondary metabolites in *Arabidopsis thaliana*. The glucosinolates., *Plant Physiol.*, 1991, **97**, 217-226.
14. MAGRATH, R., BANO, F., MORGNER, M., PARKIN, I., SHARPE, A., LISTER, C., DEAN, C., TURNER, J., LYDIATE, D., MITHEN, R., Genetics of aliphatic glucosinolates: I. Side chain elongation in *Brassica napus* and *Arabidopsis thaliana*., *Heredity*, 1994, **72**, 290-299.
15. HALKIER, B.A., Glucosinolates, *in*: Naturally Occurring Glycosides: : Chemistry, Distribution and Biological Properties (R. Ikan, ed.), John Wiley, New York. 1999, pp. 193-223
16. PICHERSKY, E., GANG, D.R., Genetics and biochemistry of secondary metabolites in plants: an evolutionary perspective., *Trends Plant Sci.*, 2000, **5**, 439-445.
17. UNDERHILL, E.W., CHISHOLM, M.D., WETTER, L.R., Biosynthesis of mustard oil glucosides 1. Administration of C14-labelled compounds to horseradish, Nasturtium, and watercress., *Can. J. Biochem. Physiol.*, 1962, **40**, 1505-1514.
18. CHISHOLM, M.D., WETTER, L.R., Biosynthesis of mustard oil glucosides 4. Administration of methionine-C14 + related compounds to horseradish., *Can. J. Biochem. Physiol.*, 1964, **42**, 1033-1040.
19. MATSUO, M., YAMAZAKI, M., Biosynthesis of sinigrin., *Chem. Pharm. Bull. Tokyo*, 1964, **12**, 1388-1389.
20. GRASER, G., SCHNEIDER, B., OLDHAM, N.J., GERSHENZON, J., The methionine chain elongation pathway in the biosynthesis of glucosinolates in *Eruca sativa* (Brassicaceae)., *Arch. Biochem. Biophys.*, 2000, **378**, 411-419.

21. CHAPPLE, C.C.S., GLOVER, J.R., ELLIS, B.E., Purification and characterization of methionine-glyoxylate aminotransferase from *Brassica carinata* and *Brassica napus.*, *Plant Physiol.,* 1990, **94**, 1887-1896.
22. CHAPPLE, C.C.S., DECICCO, C., ELLIS, B.E., Biosynthesis of 2-(2'-methylthio)ethylmalate in *Brassica carinata.*, *Phytochemistry*, 1988, **27**, 3461-3463.
23. TEXTOR S., BARTRAM, S., KROYMANN, J., FALK, K.L., HICK, A., PICKETT, J.A., GERSHENZON, J., Biosynthesis of methionine-derived glucosinolates in *Arabidopsis thaliana* : Recombinant expression and characterization of methylthioalkylmalate synthase, the condensing enzyme of the chain elongation cycle., *Planta*, 2003, in press.
24. 24. FALK K.L., VOGEL, C., TEXTOR, S., BARTRAM, S., HICK, A., PICKETT, J.A., GERSHENZON, J., Glucosinolate biosynthesis: Demonstration and characterization of the condensing enzymethe chain elongation cycle in *Eruca sativa.*, *Phytochemistry*, 2004, in press.
25. JOSEFSSON, E., JONSSON, R., Studies of variation in glucosinolate content of seed of cruciferae plants, especially in material with a high erucic acid content., *Z. Pflanzenzücht.*, 1969, **62**, 272.
26. MAGRATH, R., HERRON, C., GIAMOUSTARIS, A., MITHEN, R., The inheritance of aliphatic glucosinolates in *Brassica napus.*, *Plant Breed.*, 1993, **111**, 55-72.
27. CAMPOS DE QUIROS, H., MAGRATH, R., MCCALLUM, D., KROYMANN, J., SCHNABELRAUCH, D., MITCHELL-OLDS, T., MITHEN, R., α-Keto acid elongation and glucosinolate biosynthesis in *Arabidopsis thaliana.*, *Theor. Appl. Genet.*, 2000, **101**, 429-437.
28. CAMPOS, H., MITHEN, R., Genetic variation of aliphatic glucosinolates in *Arabidopsis thaliana* and prospects for map-based gene cloning., *Entomol. Exp. Appl.,* 1996, **80**, 202-205.
29. KLIEBENSTEIN, D.J., KROYMANN, J., BROWN, P., FIGUTH, A., PEDERSEN, D., GERSHENZON, J., MITCHELL-OLDS, T., Genetic control of natural variation in *Arabidopsis* glucosinolate accumulation., *Plant Physiol.,* 2001, **126**, 811-825.
30. KROYMANN, J., TEXTOR, S., TOKUHISA, J.G., FALK, K.L., BARTRAM, S., GERSHENZON, J., MITCHELL-OLDS, T., A gene controlling variation in *Arabidopsis* glucosinolate composition is part of the methionine chain elongation pathway., *Plant Physiol.,* 2001, **127**, 1077-1088.
31. KROYMANN, J., DONNERHACKE, S., SCHNABELRAUCH, D., MITCHELL-OLDS, T., Evolutionary dynamics of an *Arabidopsis* insect resistance quantitative trait locus., *Proc. Natl. Acad. Sci. USA*, 2003, **100**, 14587-14592.
32. BOLWELL, G.P., BOZAK, K., ZIMMERLIN, A., Plant Cytochrome P450., *Phytochemistry*, 1994, **37**, 1491-1506.
33. HALKIER, B.A., Catalytic reactivities and structure/function relationships of cytochrome P450 enzymes., *Phytochemistry*, 1996, **43**, 1-21.
34. WITTSTOCK, U., HALKIER, B.A., Glucosinolate research in the *Arabidopsis* era., *Trends Plant Sci.*, 2002, **7**, 263-270.
35. CHEN, S., ANDREASSON, E., Update on glucosinolate metabolism and transport., *Plant Physiol. Biochem.,* 2001, **39**, 743-758.

36. CHEN, S., GLAWISCHNIG, E., JØRGENSEN, K., NAUR, P., JØRGENSEN, B., OLSEN, C.E., HANSEN, C.H., RASMUSSEN, H., PICKETT, J.A., HALKIER, B.A., CYP79F1and CYP79F2 have distinct functions in the biosynthesis of aliphatic glucosinolates in *Arabidopsis.*, *Plant J.,* 2003, **33**, 923-937.
37. BAK, S., FEYEREISEN, R., The involvement of two P450 enzymes, CYP83B1 and CYP83A1, in auxin homeostasis and glucosinolate biosynthesis., *Plant Physiol.,* 2001, **127**, 108-118.
38. NAUR, P., PETERSEN, B.L., MIKKELSEN, M.D., BAK, S., RASMUSSEN, H., OLSEN, C.E., HALKIER, B.A., CYP83A1 and CYP83B1, two nonredundant cytochrome P450 enzymes metabolizing oximes in the biosynthesis of glucosinolates in *Arabidopsis.*, *Plant Physiol.,* 2003, **133**, 63-72.
39. HEMM, M.R., RUEGGER, M.O., CHAPPLE, C., The *Arabidopsis ref2* mutant is defective in the gene encoding CYP83A1 and shows both phenylpropanoid and glucosinolate phenotypes., *Plant Cell,* 2003, **15**, 179-194.
40. POULTON, J.E., MØLLER, B.L., Glucosinolates, *in*: Enzymes of Secondary Metabolism (P.J. Lea, ed.), Academic Press, New York. 1993, pp. 209-237
41. GROOTWASSINK, J.W.D., BALSEVICH, J.J., KOLENOVSKY, A.D., Formation of sulfatoglucosides from exogenous aldoximes in plant cell cultures and organs., *Plant Sci.,* 1990, **66**, 11-20.
42. KIDDLE, G.A., BENNETT, R.N., HICK, A.J., WALLSGROVE, R.M., C-S lyase activities in leaves of crucifers and non-crucifers, and the characterization of three classes of C-S lyase activities from oilseed rape (*Brassica napus* L.)., *Plant Cell Environ.*, 1999, **22**, 433-445.
43. BROWN, P.D., TOKUHISA, J.G., REICHELT, M., GERSHENZON, J., Variation of glucosinolate accumulation among different organs and developmental stages of *Arabidopsis thaliana.*, *Phytochemistry*, 2003, **62**, 471-481.
44. PETERSEN, B.L., CHEN, S.X., HANSEN, C.H., OLSEN, C.E., HALKIER, B.A., Composition and content of glucosinolates in developing *Arabidopsis thaliana.*, *Planta*, 2002, **214**, 562-571.
45. GUSTAVSSON, N., KOKKE, B.P., HARNDAHL, U., SILOW, M., BECHTOLD, U., POGHOSYAN, Z., MURPHY, D., BOELENS, W.C., SUNDBY, C., A peptide methionine sulfoxide reductase highly expressed in photosynthetic tissue in *Arabidopsis thaliana* can protect the chaperone-like activity of a chloroplast-localized small heat shock protein., *Plant J.,* 2002, **29**, 545-553.
46. MITHEN, R., CLARKE, J., LISTER, C., DEAN, C., Genetics of aliphatic glucosinolates. III. Side chain structure of aliphatic glucosinolates in *Arabidopsis* thaliana., *Heredity*, 1995, **74**, 210-215.
47. GIAMOUSTARIS, A., MITHEN, R., Genetics of aliphatic glucosinolates 4. Side-chain modification in *Brassica oleracea.*, *Theor. Appl. Genet.,* 1996, **93**, 1006-1010.
48. HALL, C., MCCALLUM, D., PRESCOTT, A., MITHEN, R., Biochemical genetics of glucosinolate modification in *Arabidopsis* and *Brassica.*, *Theor. Appl. Genet.,* 2001, **102**, 369-374.
49. KLIEBENSTEIN, D.J., LAMBRIX, V.M., REICHELT, M., GERSHENZON, J., MITCHELL-OLDS, T., Gene duplication in the diversification of secondary

metabolism: Tandem 2-oxoglutarate-dependent dioxygenases control glucosinolate biosynthesis in *Arabidopsis.*, *Plant Cell*, 2001, **13**, 681-693.

50. PRESCOTT, A., Two-oxoacid-dependent dioxygenases: Inefficient enzymes or evolutionary driving force?, *in*: Evolution of Metablic Pathways (J.T. Romeo, R. Ibrahim, L. Varin, V. De Luca, eds.), Pergamon, New York. 2001, pp. 249-284
51. GRASER, G., OLDHAM, N.J., BROWN, P.D., TEMP, U., GERSHENZON, J., The biosynthesis of benzoic acid glucosinolate esters in *Arabidopsis thaliana.*, *Phytochemistry*, 2001, **57**, 23-32.
52. ZHAO, Y.D., HULL, A.K., GUPTA, N.R., GOSS, K.A., ALONSO, J., ECKER, J.R., NORMANLY, J., CHORY, J., CELENZA, J.L., Trp-dependent auxin biosynthesis in *Arabidopsis*: Involvement of cytochrome P450s CYP79B2 and CYP79B3., *Genes Dev.,* 2002, **16**, 3100-3112.
53. HANSEN, C.H., WITTSTOCK, U., OLSEN, C.E., HICK, A.J., PICKETT, J.A., HALKIER, B.A., Cytochrome P450 CYP79F1 from *Arabidopsis* catalyzes the conversion of dihomomethionine and trihomomethionine to the corresponding aldoximes in the biosynthesis of aliphatic glucosinolates., *J. Biol. Chem.,* 2001, **276**, 11078-11085.
54. REINTANZ, B., LEHNEN, M., REICHELT, M., GERSHENZON, J., KOWALCZYK, M., SANDBERG, G., GODDE, M., UHL, R., PALME, K., *bus*, a bushy *Arabidopsis CYP79F1* knockout mutant with abolished synthesis of short-chain aliphatic glucosinolates., *Plant Cell*, 2001, **13**, 351-367.
55. TANTIKANJANA, T., YONG, J.W.H., LETHAM, D.S., GRIFFITH, M., HUSSAIN, M., LJUNG, K., SANDBERG, G., SUNDARESAN, V., Control of axillary bud initiation and shoot architecture in *Arabidopsis* through the *SUPERSHOOT* gene., *Genes Dev.,* 2001, **15**, 1577-1588.

Chapter Three

THE PHENYLPROPANOID PATHWAY IN ARABIDOPSIS: LESSONS LEARNED FROM MUTANTS IN SINAPATE ESTER BIOSYNTHESIS

Jake Stout and Clint Chapple*

Department of Biochemistry
Purdue University
West Lafayette, IN 47907, USA

**Author for correspondence:* chapple@purdue.edu

INTRODUCTION

Over the past three decades, phytochemistry has been progressing from the identification of individual compounds to the elucidation of the structural and regulatory elements of metabolic networks. Although *Arabidopsis* accumulates only a subset of the natural products known in the plant kingdom, it produces a range of secondary metabolites representative of several structural classes, including glucosinolates, indole phytoalexins, and terpenoids, as well as phenylpropanoids including flavonoids, sinapate esters, and lignin.[1,2] The structural and regulatory elements of the pathways responsible for the production of these metabolites are rapidly being elucidated using the genetic and genomic tools available to *Arabidopsis* researchers. The knowledge gained from these studies will not only further our understanding of these pathways in *Arabidopsis* and other species, but will also facilitate research on the catalysts and regulatory factors involved in the synthesis of compounds not found in *Arabidopsis*. The phenylpropanoid pathway has been particularly amenable for study in *Arabidopsis* due to the accumulation of readily observable end-products produced from different branches. The goal of this review is to outline the analysis of mutants impaired in the accumulation of one class of these end-products, the sinapate esters. These mutants have improved our understanding of the enzymes and metabolites involved in the phenylpropanoid pathway, have demonstrated interactions between pathways of secondary metabolism, and have provided a glimpse into their evolution.

PHYSIOLOGICAL ROLES OF PHENYLPROPANOIDS

The phenylpropanoid pathway (Fig. 3.1) is responsible for the production of many natural products that are of interest in the context of plant growth and development, human health, and ecology. For example, flavonoids are necessary for pollen viability in maize and petunia,[3-5] and have been suggested to play a role in directed auxin transport.[6,7] Flavonoids and sinapate esters have been found to be important UV-protectants in many species, including *Arabidopsis*.[8, 9] Furthermore, wall-bound phenolics are thought to impart control over cell wall expansion,[10, 11] and hydroxycinnamic acids are an important structural component of the hydrophobic barrier polymer suberin.[12, 13] Finally, lignin is a phenylpropanoid polymer ubiquitous in higher plants, which is necessary for mechanical support and water transport.[14]

From the perspective of human health, phenylpropanoids such as resveratrol, steryl ferulate, and isoflavones have been implicated in reducing the risk of heart disease[15-17] and certain cancers.[18-21] Recently, it has been suggested that resveratrol may also increase longevity by inducing a signal cascade normally associated with a calorie reduced diet.[22] Finally, phenylpropanoids have been found to play diverse roles in ecology. A host of compounds, including the phenylpropanoid methylbenzoate, is volatilized by the reproductive organs of various species to attract pollinating insects.[23] It has also been shown that plants produce phenylpropanoids that inhibit herbivory[24] and serve as allelopathic agents that inhibit the growth of competing plants.[25, 26] Furthermore, lignin is relevant in an ecological context as the second most abundant polymer in Nature, providing a sink for over 4 X 10^{11} kg of carbon annually.[27]

Arabidopsis ***as a Model for Understanding Phenylpropanoid Metabolism***

Arabidopsis has become the model system of choice in which to study many aspects of plant growth, development, and metabolism, including the biosynthesis of phenylpropanoid natural products. This is, in part, because *Arabidopsis* accumulates two classes of phenylpropanoid end products that are good targets for mutant screens. For example, many screens have identified mutants defective in flavonoid biosynthesis. Defects in this pathway in *Arabidopsis* lead to *transparent testa* (*tt*) and *transparent testa glabrous* (*ttg*) phenotypes that result from decreases in the condensed tannins found in the seed coat. These mutants have already been exhaustively reviewed,[28, 29] and hence will not be covered here.

Although *tt* and *ttg* mutants can easily be identified because of the obvious visible phenotype associated with defects in flavonoid biosynthesis, the branch of the phenylpropanoid pathway leading to lignin precursors does not lead to the production of colored end products. Fortunately, members of Brassicaceae including *Arabidopsis* accumulate sinapate esters that fluoresce when illuminated with ultraviolet light.[30, 31] These compounds include sinapoylmalate, which accumulates in the adaxial leaf epidermis, and sinapoylcholine, the major sinapate ester found in seeds, which serves as a reserve of choline and sinapate for the developing seedling.[32, 33] The UV-fluorescent nature of these compounds has formed the foundation of a number of mutant screens. Many have been identified following TLC analysis of methanolic tissue extracts; however, the most comprehensive screens have taken advantage of the fact that sinapoylmalate causes leaves of *Arabidopsis* to fluoresce blue-green when observed under UV-light. Mutants identified from such screens exhibit a *reduced epidermal fluorescence (ref)* phenotype.[34] In total, eight independently segregating *ref* loci and bright trichomes (*brt1*), a mutant with hyperflourescent trichomes, have been identified (Table 3.1).

All of the mutants identified in these screens accumulate less sinapoylmalate and/or sinapoylcholine than the wild type.[34] In addition, because the sinapic acid moiety of sinapate esters is derived from the same pathway that generates lignin monomers, some mutants also exhibit alterations in lignin quality and quantity. Furthermore, the *ref3*, *ref4*, and *ref8* mutants exhibit aberrations in morphology, indicating that alterations in phenylpropanoid synthesis can have unexpected effects on plant growth and development. The characterizations of these mutants have been instrumental in unraveling the complexity of the phenylpropanoid pathway, and have afforded many surprises along the way.

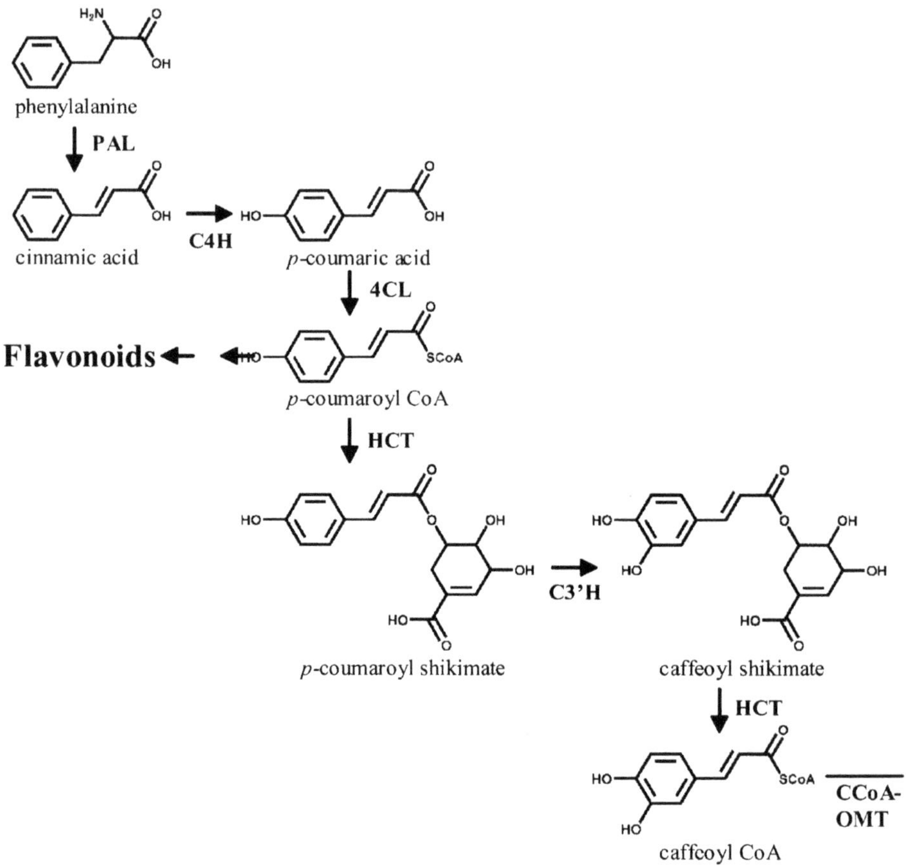

Figure 3.1: Primary flux of carbon through phenylpropanoid pathway in *Arabidopsis*. PAL, phenylalanine ammonia-lyase; 4CL, 4-(hydroxy)cinnamoyl CoA ligase; C4H, cinnamate 4-hydroxylase; HCT, hydroxycinnamoyl-CoA shikimate/quinate hydroxycinnamoyltransferase; C3'H, *p*-coumaroylshikimate 3'-hydroxylase; CCoAOMT, caffeoyl CoA *O*-methyltransferase; F5H, ferulate 5-hydroxylase; COMT, caffeic acid/5-hydroxyferulic acid o-methyltransferase; CCR, cinnamoyl CoA reductase; CAD, cinnamyl alcohol dehydrogenase. Not depicted is the HCT catalyzed synthesis of *p*-coumaroyl quinate.

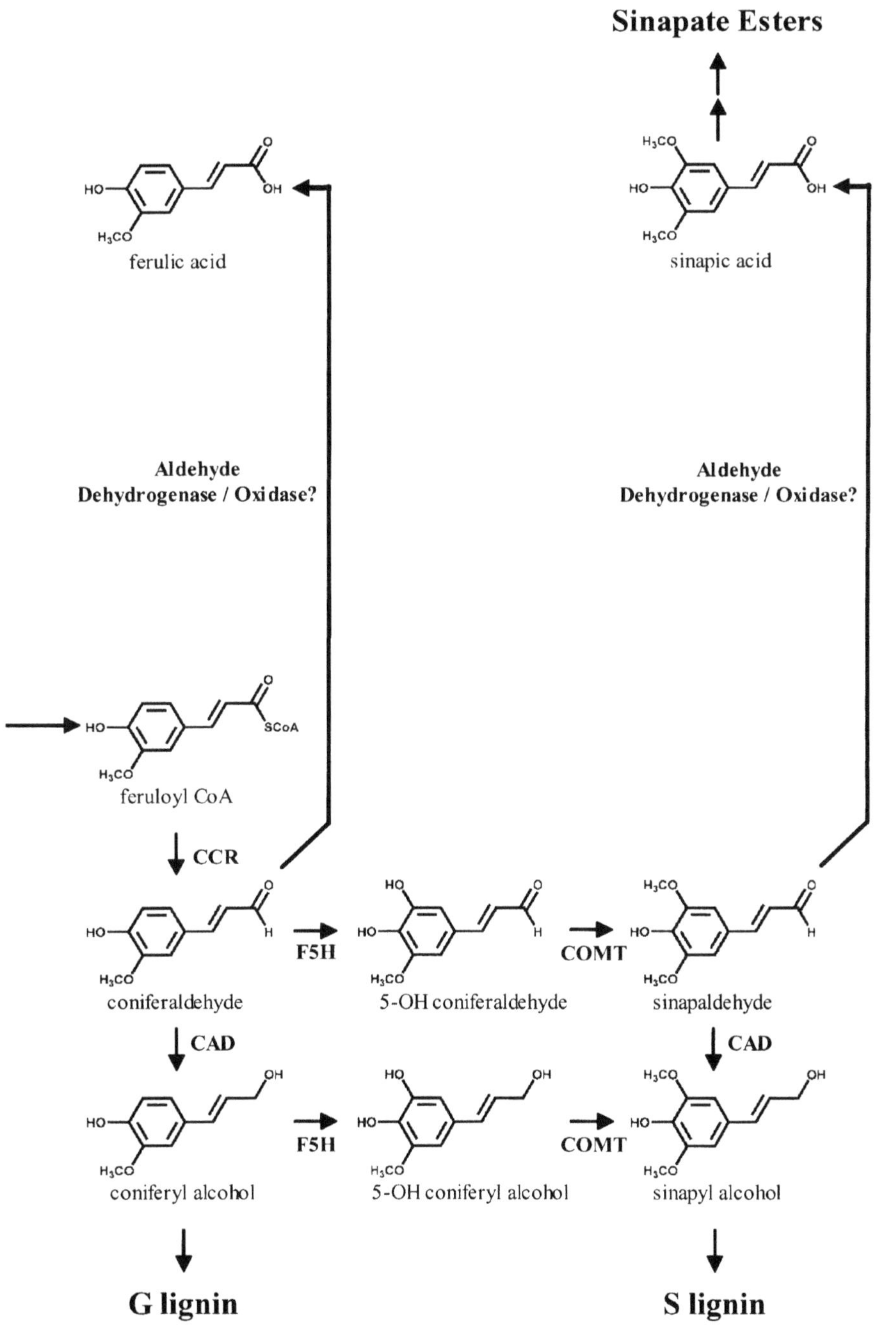

Sinapate Esters
ferulic acid
sinapic acid
Aldehyde
Dehydrogenase / Oxidase?
Aldehyde
Dehydrogenase / Oxidase?
feruloyl CoA
CCR
F5H
COMT
coniferaldehyde
5-OH coniferaldehyde
sinapaldehyde
CAD
CAD
F5H
COMT
coniferyl alcohol
5-OH coniferyl alcohol
sinapyl alcohol
G lignin
S lignin

Table 3.1: *Arabidopsis* Mutants Affected in Sinapate Ester or Lignin Biosynthesis.

			Phenotype[1]				
Mutant	**Enzyme**	**Locus**	**Growth**	**Sinapate Ester Content**	**Lignin Quantity**	**Lignin Quality**	**Other**
fah1	F5H	At4g36220	wild-type	none	wild-type	no S lignin	
ref1	unknown						
ref2	CYP83A1	At4g13770	wild-type	reduced SM	wild-type	reduced S lignin	reduced methionine-derived glucosinolates
ref3	C4H	At2g30490	dwarfed	severely reduced	severely reduced	wild-type	reduced seed tannin content
ref4	unknown		dwarfed	reduced	reduced	wild-type	reduced seed tannin content
ref5	unknown						
ref8	C3H	At2g40890	dwarfed	none	severely reduced	H lignin only	accumulates p-coumarate esters
irx4	CCR	At1g15950			reduced	variable	lignin quality dependant on growth conditions
AtOMT1	COMT	At5g54160		reduced	wild-type	deposits 5-OH G	accumulates 5-OH FM
sng1	SMT	At2g22990	wild-type	no leaf SM			
sng2	SCT	At5g09640	wild-type	no seed SC			
brt1	unknown		wild-type	reduced SM	wild-type	wild-type	hyperfluorescent trichomes

1. Phenotype of the most severe allele described.

MUTANTS AFFECTING MONOLIGNOL BIOSYNTHESIS

Lignin Biosynthesis and Deposition

Although the phenylpropanoid pathway produces many compounds of interest, a major goal of research on the pathway has been to improve our understanding of lignin biosynthesis. The extraction of lignin during the pulping process is both costly and damaging to the environment.[35] Hence, the production of plants with more readily extractable lignin would be beneficial for both economic gain and for long-term environmental sustainability.[36] Furthermore, the quantity and quality of lignin in forage species has been found to impact negatively their digestibility in ruminant animals,[37-39] thus, the application of similar strategies to crops used as animal feedstocks would be expected to lead to comparable gains.

Ubiquitous in higher plants, lignin imparts structural support to the stem, contributes to the hydrophobicity of vascular elements, and provides reinforcement to the xylem, thus preventing cavitation during water transport.[40-42] The lignin heteropolymer is produced via the oxidative coupling of *p*-coumaryl alcohol, coniferyl alcohol, and sinapyl alcohol subunits (collectively termed monolignols) by both peroxidases and laccases *in muro*.[43] The polymerization of these subunits leads to the formation of *p*-hydroxyphenyl (H), guaiacyl (G), and syringyl (S) lignin, respectively. The degree to which G and S lignin is deposited (commonly denoted as the S:G ratio) varies widely among species, tissue types, and even within an individual cell wall.[44, 45] For example, in the rachis (stem) of *Arabidopsis*, guaiacyl lignin is deposited in the cell walls of the vascular bundles; whereas, syringyl lignin is deposited at high levels in the adjacent sclerified parenchyma.[31] This cell type specificity indicates that there exists in plants a high degree of control in monolignol biosynthesis. The analysis of the first sinapate-ester deficient mutant of *Arabidopsis* helped to elucidate the mechanism by which this specificity is regulated.

fah1

The *ferulic acid hydroxylase-1* (*fah1*) mutant was isolated by using thin layer chromatography to screen an ethyl methanesulfonate-mutagenized population of seedlings for individuals that lacked sinapoylmalate.[31] Characterization of the *fah1* mutants demonstrated that, in addition to severe reductions in sinapoylmalate content in leaf tissues, sinapoylcholine was below detectable limits in seeds. Furthermore, nitrobenzene oxidation of rachis tissue showed that the *fah1* mutant does not deposit S lignin. In conjunction with radiotracer feeding studies, these data suggested that the *fah1* mutants were compromised in a step common to both sinapic acid and syringyl lignin biosynthesis.

Following the laborious TLC screen, it was found that the *fah1* mutant had the obvious reduction in epidermal fluorescence under long wave UV light that later served as the basis for the *ref* mutant screen described above.[34] This *fah1* mutant phenotype was then used to isolate the *fah1-9* allele from a T-DNA mutagenized population.[46] Using this allele, the gene corresponding to the *FAH1* locus was cloned and found to encode a cytochrome P450-dependent monooxygenase (P450) sufficiently divergent from previously known plant P450s to qualify as the first new member of a new subfamily, designated CYP84.

The cloning of the *F5H* gene led to a number of investigations into the role of F5H in both lignin and sinapate ester synthesis. The observation that syringyl lignin deposition is blocked in the *fah1* mutant led to the hypothesis that the tissue specificity of F5H expression is the key determinant of syringyl lignin deposition patterns. To test this hypothesis, *fah1* plants were transformed with constructs in which *F5H* expression was driven by the CaMV 35S promoter (35S-F5H).[47] In these transformants, deposition of syringyl lignin was observed in the vascular bundle cell walls in addition to the adjacent sclerified parenchyma, demonstrating that syringyl lignin accumulation is regulated at the level of F5H expression.

Interestingly, the lignin of lines carrying the 35S-F5H construct was still dominated by guaiacyl subunits.[47] Although the CaMV 35S promoter generally leads to strong, constitutive expression, the limited efficacy of the 35S-F5H construct was not inconsistent with previous reports that the promoter leads to only weak transgene expression in certain tissues and/or cell types. Thus, a lignification-specific promoter might be required to ensure the conversion of a higher percentage of guaiacyl subunits to syringyl monomers.

Previous experiments had shown that the cinnamate 4-hydroxylase (C4H) promoter conferred high expression of a GUS reporter gene in lignifying tissues.[48] Furthermore, it had also been shown that transcription of *C4H* is evident in tissues at the earliest stages of lignification.[47] These data suggested that the *C4H* promoter would be an appropriate choice for subsequent experiments. Thus, a chimeric C4H-F5H transgene was generated and introduced into *fah1* plants in order to test whether targeted overexpression of *F5H* could substantially increase the lignin S:G ratio. Surprisingly, plants carrying this construct were found to deposit lignin with an S monomer content much higher than the 35S-F5H transgenics. Indeed, the lignin syringyl monomer content of some of the plants exceeded 95%. NMR analysis confirmed that lignin within these transformants mostly contained linkages associated with S lignin.[49] These data further supported the critical role of *F5H* expression in the regulation of lignin monomer content in *Arabidopsis*, and also demonstrated the plasticity of lignin monomer composition, and the feasibility of generating S-rich lignins that may be of utility in agriculture and forestry.

Although these experiments showed that F5H is a critical player in syringyl lignin deposition, it was found that ectopic *F5H* expression is not sufficient for the accumulation of other sinapate derived metabolites in *Arabidopsis*.[50] As previously discussed, wild-type plants accumulate sinapoylmalate in the adaxial epidermis. Overexpression of F5H with both the 35S-F5H and C4H-F5H transgenes did not lead to the accumulation of sinapoylmalate in other leaf cell types, nor did it lead to increases in overall sinapoylmalate content. Furthermore, these transgenic plants did not over-accumulate sinapoylcholine in developing embryos. These data indicate that, unlike the deposition of syringyl lignin, the biosynthesis of sinapate esters is not regulated by the transcription of *F5H*.

The phenylpropanoid pathway has undergone numerous revisions as new data concerning its intermediates and catalysts have emerged.[51] The "classic" model of the lignin biosynthetic pathway postulated a series of ring hydroxylation and *O*-methylation reactions that occurred at the level of the free acids. Ferulic acid and sinapic acid were then thought to be reduced to their corresponding alcohols and polymerized.

An alternate pathway to guaiacyl lignin was later proposed following the characterization of *S*-adenosyl-L-methionine:*trans*-caffeoyl-coenzyme A 3-*O*-

methyltranserase (CCoAOMT) activity in parsley and carrot cell cultures,[52-54] and in lignifying stem tissue.[55-57] The presence of this shunt in *Arabidopsis* made it difficult to reconcile the finding that overexpression of F5H in *Arabidopsis* can lead to the deposition of primarily syringyl lignin. Given that the so-called "alternative pathway" provides a route to G lignin that does not include ferulic acid, and assuming this route is quantitatively important, how could overexpression of F5H redirect virtually all flux toward syringyl monomer biosynthesis? Similarly, in the "classic" model of the phenylpropanoid pathway, conjugation of the free hydroxycinnamic acids to CoA by 4-coumarate:Coenzyme A ligase (4CL) activity was thought to be required for the reduction of the phenylpropane side chain to the corresponding aldehydes and alcohols. This model conflicted with the observation that recombinant 4CL from *Arabidopsis* and other species exhibits negligible activity towards sinapic acid.[58-60] These findings cast further doubt on the pathway by which syringyl lignin is synthesized. If F5H functions in the synthesis of sinapic acid, but sinapoyl-CoA cannot be made by plants, how are sinapaldehyde and sinapyl alcohol produced? Analysis of F5H expressed in *Saccharomyces cerevisiae* resolved these apparent conflicts.[61, 62]

The only previous report of F5H activity used poplar xylem extracts to demonstrate hydroxylation of ferulic acid.[63] Surprisingly, when F5H from *Arabidopsis* was expressed in yeast and used in standard kinetic analyses, the enzyme exhibited a Km for ferulic acid of 1 mM, a value that is very high when compared to other pathway enzymes and their substrates. This finding suggested that other guaiacyl-substituted intermediates of the phenylpropanoid pathway were more likely to be the true substrates for F5H. Indeed, assays using coniferaldehyde and coniferyl alcohol demonstrated that F5H exhibited Km values for these substrates in the low micromolar range.[61, 62] Further, experiments with caffeic acid / 5-hydroxyferulic acid O-methyltransferase (COMT) showed that the corresponding 5-hydroxylated F5H products were preferred substrates for the enzyme.[61-62,64] These data strongly suggested that *in vivo*, both F5H and COMT function later in the pathway than had previously been suggested, downstream of the proposed "alternative pathway". This repositioning reconciled the proposed existence of the ferulate-independent "alternative pathway" with the efficacy of F5H overexpression. This new pathway model also explained why transgenic tobacco with reduced CCoAOMT activity exhibit a reduction in both G and S lignin.[65] Finally, F5H activity towards coniferaldehyde and coniferyl alcohol obviated the need for 4CL activity towards sinapic acid.

ref8

We now know that the three hydroxylation steps necessary for the production of sinapyl alcohol and sinapic acid are catalyzed by cytochrome P450s. Their membrane-bound nature, instability, and low abundance make plant P450s difficult to isolate and characterize via classical biochemical techniques. Despite these technical obstacles, the *C4H* gene was cloned after purification of the protein,[66-69] and as mentioned previously, *F5H* was cloned via T-DNA tagging. In contrast, the gene encoding the 3-hydroxylase of the pathway proved to be a more elusive target. Early studies had reported that *p*-coumarate 3-hydroxylase (C3H) was either an ascorbate-, NADPH-, or flavin-dependent mixed function oxidase,[70-73] a plastidic enzyme that uses plastoquinone or ferredoxin as an electron donor,[74] or a phenolase that also oxidizes dihydroxyphenols to their corresponding orthoquinones.[71] Despite these early efforts, the enzyme remained uncharacterized until it was identified as the cytochrome P450 CYP98A3 using parallel genetic and bioinformatics approaches in *Arabidopsis*.[75-77]

The *ref8* mutant was one of the first *ref* mutants studied in detail because radiotracer feeding experiments and phenotypic characterization suggested that it was blocked early in the phenylpropanoid pathway, possibly at C3H.[76] The *REF8* gene was isolated through a combination of positional cloning and candidate gene approaches.[76] Concurrently, the completed sequence of the *Arabidopsis* genome made it possible for two other groups to identify the gene encoding C3H based upon its limited similarity to C4H and the pattern of its expression.[75, 77]

The kinetic analysis of C3H necessitated further revisions to the monolignol biosynthetic pathway. Hydroxylase activity measured in a yeast expression system,[77] or from prepared yeast microsomes,[76] was found to be extremely low towards free *p*-coumaric acid and *p*-coumaraldehyde, and activity towards *p*-coumaryl alcohol was below detectable limits. The fact that the enzyme's Km for these compounds was well above reasonable physiological concentrations excluded them as potential substrates *in vivo*. Fortunately, previous reports of 3'-hydroxylase activity on *p*-coumaroyl shikimate and *p*-coumaroyl quinate in carrot[78] and parsley cell cultures[79] led to the examination of these compounds as substrates for *Arabidopsis* CYP98A3 activity.[75] These *p*-coumaroyl esters were indeed found to be excellent substrates for C3H [now more properly called p-coumaroyl shikimate/quinate 3'-hydroxylase (C3'H)], suggesting that one or both are *bona fide* intermediates in the monolignol biosynthesis. The enzyme that catalyzes both their production and the conversion of the 3'-hydroxylated caffeoyl products back to the corresponding CoA-esters, hydroxycinnamoyl-CoA: shikimate/quinate hydroxycinnamoyltransferase (HCT), has recently been cloned and characterized in tobacco.[80]

The role of CYP98A3 in phenylpropanoid biosynthesis was validated by the analysis of the *ref8* mutant.[76, 81] As a result of the lack of C3'H activity, and as would be expected based upon its *ref* phenotype, the *ref8* mutant lacks sinapoylmalate. Further, saponification of leaf extracts yielded high levels of *p*-coumaric acid, indicating that *ref8* accumulates *p*-coumaric acid esters that are not normally found in wild-type plants. The presence of these novel compounds demonstrates a degree of plasticity in hydroxycinnamic acid ester synthesis in

Arabidopsis, which may be accounted for by the broad substrate specificity of the enzymes that catalyze their formation.[82, 83]

The deposition of lignin is also altered in the *ref8* mutant.[81] Most strikingly, *ref8* deposits very little G and S lignin, and instead deposits H lignin derived from *p*-coumaryl alcohol. Although many plants deposit trace amounts of H lignin, the *ref8* mutant is the first plant described in which H monomers are the dominant subunits. Quantitatively, *ref8* accumulates only 20-40% of lignin normally found in the wild type.[81] This reduction in lignin content may result from *p*-hydroxyphenyl intermediates being poor substrates for downstream enzymes such as (hydroxy)cinnamoyl CoA reductase (CCR) and (hydroxy)cinnamyl alcohol dehydrogenase (CAD), or those involved in lignin polymerization. In this context, it is interesting to note that the *ref8* mutant is severely dwarfed, and exhibits collapsed vasculature.[81] It is currently unclear whether the vascular collapse observed in *ref8* is due to the decreased amount of deposited lignin, or because the novel lignin deposited in *ref8* is mechanically inferior to the wild-type mixed G and S lignin. Taken together, these data unequivocally demonstrated the role of CYP98A3 in phenylpropanoid metabolism, and showed once again that earlier models of the pathway were incorrect.

irx4

Alterations in either lignin (*e.g.*, *ref8*[81]) or cell wall carbohydrate polymers[84] can lead to changes in the physical properties of cell walls, which often result in similar phenotypic consequences. Such was the case with the *irregular xylem (irx)* mutants that exhibit collapsed tracheary elements.[85] These mutants were isolated by microscopic inspection of stem hand sections. The *irx1, irx2, and irx3* mutants exhibited reductions in cellulose synthesis and/or deposition. The *irx3* mutant was later characterized to encode a cellulose synthase.[86]

Unlike the other *irx* mutants, *irx4* was found to be a null allele of CCR (AtCCR1).[87] The mutant contained only 50% of the lignin found in the wild type, as measured by thioglycolic acid (TGA) assays. Furthermore, the interfasicular cell walls in the mutant were abnormally thick, and the stems were reduced in tensile strength and stiffness. Similar results were also reported following antisense inhibition of *CCR* expression in *Arabidopsis*.[88] It should be noted that there are at least ten putative CCR genes in *Arabidopsis*.[87, 89] Thus, although the phenotype of the *irx4* mutant makes it clear that AtCCR1 is one of the quantitatively most important members of the CCR gene family in *Arabidopsis*, specific CCR isoforms may be partially redundant with AtCCR1 or may perform this enzymatic function in a substrate- and/or cell-specific manner.

AtOMT1

Recently, a COMT-deficient *Arabidopsis* mutant was identified using the Versailles β-glucuronidase promoter trap T-DNA collection by screening for GUS staining in root vascular tissues.[90] The sinapoylmalate content of the *AtOMT1* mutant was approximately 50% that of wild type, and in its place, the mutant accumulates low levels of 5-hydroxyferuloylmalate and 5-hydroxyferuloylglucose, neither of which are observed in the wild type.

Consistent with the repositioning of COMT described above, GC-MS analysis of lignin thioacidolysis products revealed that the mutant deposits almost no S lignin. Rather, 5-hydroxy guaiacyl (5HG) units were observed that are not found in the wild type. 5HG units have also been observed in a poplar mutant that is deficient in COMT,[91-93] and in the maize *bm3* mutant,[94] as well as in plants down-regulated in COMT transcription,[95-97] but were not observed in the COMT-deficient sorghum *bmr3* mutant.[98] The incorporation of the 5HG units into lignin creates a novel benzodioxane linkage.,[92] although the mechanism by which it is formed is currently a matter of debate.[99, 100] A COMT gene from poplar complemented the *AtOMT1* mutant, but its over-expression did not lead to an increase in S lignin deposition.[90] These data indicate that, unlike F5H, COMT is not a major control point in S lignin biosynthesis. On the other hand, C4H-F5H plants do incorporate 5HG units into their lignin,[93] indicating that COMT does become a rate limiting step in S lignin biosynthesis when F5H is over-expressed.

ref2

Research into secondary metabolism is often focused on individual pathways analyzed in isolation. Only with the advent of metabolomic tools is it becoming possible to study the effect of perturbations in specific pathways within the context of whole plant metabolism. Recent research into the *ref2* mutant[101] highlighted the interactions that can occur in plant metabolic networks, and the need to consider metabolism as a whole, rather than as an array of isolated pathways.Other than its *ref* phenotype, the four alleles of the *ref2* mutant do not exhibit any deviations from wild-type morphology. Although lignin content was found to be at wild-type levels in *ref2* plants, S monomer content was lower in the mutant.[101] These data indicated that the *ref2* mutant is compromised in its ability to synthesize sinapyl alcohol, but not coniferyl alcohol, suggesting that either F5H or COMT activity is decreased in the mutant.

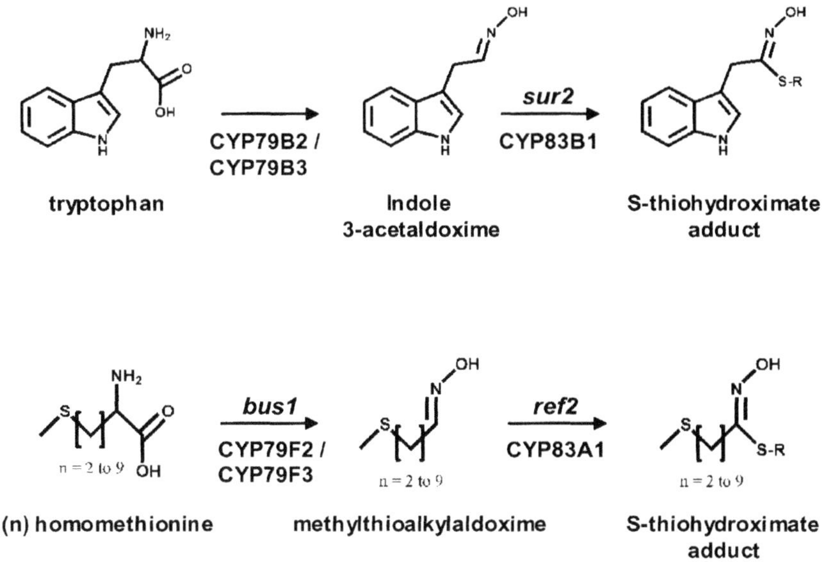

Figure 3.2: P450-mediated reactions involved in the formation of indole and methionine-derived glucosinolates in *Arabidopsis*. Enzymes catalyzing each step are indicated below the reaction arrows. *Arabidopsis* mutants blocked in the corresponding reaction are indicated above.

A combination of map based cloning, complementation analysis, and DNA sequencing revealed that the *REF2* locus encodes the cytochrome P450 CYP83A1. This in itself was unexpected, in that the genes necessary for ring hydroxylations within the phenylpropanoid pathway were already accounted for. Further, previous work with the *sur2 Arabidopsis* mutants had shown that CYP83B1, the closest homolog to CYP83A1, oxidizes indole 3-acetaldoxime during indole glucosinolate biosynthesis.[102, 103] The close homology between these P450s prompted the analysis of glucosinolate levels in the *ref2* mutant. These experiments revealed that the level of all methionine-derived glucosinolates was reduced in *ref2* mutants, suggesting that CYP83A1 oxidizes methylthioalkylaldoximes, a reaction analogous to the role of CYP83B1 in indole glucosinolate biosynthesis (Fig. 3.2). This hypothesis has since been confirmed by *in vitro* analysis of the REF2 protein.[104]

A genetic approach was first used to address how a defect in glucosinolate biosynthesis results in a decrease in sinapoylmalate and syringyl lignin accumulation. First, quantification of the sinapoylmalate levels of the *sur2-1* mutant revealed that they were decreased compared to wild type.[101] In contrast, wild-type levels of sinapoylmalate were observed in *bus1-1f*, a mutant defective in the glucosinolate biosynthetic enzyme immediately upstream of *REF2*.[101, 105] It thus appeared that the decrease in sinapoylmalate accumulation could be attributed to a block in aldoxime oxidization by either CYP83A1 or CYP83B1, rather than to a decrease in glucosinolate biosynthesis.

These observations led to the hypothesis that a defect in aldoxime oxidization could lead to the inhibition of F5H or COMT. Although genetic evidence suggested that F5H activity was unaffected in *ref2* plants, the addition of *ref2* leaf extracts to *in vitro* COMT assays led to the inhibition of enzyme activity. The addition of 3-nitrobenzaldoxime, a commercially available aldoxime, produced a similar inhibition of COMT activity.[101] These data supported the hypothesis that aldoximes play a role in the phenylpropanoid phenotypes of *ref2* and *sur2*.

This finding provides an example of a defect in one pathway having an impact on another even though the two normally function independently in wild-type plants. This suggests that the evolution of pathways may be constrained by other, apparently unrelated, areas of metabolism. For example, although extensive allelic variation exists for many glucosinolate biosynthetic loci in *Arabidopsis* ecotypes,[106] none has yet been reported for CYP83A1. Considering that sinapoylmalate affords UV protection to *Arabidopsis*, mutations in CYP83A1 may have been eliminated from natural populations due to UV-induced decreases in plant fitness.

MUTANTS AFFECTING THE FINAL STAGES OF SINAPATE ESTER SYNTHESIS

sng1 and *sng2*

The vast number of plant secondary metabolites isolated to date implies that there exists a correspondingly large number of enzymes that are required for their synthesis. The creation of large sets of sequence data from EST and genome sequencing initiatives has allowed for the comparison of gene families, an undertaking that may ultimately help to explain the evolutionary origin of the catalytic diversity observed in the plant kingdom. The analysis of the *Arabidopsis sinapoylglucose accumulator* (*sng*) mutants has led to insights into how a small portion of this diversity may have arisen.

The initial step in sinapoylglucose synthesis is the conjugation of sinapic acid and UDPG to form sinapoylglucose, a reaction catalyzed by sinapic acid: UDPG sinapoyltransferase[81] (SGT; Fig. 3.3). The 1-*O*-glucose ester bond of sinapoylglucose has a high free energy of hydrolysis, making the compound a suitable sinapate donor in subsequent transacylation reactions.[107] One such reaction occurs in leaves, where the sinapate moiety of SG is transferred to malate to form

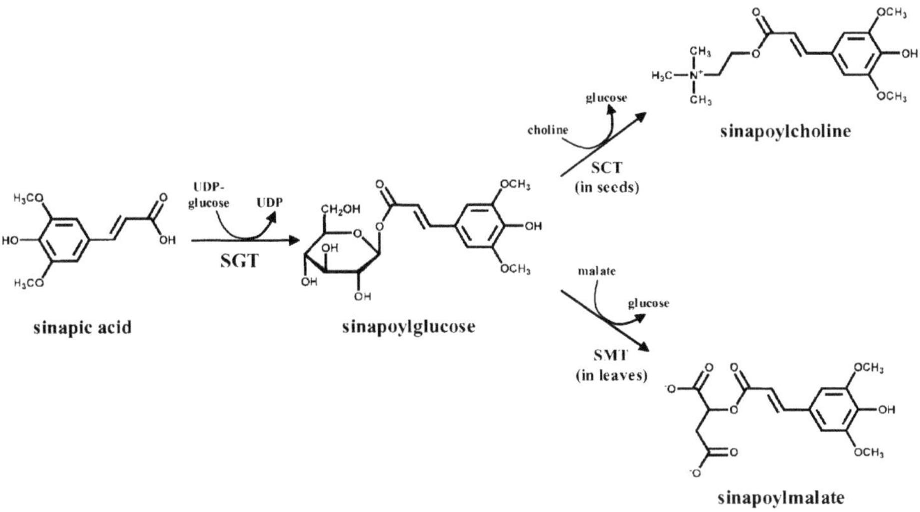

Figure 3.3: Synthesis of sinapate esters in *Arabidopsis*. SGT, sinapic acid: UDPG sinapoyltransferase; SMT, sinapoylglucose: sinapoylmalate sinapate transferase; SCT, sinapoylglucose: sinapoylcholine sinapate transferase.

sinapoylmalate in a reaction catalyzed by sinapoylglucose: sinapoylmalate sinapoyltransferase (SMT).[108] An analogous reaction catalyzed by sinapoylglucose: sinapoylcholine sinapoyltransferase (SCT) occursin seeds to produce sinapoylcholine using choline as a sinapate acceptor.[109] During germination, sinapoylcholine is hydrolyzed by sinapoylcholinesterse (SCE). The liberated choline is subsequently used for membrane lipid biosynthesis,[32] whereas the sinapic acid moiety is used for sinapoylmalate synthesis in the developing cotyledons.[30]

A TLC-based screen was used to identify the *Arabidopsis sng1* mutant. The leaves of the mutant contain sinapoylglucose in place of sinapoylmalate as a result of a block in SMT activity.[110] Unexpectedly, although *sng1* leaves accumulate sinapoylglucose to levels that are comparable to those of sinapoylmalate found in the wild type, the leaves of the mutant show a diminished fluorescence under UV light. This fluorescence phenotype was used to identify a T-DNA tagged *sng1* allele, which was subsequently used to clone the gene encoding SMT.[111]

A similar approach was taken to isolate a mutant in SCT. A TLC -based screen of seed extracts from 3000 EMS-mutagenized M2 seed was conducted to

A. Peptidase Activity **B.** Hypothetical Acyltransferase Activity

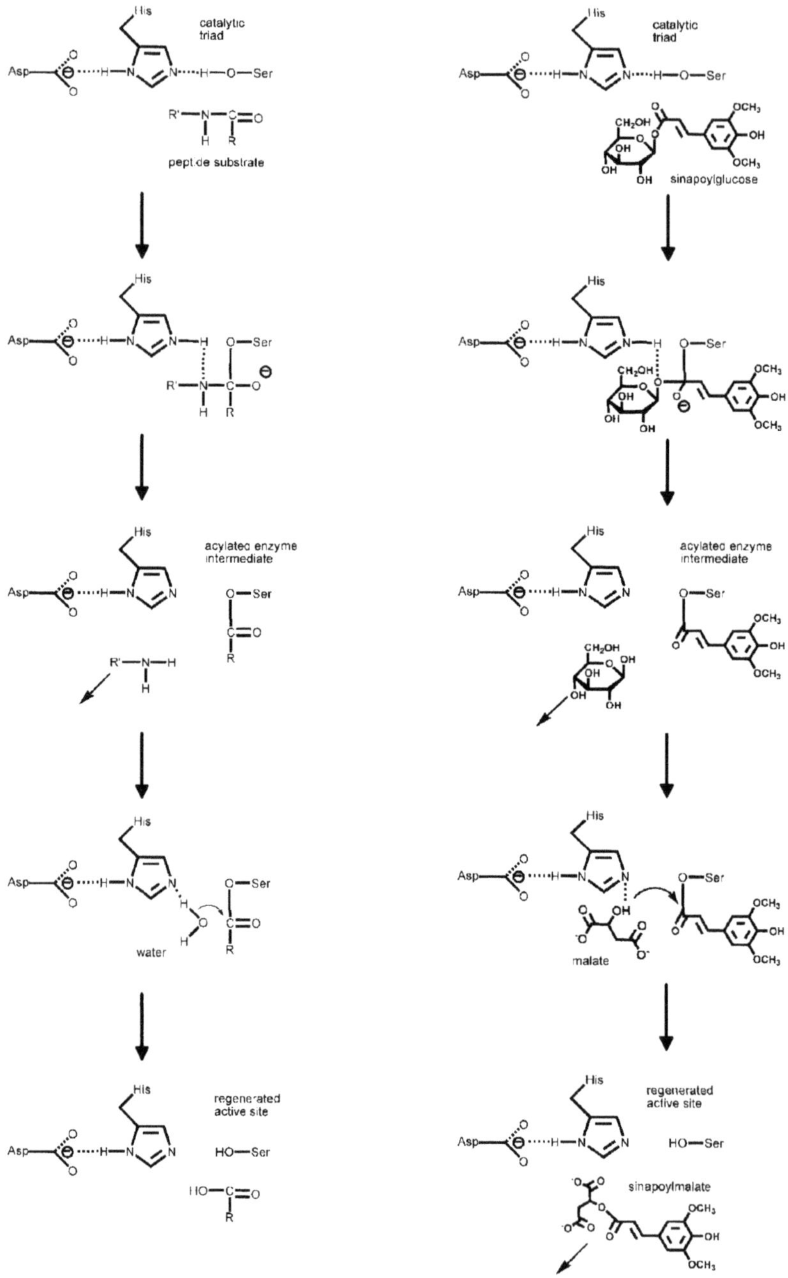

identify plants that accumulate sinapoylglucose, rather than sinapoylcholine, in their seed. One such mutant, designated *sng2*, was isolated,[112] and a positional cloning effort was used to isolate the *SCT* gene. Protein produced by expressing this gene in *E. coli* was able to catalyze the formation of sinapoylcholine from choline and sinapoylglucose, providing conclusive evidence that the gene encoded SCT.[112]

The inferred amino acid sequences of SMT and SCT were found to share significant identity with serine carboxypeptidases from yeast and plants. Members of this enzyme family have been shown to play diverse roles in protein processing and turnover in a wide variety of eukaryotic organisms (for example[113-116]). Serine carboxypeptidases remove the terminal amino acid from their protein substrates through the action of a catalytic triad of serine, histidine, and aspartic acid residues.[117-119] SMT and SCT also contain these conserved catalytic residues, as do other serine carboxypeptidase-like (SCPL) proteins involved in other aspects of plant secondary metabolism, such as the SCPL hydroxynitrile lyase involved in cyanogenic glycoside degradation,[120] and SCPL acyltransferases that catalyze the formation of isobutyryl glucose polyesters in tomato.[121] The completed genome sequence of *Arabidopsis* revealed that *SMT* and *SCT* belong to an SCPL gene family of over 50 members.

The conservation of catalytic residues between carboxypeptidases and SCPL acyltransferases led to the hypothesis that SCPL proteins may carry out their catalytic function through reaction mechanisms similar to that used by genuine carboxypeptidases. During carboxypeptidase-mediated peptide bond hydrolysis, the catalytic serine performs a nucleophilic attack on carbonyl carbon of the peptide backbone, forming an acyl-enzyme intermediate (Fig. 3.4a). This intermediate is rapidly hydrolyzed, regenerating the serine residue and releasing the newly cleaved products. Although the mechanism of SCPL acyltranferases has not yet been elucidated, the acyl acceptor (*e.g.*, malate in the case of SMT), may be activated to perform the degradation of a similar acyl-enzyme intermediate (*i.e.*, a sinapoylated enzyme in the case of SMT; Fig. 3.4b). It is interesting to note that SCPL acyltransferases must have been modified throughout evolution such that they catalyze acyltransferase rather than a hydrolysis reactions. These changes may include the ability to exclude water from the active site, or the ability to adopt a catalytically inactive conformation in the absence of the acyl acceptor.

In light of these findings, it appears that enzymes involved in primary metabolism, in this case the turnover and processing of proteins, have be co-opted to perform reactions on small molecules within secondary metabolic pathways. If this

Figure 3.4: Catalytic mechanism of yeast carboxypeptidase Y mediated peptide hyrolysis and a model for the acyltransferase activity of SMT.

is indeed the case, how many *Arabidopsis* genes that are annotated as encoding enzymes of primary metabolism are actually involved in the production of secondary metabolites? The identification of SMT, SCT, and other plant acyltransferases as SCPL proteins demonstrates that, even in this time of systems biology, gene and protein function must always be empirically verified.

SUMMARY AND FUTURE DIRECTIONS

The analysis of mutants of the phenylpropanoid pathway in *Arabidopsis*, as outlined in this review, has led to numerous revisions of the pathway over the past decade. The presently accepted pathway clarifies some of the contradictory data of the past, but also poses new questions for which we do not yet have answers. For example, a growing body of evidence suggests that neither ferulic acid nor sinapic acid are intermediates in phenylpropanoid biosynthesis. This is problematic in that many plant cell walls contain esterified ferulic acid,[10,11] and sinapic acid esters are major soluble secondary metabolites in *Arabidopsis* leaves and seeds.[31] If the most current model of the pathway is correct, how are these molecules synthesized?

Another challenge will be to assign function to individual members of enzymes that belong to gene families, which include CAD, CCR, and 4CL.[122] Different isoforms may exhibit specific spatial or temporal expression during development. Alternatively, individual members of a gene family may possess different substrate specificities towards intermediates of the pathway, which in turn may control the flux of the pathway towards different phenylpropanoid end products. The analysis of mutants with null alleles of these isoforms, either from publicly available T-DNA insertion lines or developed utilizing RNAi, will be necessary to elucidate their roles.

Evidence that supports the assembly of multi-enzyme complexes responsible for the metabolic channeling of intermediates during flavonoid biosynthesis has been described in *Arabidopsis*.[123, 124] Multi-enzyme assemblies, or "metabolons", would concentrate substrate pools for each reaction, leading to an overall more efficient production of final products. Such a complex has recently been proposed to operate in the production of monolignols,[125] in which P450s would provide an anchor to which the soluble enzymes of the pathway would be tethered via protein/protein interactions.[126] It has been further suggested that these metabolons may be differentially assembled for the production of either H, G, or S monolignols. If this proves to be the case, it will provide significant new opportunities for the study of phenylpropanoid biosynthetic regulation.

To date, most of the phenylpropanoid pathway genes isolated from *Arabidopsis* using genetic approaches encode enzymes. In contrast, little is known regarding the transcriptional regulatory elements of monolignol and sinapate ester biosynthesis.[127] This is in stark contrast to our understanding of the regulation of flavonoid and anthocyanin biosynthesis, which has been elucidated in detail through the analysis of maize and petunia mutants.[128] Recently, a number of *Arabidopsis* flavonoid regulatory mutants and their corresponding genes have been described.[129-134] In contrast, the sole regulatory element shown to be required for sinapate ester and monolignol biosynthesis is AtMyb4, an ortholog of the *Antirrhinum majus* gene *Am*MYB308,[135] which represses *C4H* transcription in response to low UV levels.[136,137] Only a few MYB regulatory proteins are found in yeast and animals, whereas the *Arabidopsis* genome contains at least 123 MYBs.[138] It seems clear that this class of proteins has evolved to regulate an array of functions in plants, including secondary metabolism.[139] The assignment of function to this class of proteins may, thus, shed further light onto the regulation of secondary metabolism in plants.

Finally, further research into the structural and regulatory aspects of phenylpropanoid biosynthesis in *Arabidopsis* may lead to interesting insights into the evolution of land plants. It is generally accepted that lignin biosynthesis was crucial for the colonization of land by plants.[140, 141] The knowledge gained by studies in *Arabidopsis* will permit the isolation and functional characterization of enzymes and regulatory factors from a wide array of genera, including pteridophytes and lycophytes, that arose before seed plants. These studies will reveal the similarities and differences in phenylpropanoid biosynthesis and its regulation that have arisen over the past 400 million years. In doing so, we may gain further appreciation for ancient evolutionary events that allowed for the spectacular diversity in plant life that we see today.

ACKNOWLEDGEMENTS

This work was supported by a grant from the National Science Foundation. This is journal paper number XXXXX of the Purdue University Agricultural Experiment Station.

REFERENCES

1. CHAPPLE, C.S., SHIRLEY, B.W., ZOOK, M., HAMMERSCHMIDT, R., SOMERVILLE, C.S., Secondary Metabolism in *Arabidopsis, in*: *Arabidopsis*, (E.M. Meyerowitz and.C.R. Somerville, eds), Cold Spring Harbor Press , Plainview, NY. 1994, pp. 989-1030.
2. CHEN, F., THOLL, D., D'AURIA, J.C., FAROOQ, A., PICHERSKY, E., GERSHENZON, J., Biosynthesis and emission of terpenoid volatiles from *Arabidopsis* flowers, *Plant Cell*, 2003, 15, 481-494.
3. COE, E.H., MCCORMICK, S.M., MODENA, S.A., White Pollen in Maize, *J. Hered.*, 1981, 72, 318-320.
4. TAYLOR, L.P., JORGENSEN, R., Conditional Male-Fertility in Chalcone Synthase-Deficient Petunia, *J. Hered.*, 1992, 83, 11-17.

5. VANDERMEER, I.M., STAM, M.E., VANTUNEN, A.J., MOL, J.N.M., STUITJE, A.R., Antisense Inhibition of Flavonoid Biosynthesis in Petunia Anthers Results in Male-Sterility, *Plant Cell*, 1992, 4, 253-262.
6. MATHESIUS, U., SCHLAMAN, H.R.M., SPAINK, H.P., SAUTTER, C., ROLFE, B.G., DJORDJEVIC, M.A., Auxin transport inhibition precedes root nodule formation in white clover roots and is regulated by flavonoids and derivatives of chitin oligosaccharides, *Plant J.*, 1998, 14, 23-34.
7. BROWN, D.E., RASHOTTE, A.M., MURPHY, A.S., NORMANLY, J., TAGUE, B.W., PEER, W.A., TAIZ, L., MUDAY, G.K., Flavonoids act as negative regulators of auxin transport in vivo in *Arabidopsis*, *Plant Physiol.*, 2001, 126, 524-535.
8. LI, J.Y., OULEE, T.M., RABA, R., AMUNDSON, R.G., LAST, R.L., *Arabidopsis* Flavonoid Mutants Are Hypersensitive to Uv-B Irradiation, *Plant Cell*, 1993, 5, 171-179.
9. LANDRY, L.G., CHAPPLE, C.C.S., LAST, R.L., *Arabidopsis* Mutants Lacking Phenolic Sunscreens Exhibit Enhanced Ultraviolet-B Injury and Oxidative Damage, *Plant Physiol.*, 1995, 109, 1159-1166.
10. FRY, S.C., Phenolic Components of the Primary-Cell Wall and Their Possible Role in the Hormonal-Regulation of Growth, *Planta*, 1979, 146, 343-351.
11. YANG, J.G., UCHIYAMA, T., Hydroxycinnamic acids and their dimers involved in the cessation of cell elongation in *Mentha* suspension culture, *Biosci. Biotech. Biochem.*, 2000, 64, 1572-1579.
12. BERNARDS, M.A., RAZEM, F.A., The poly(phenolic) domain of potato suberin: a non-lignin cell wall bio-polymer, *Phytochemistry*, 2001, 57, 1115-1122.
13. BERNARDS, M.A., Demystifying suberin, *Can. J. Botany*, 2002, 80, 227-240.
14. WHETTEN, R., SEDEROFF, R., Lignin Biosynthesis, *Plant Cell*, 1995, 7, 1001-1013.
15. HAKALA, P., LAMPI, A.M., OLLILAINEN, V., WERNER, U., MURKOVIC, M., WAHALA, K., KARKOLA, S., PIIRONEN, V., Steryl phenolic acid esters in cereals and their milling fractions, *J. Agri. Food Chem.*, 2002, 50, 5300-5307.
16. HERMANSEN, K., DINESEN, B., HOIE, L.H., MORGENSTERN, E., GRUENWALD, J., Effects of soy and other natural products on LDL : HDL ratio and other lipid parameters: A literature review, *Adv. Ther.*, 2003, 20, 50-78.
17. OSTLUND, R.E., RACETTE, S.B., STENSON, W.F., Inhibition of cholesterol absorption by phytosterol-replete wheat germ compared with phytosterol-depleted wheat germ, *Am. J. Clin. Nutr.*, 2003, 77, 1385-1389.
18. MIDDLETON, E., KANDASWAMI, C., THEOHARIDES, T.C., The effects of plant flavonoids on mammalian cells: Implications for inflammation, heart disease, and cancer, *Pharmacol. Rev.*, 2000, 52, 673-751.
19. DABROSIN, C., CHEN, J.M., WANG, L., THOMPSON, L.U., Flaxseed inhibits metastasis and decreases extracellular vascular endothelial growth factor in human breast cancer xenografts, *Canc. Lett.*, 2002, 185, 31-37.
20. GIRIDHARAN, P., SOMASUNDARAM, S.T., PERUMAL, K., VISHWAKARMA, R.A., KARTHIKEYAN, N.P., VELMURUGAN, R., BALAKRISHNAN, A., Novel substituted methylenedioxy lignan suppresses

proliferation of cancer cells by inhibiting telomerase and activation of c-myc and caspases leading to apoptosis, *Br. J. Canc.*, 2002, 87, 98-105.

21. HEDLUND, T.E., JOHANNES, W.U., MILLER, G.J., Soy isoflavonoid equol modulates the growth of benign and malignant prostatic epithelial cells in vitro, *Prostate*, 2003, 54, 68-78.
22. HOWITZ, K.T., BITTERMAN, K.J., COHEN, H.Y., LAMMING, D.W., LAVU, S., WOOD, J.G., ZIPKIN, R.E., CHUNG, P., KISIELEWSKI, A., ZHANG, L.L., SCHERER, B., SINCLAIR, D.A., Small molecule activators of sirtuins extend *Saccharomyces cerevisiae* lifespan, *Nature*, 2003, 425, 191-196.
23. DUDAREVA, N., MURFITT, L.M., MANN, C.J., GORENSTEIN, N., KOLOSOVA, N., KISH, C.M., BONHAM, C., WOOD, K., Developmental regulation of methyl benzoate biosynthesis and emission in snapdragon flowers, *Plant Cell*, 2000, 12, 949-961.
24. GANG, D.R., WANG, J.H., DUDAREVA, N., NAM, K.H., SIMON, J.E., LEWINSOHN, E., PICHERSKY, E., An investigation of the storage and biosynthesis of phenylpropenes in sweet basil, *Plant Physiol.*, 2001, 125, 539-555.
25. WU, H., PRATLEY, J., LEMERLE, D., HAIG, T., Crop cultivars with allelopathic capability, *Weed Res.*, 1999, 39, 171-180.
26. BAIS, H.P., VEPACHEDU, R., GILROY, S., CALLAWAY, R.M., VIVANCO, J.M., Allelopathy and exotic plant invasion: From molecules and genes to species interactions, *Science*, 2003, 301, 1377-1380.
27. BATTLE, M., BENDER, M.L., TANS, P.P., WHITE, J.W.C., ELLIS, J.T., CONWAY, T., FRANCEY, R.J., Global carbon sinks and their variability inferred from atmospheric O-2 and delta C-13, *Science*, 2000, 287, 2467-2470.
28. WINKEL-SHIRLEY, B., Flavonoid biosynthesis. A colorful model for genetics, biochemistry, cell biology, and biotechnology, *Plant Physiol.*, 2001, 126, 485-493.
29. WINKEL-SHIRLEY, B., Biosynthesis of flavonoids and effects of stress, *Curr. Opin. Plant Biol.*, 2002, 5, 218-223.
30. STRACK, D., Sinapic acid ester fluctuations in cotyledons of *Raphanus sativus.*, *Z. Pflanzenphysiol.*, 1977, 84, 139-154.
31. CHAPPLE, C.C.S., VOGT, T., ELLIS, B.E., SOMERVILLE, C.R., An *Arabidopsis* Mutant Defective in the General Phenylpropanoid Pathway, *Plant Cell*, 1992, 4, 1413-1424.
32. STRACK, D., Sinapine as a supply of choline for the biosynthesis of phosphotidylcholine in *Raphanus sativus* seedlings, *Z. Naturforsch.*, 1981, 36c, 215-221.
33. RUEGGER, M., MEYER, K., CUSUMANO, J.C., CHAPPLE, C., Regulation of ferulate-5-hydroxylase expression in *Arabidopsis* in the context of sinapate ester biosynthesis, *Plant Physiol.*, 1999, 119, 101-110.
34. RUEGGER, M., CHAPPLE, C., Mutations that reduce sinapoylmalate accumulation in *Arabidopsis thaliana* define loci with diverse roles in phenylpropanoid metabolism, *Genetics*, 2001, 159, 1741-1749.
35. CHEN, C.Y., BAUCHER, M., HOLST CHRISTENSEN, J., BOERJAN, W., Biotechnology in trees: Towards improved paper pulping by lignin engineering, *Euphytica*, 2001, 118, 185-195.

36. HUNTLEY, S.K, ELLIS, D., GILBERT, M., CHAPPLE, C.S., MANSFIELD, S.D. Significant Increases in Pulping Efficiency in C4H-F5H-Transformed Poplars: Improved Chemical Savings and Reduced Environmental Toxins, *J. Agric. Food Chem.*, 2003, 51, 6178 -6183.
37. SEWALT, V.J.H., DEOLIVEIRA, W., GLASSER, W.G., FONTENOT, J.P., Lignin impact on fibre degradation .2. A model study using cellulosic hydrogels, *J. Sci. Food Agr.*, 1996, 71, 204-208.
38. JUNG, H.G., MERTENS, D.R., PAYNE, A.J., Correlation of acid detergent lignin and Klason lignin with digestibility of forage dry matter and neutral detergent fiber, *J. Dairy Sci.*, 1997, 80, 1622-1628.
39. BARRIERE, Y., GUILLET, C., GOFFNER, D., PICHON, M., Genetic variation and breeding strategies for improved cell wall digestibility in annual forage crops. A review, *An. Res.*, 2003, 52, 193-228.
40. LEWIS, N.G., YAMAMOTO, E., Lignin - Occurrence, Biogenesis and Biodegradation, *Annu. Rev. Plant Physiol. Plant Molec. Biol.*, 1990, 41, 455-496.
41. CAMPBELL, M.M., SEDEROFF, R.R., Variation in lignin content and composition - Mechanism of control and implications for the genetic improvement of plants, *Plant Physiol.*, 1996, 110, 3-13.
42. WHETTEN, R.W., MACKAY, J.J., SEDEROFF, R.R., Recent advances in understanding lignin biosynthesis, *Annu. Rev. Plant Physiol. Plant Molec. Biol.*, 1998, 49, 585-609.
43. BOUDET, A.M., CHABANNES, M., Gains achieved by molecular approaches in the area of lignification, *Pure Appl. Chem.*, 2001, 73, 561-566.
44. DONALDSON, L.A., Lignification and lignin topochemistry - an ultrastructural view, *Phytochemistry*, 2001, 57, 859-873.
45. SINGH, A., DANIEL, G., NILSSON, T., Ultrastructure of the S-2 layer in relation to lignin distribution in Pinus radiata tracheids, *J. Wood Sci.*, 2002, 48, 95-98.
46. MEYER, K., CUSUMANO, J.C., SOMERVILLE, C., CHAPPLE, C.C.S., Ferulate-5-hydroxylase from *Arabidopsis thaliana* defines a new family of cytochrome P450-dependent monooxygenases, *Proc. Natl. Acad. Sci. U. S. A.*, 1996, 93, 6869-6874.
47. MEYER, K., SHIRLEY, A.M., CUSUMANO, J.C., BELL-LELONG, D.A., CHAPPLE, C., Lignin monomer composition is determined by the expression of a cytochrome P450-dependent monooxygenase in *Arabidopsis*, *Proc. Natl. Acad. Sci. U. S. A.*, 1998, 95, 6619-6623.
48. BELL-LELONG, D.A., CUSUMANO, J.C., MEYER, K., CHAPPLE, C., Cinnamate-4-hydroxylase expression in *Arabidopsis* - Regulation in response to development and the environment, *Plant Physiol.*, 1997, 113, 729-738.
49. MARITA, J.M., RALPH, J., HATFIELD, R.D., CHAPPLE, C., NMR characterization of lignins in *Arabidopsis* altered in the activity of ferulate 5-hydroxylase, *Proc. Natl. Acad. Sci. U. S. A.*, 1999, 96, 12328-12332.
50. RUEGGER, M., MEYER, K., CUSUMANO, J.C., CHAPPLE, C., Regulation of ferulate-5-hydroxylase expression in *Arabidopsis* in the context of sinapate ester biosynthesis, *Plant Physiol.*, 1999, 119, 101-110.
51. HUMPHREYS, J.M., CHAPPLE, C., Rewriting the lignin roadmap, *Curr. Opin. Plant Biol.*, 2002, 5, 224-229.

52. MARTERN, U., WENDORFF, H., HAMERSKI, D., PAKUSCH, A.E., KNEUSEL, R.E., Elicitor-induced phenylpropanoid synthesis in *Apiaceae* cell cultures, *Bull. Liaison Group Polyphenols*, 1988, 14, 173-184.
53. KUHNL, T., KOCH, U., HELLER, W., WELLMANN, E., Elicitor Induced S-Adenosyl-L-Methionine - Caffeoyl-Coa 3-O-Methyltransferase from Carrot Cell-Suspension Cultures, *Plant Science*, 1989, 60, 21-25.
54. PAKUSCH, A.E., KNEUSEL, R.E., MATERN, U., S-Adenosyl-L-Methionine - Trans-Caffeoyl-Coenzyme-a 3-O-Methyltransferase from Elicitor-Treated Parsley Cell-Suspension Cultures, *Arch. Biochem. Biophys.*, 1989, 271, 488-494.
55. YE, Z.H., KNEUSEL, R.E., MATERN, U., VARNER, J.E., An Alternative Methylation Pathway in Lignin Biosynthesis in Zinnia, *Plant Cell*, 1994, 6, 1427-1439.
56. YE, Z.H., VARNER, J.E., Differential Expression of 2 O-Methyltransferases in Lignin Biosynthesis in Zinnia-Elegans, *Plant Physiol.*, 1995, 108, 459-467.
57. YE, Z.H., Association of caffeoyl coenzyme A 3-O-methyltransferase expression with lignifying tissues in several dicot plants, *Plant Physiol.*, 1997, 115, 1341-1350.
58. LEE, D., MEYER, K., CHAPPLE, C., DOUGLAS, C.J., Antisense suppression of 4-coumarate:coenzyme A ligase activity in *Arabidopsis* leads to altered lignin subunit composition, *Plant Cell*, 1997, 9, 1985-1998.
59. ALLINA, S.M., PRI-HADASH, A., THEILMANN, D.A., ELLIS, B.E., DOUGLAS, C.J., 4-coumarate : coenzyme A ligase in hybrid poplar - Properties of native enzymes, cDNA cloning, and analysis of recombinant enzymes, *Plant Physiol.*, 1998, 116, 743-754.
60. EHLTING, J., BUTTNER, D., WANG, Q., DOUGLAS, C.J., SOMSSICH, I.E., KOMBRINK, E., Three 4-coumarate : coenzyme A ligases in *Arabidopsis thaliana* represent two evolutionarily divergent classes in angiosperms, *Plant J.*, 1999, 19, 9-20.
61. HUMPHREYS, J.M., HEMM, M.R., CHAPPLE, C., New routes for lignin biosynthesis defined by biochemical characterization of recombinant ferulate 5-hydroxylase, a multifunctional cytochrome P450-dependent monooxygenase, *Proc. Natl. Acad. Sci. U. S. A.*, 1999, 96, 10045-10050.
62. OSAKABE, K., TSAO, C.C., LI, L.G., POPKO, J.L., UMEZAWA, T., CARRAWAY, D.T., SMELTZER, R.H., JOSHI, C.P., CHIANG, V.L., Coniferyl aldehyde 5-hydroxylation and methylation direct syringyl lignin biosynthesis in angiosperms, *Proc. Natl. Acad. Sci. U. S. A.*, 1999, 96, 8955-8960.
63. GRAND, C., Ferulic Acid 5-Hydroxylase - a New Cytochrome P-450-Dependent Enzyme from Higher-Plant Microsomes Involved in Lignin Synthesis, *FEBS Lett.*, 1984, **169**, 7-11.
64. LI, L.G., POPKO, J.L., UMEZAWA, T., CHIANG, V.L., 5-Hydroxyconiferyl aldehyde modulates enzymatic methylation for syringyl monolignol formation, a new view of monolignol biosynthesis in angiosperms, *J. Biol. Chem.*, 2000, 275, 6537-6545.
65. ZHONG, R.Q., MORRISON, W.H., NEGREL, J., YE, Z.H., Dual methylation pathways in lignin biosynthesis, *Plant Cell*, 1998, 10, 2033-2045.

66. GABRAIC, B., WERCK-REICHHART, D., TEUTSCH, H., DURST, F., Purification and immunocharacterization of a plant cytochrome P450: the cinnamic acid 4-hydroxylase, *Arch. Biochem. Biophys.*, 1991, 288, 302-309.
67. MIZUTANI, M., OHTA, D., SATO, R., Purification and Characterization of a Cytochrome-P450 (Trans-Cinnamic Acid 4-Hydroxylase) from Etiolated Mung Bean Seedlings, *Plant Cell Physiol.*, 1993, 34, 481-488.
68. MIZUTANI, M., WARD, E., DIMAIO, J., OHTA, D., RYALS, J., SATO, R., Molecular-Cloning and Sequencing of a Cdna-Encoding Mung Bean Cytochrome-P450 (P450c4h) Possessing Cinnamate 4-Hydroxylase Activity, *Biochem. Biophys. Res. Commun.*, 1993, 190, 875-880.
69. TEUTSCH, H.G., HASENFRATZ, M.P., LESOT, A., STOLTZ, C., GARNIER, J.M., JELTSCH, J.M., DURST, F., WERCKREICHHART, D., Isolation and Sequence of a Cdna-Encoding the Jerusalem-Artichoke Cinnamate 4-Hydroxylase, a Major Plant Cytochrome-P450 Involved in the General Phenylpropanoid Pathway, *Proc. Natl. Acad. Sci. U. S. A.*, 1993, 90, 4102-4106.
70. VAUGHAN, P.F.T., BUTT, V.S., The action of *o*-dihydric phenols in the hydroxylation of *p*-coumaric acid by a phenolase from leaves of spinach beet (*Beta vulgaris* L.), *Biochem. J.*, 1970, 119, 89-94.
71. STAFFORD, H.A., DRESLER, S., 4-hydroxycinnamic acid hydroxylase and polyphenolase activities in *Sorghum vulgare*, *Plant Physiol.*, 1972, 49, 151-157.
72. BONIWELL, J.M., BUTT, V.S., Flavin Nucleotide-Dependent 3-Hydroxylation of 4-Hydroxyphenylpropanoid Carboxylic-Acids by Particulate Preparations from Potato-Tubers, *Z. Naturforsch. Sect. C*, 1986, 41, 56-60.
73. KOJIMA, M., TAKEUCHI, W., Detection and Characterization of Para-Coumaric Acid Hydroxylase in Mung Bean, Vigna-Mungo, Seedlings, *J. Biochem.*, 1989, 105, 265-270.
74. BARTLETT, D.J., ARLOTTO, M.P., WATERMAN, M.R., Hydroxylation of *p*-coumaric acid by illuminated chloroplasts from spinach beet leaves, *FEBS Lett.*, 1972, 23, 265-267.
75. SCHOCH, G., GOEPFERT, S., MORANT, M., HEHN, A., MEYER, D., ULLMANN, P., WERCK-REICHHART, D., CYP98A3 from *Arabidopsis thaliana* is a 3 '-hydroxylase of phenolic esters, a missing link in the phenylpropanoid pathway, *J. Biol. Chem.*, 2001, 276, 36566-36574.
76. FRANKE, R., HUMPHREYS, J.M., HEMM, M.R., DENAULT, J.W., RUEGGER, M.O., CUSUMANO, J.C., CHAPPLE, C., The *Arabidopsis* REF8 gene encodes the 3-hydroxylase of phenylpropanoid metabolism, *Plant J.*, 2002, 30, 33-45.
77. NAIR, R.B., XIA, Q., KARTHA, C.J., KURYLO, E., HIRJI, R.N., DATLA, R., SELVARAJ, G., *Arabidopsis* CYP98A3 mediating aromatic 3-hydroxylation. Developmental regulation of the gene, and expression in yeast, *Plant Physiol.*, 2002, 130, 210-220.
78. HELLER, W., KUHNL, T., Elicitor Induction of a Microsomal 5-O-(4-Coumaroyl)Shikimate 3'-Hydroxylase in Parsley Cell-Suspension Cultures, *Arch. Biochem. Biophys.*, 1985, 241, 453-460.
79. KUHNL, T., KOCH, U., HELLER, W., WELLMANN, E., Chlorogenic Acid Biosynthesis - Characterization of a Light-Induced Microsomal 5-O-(4-Coumaroyl)-

D-Quinate/Shikimate 3'-Hydroxylase from Carrot (Daucus-Carota L) Cell-Suspension Cultures, *Arch. Biochem. Biophys.*, 1987, 258, 226-232.

80. HOFFMANN, L., MAURY, S., MARTZ, F., GEOFFROY, P., LEGRAND, M., Purification, cloning, and properties of an acyltransferase controlling shikimate and quinate ester intermediates in phenylpropanoid metabolism, *J. Biol. Chem.*, 2003, 278, 95-103.
81. FRANKE, R., MCMICHAEL, C.M., MEYER, K., SHIRLEY, A.M., CUSUMANO, J.C., CHAPPLE, C., Modified lignin in tobacco and poplar plants over-expressing the *Arabidopsis* gene encoding ferulate 5-hydroxylase, *Plant J.*, 2000, 22, 223-234.
82. STRACK, D., Enzymatic synthesis of 1-sinapoylglucose from free sinapic acid and UDP-glucose by a cell-free system from *Raphanus sativus* seedlings, *Z. Naturforsch.*, 1980, 35c, 204-208.
83. GRAWE, W., BACHHUBER, P., MOCK, H.P., STRACK, D., Purification and Characterization of Sinapolyglucose - Malate Sinapoyltransferase from Raphanus-Sativus L, *Planta*, 1992, 187, 236-241.
84. REITER, W.D., CHAPPLE, C., SOMERVILLE, C.R., Mutants of *Arabidopsis* thaliana with altered cell wall polysaccharide composition, *Plant J.*, 1997, 12, 335-345.
85. TURNER, S.R., SOMERVILLE, C.R., Collapsed xylem phenotype of *Arabidopsis* identifies mutants deficient in cellulose deposition in the secondary cell wall, *Plant Cell*, 1997, 9, 689-701.
86. TAYLOR, N.G., SCHEIBLE, W.R., CUTLER, S., SOMERVILLE, C.R., TURNER, S.R., The irregular xylem3 locus of *Arabidopsis* encodes a cellulose synthase required for secondary cell wall synthesis, *Plant Cell*, 1999, 11, 769-779.
87. JONES, L., ENNOS, A.R., TURNER, S.R., Cloning and characterization of irregular xylem4 (irx4): a severely lignin-deficient mutant of *Arabidopsis*, *Plant J.*, 2001, 26, 205-216.
88. GOUJON, T., FERRET, V., MILA, I., POLLET, B., RUEL, K., BURLAT, V., JOSELEAU, J.P., BARRIERE, Y., LAPIERRE, C., JOUANIN, L., Down-regulation of the AtCCR1 gene in *Arabidopsis thaliana*: effects on phenotype, lignins and cell wall degradability, *Planta*, 2003, 217, 218-228.
89. LAUVERGEAT, V., LACOMME, C., LACOMBE, E., LASSERRE, E., ROBY, D., GRIMA-PETTENATI, J., Two cinnamoyl-CoA reductase (CCR) genes from *Arabidopsis* thaliana are differentially expressed during development and in response to infection with pathogenic bacteria, *Phytochemistry*, 2001, 57, 1187-1195.
90. GOUJON, T., SIBOUT, R., POLLET, B., MABA, B., NUSSAUME, L., BECHTOLD, N., LU, F.C., RALPH, J., MILA, I., BARRIERE, Y., LAPIERRE, C., JOUANIN, L., A new *Arabidopsis thaliana* mutant deficient in the expression of O-methyltransferase impacts lignins and sinapoyl esters, *Plant Mol. Biol.*, 2003, 51, 973-989.
91. JOUANIN, L., GOUJON, T., DE NADAI, V., MARTIN, M.T., MILA, I., VALLET, C., POLLET, B., YOSHINAGA, A., CHABBERT, B., PETIT-CONIL, M., LAPIERRE, C., Lignification in transgenic poplars with extremely reduced caffeic acid O-methyltransferase activity, *Plant Physiol.*, 2000, 123, 1363-1373.

92. RALPH, J., LAPIERRE, C., LU, F.C., MARITA, J.M., PILATE, G., VAN DOORSSELAERE, J., BOERJAN, W., JOUANIN, L., NMR evidence for benzodioxane structures resulting from incorporation of 5-hydroxyconiferyl alcohol into lignins of O-methyltransferase-deficient poplars, *J. Agric. Food Chem.*, 2001, 49, 86-91.
93. RALPH, J., LAPIERRE, C., MARITA, J.M., KIM, H., LU, F.C., HATFIELD, R.D., RALPH, S., CHAPPLE, C., FRANKE, R., HEMM, M.R., VAN DOORSSELAERE, J., SEDEROFF, R.R., O'MALLEY, D.M., SCOTT, J.T., MACKAY, J.J., YAHIAOUI, N., BOUDET, A.M., PEAN, M., PILATE, G., JOUANIN, L., BOERJAN, W., Elucidation of new structures in lignins of CAD- and COMT-deficient plants by NMR, *Phytochemistry*, 2001, 57, 993-1003.
94. MARITA, J.M., VERMERRIS, W., RALPH, J., HATFIELD, R.D., Variations in the cell wall composition of maize brown midrib mutants, *J. Agric. Food Chem.*, 2003, 51, 1313-1321.
95. ATANASSOVA, R., FAVET, N., MARTZ, F., CHABBERT, B., TOLLIER, M.T., MONTIES, B., FRITIG, B., LEGRAND, M., Altered Lignin Composition in Transgenic Tobacco Expressing O-Methyltransferase Sequences in Sense and Antisense Orientation, *Plant J.*, 1995, 8, 465-477.
96. VAILHE, M.A.B., MIGNE, C., CORNU, A., MAILLOT, M.P., GRENET, E., BESLE, J.M., ATANASSOVA, R., MARTZ, F., LEGRAND, M., Effect of modification of the O-methyltransferase activity on cell wall composition, ultrastructure and degradability of transgenic tobacco, *J. Sci. Food Agric.*, 1996, 72, 385-391.
97. TSAI, C.J., POPKO, J.L., MIELKE, M.R., HU, W.J., PODILA, G.K., CHIANG, V.L., Suppression of O-methyltransferase gene by homologous sense transgene in quaking aspen causes red-brown wood phenotypes, *Plant Physiol.*, 1998, 117, 101-112.
98. BOUT, S., VERMERRIS, W., A candidate-gene approach to clone the sorghum Brown midrib gene encoding caffeic acid O-methyltransferase, *Mol. Genet. Genomics*, 2003, 269, 205-214.
99. SEDEROFF, R.R., MACKAY, J.J., RALPH, J., HATFIELD, R.D., Unexpected variation in lignin, *Curr. Opin. Plant Biol.*, 1999, 2, 145-152.
100. ANTEROLA, A.M., LEWIS, N.G., Trends in lignin modification: a comprehensive analysis of the effects of genetic manipulations/mutations on lignification and vascular integrity, *Phytochemistry*, 2002, 61, 221-294.
101. HEMM, M.R., RUEGGER, M.O., CHAPPLE, C., The *Arabidopsis* ref2 mutant is defective in the gene encoding CYP83A1 and shows both phenylpropanoid and glucosinolate phenotypes, *Plant Cell*, 2003, 15, 179-194.
102. BAK, S., FEYEREISEN, R., The involvement of two P450 enzymes, CYP83B1 and CYP83A1, in auxin homeostasis and glucosinolate biosynthesis, *Plant Physiol.*, 2001, 127, 108-118.
103. BAK, S., TAX, F.E., FELDMANN, K.A., GALBRAITH, D.W., FEYEREISEN, R., CYP83B1, a cytochrome P450 at the metabolic branch paint in auxin and indole glucosinolate biosynthesis in *Arabidopsis*, *Plant Cell*, 2001, 13, 101-111.

104. NAUR, P., PETERSEN, B.L., MIKKELSEN, M.D., BAK, S., RASMUSSEN, H., OLSEN, C.E., HALKIER, B.A., CYP83A1 and CYP83B1, two nonredundant cytochrome P450 enzymes metabolizing oximes in the biosynthesis of glucosinolates in *Arabidopsis*, *Plant Physiol.*, 2003, 133, 63-72.
105. REINTANZ, B., LEHNEN, M., REICHELT, M., GERSHENZON, J., KOWALCZYK, M., SANDBERG, G., GODDE, M., UHL, R., PALME, K., Bus, a bushy *Arabidopsis* CYP79F1 knockout mutant with abolished synthesis of short-chain aliphatic glucosinolates, *Plant Cell*, 2001, **13**, 351-367.
106. KLIEBENSTEIN, D., PEDERSEN, D., BARKER, B., MITCHELL-OLDS, T., Comparative analysis of quantitative trait loci controlling glucosinolates, myrosinase and insect resistance in Arabidapsis thaliana, *Genetics*, 2002, 161, 325-332.
107. MOCK, H.P., STRACK, D., Energetics of the Uridine 5'-Diphosphoglucose - Hydroxy-Cinnamic Acid Acyl-Glucosyltransferase Reaction, *Phytochemistry*, 1993, 32, 575-579.
108. STRACK, D., Development of 1-*O*-sinapoyl-β-D-glucose: L-malate sinapoyltransferase activity in cotyledons of red raddish (*Raphanus sativus* L. var. *sativus*), *Planta*, 1982, 155, 31-36.
109. STRACK, D., KNOGGE, W., DAHLBENDER, B., Enzymatic-Synthesis of Sinapine from 1-0-Sinapoyl-Beta-D- Glucose and Choline by a Cell-Free System from Developing Seeds of Red Radish (*Raphanus Sativus* L. var. *Sativus*), *Z. Naturforsch. Sect. C*, 1983, 38, 21-27.
110. LORENZEN, M., RACICOT, V., STRACK, D., CHAPPLE, C., Sinapic acid ester metabolism in wild type and a sinapoylglucose-accumulating mutant of *Arabidopsis*, *Plant Physiol.*, 1996, 112, 1625-1630.
111. LEHFELDT, C., SHIRLEY, A.M., MEYER, K., RUEGGER, M.O., CUSUMANO, J.C., VIITANEN, P.V., STRACK, D., CHAPPLE, C., Cloning of the SNG1 gene of *Arabidopsis* reveals a role for a serine carboxypeptidase-like protein as an acyltransferase in secondary metabolism, *Plant Cell*, 2000, 12, 1295-1306.
112. SHIRLEY, A.M., MCMICHAEL, C.M., CHAPPLE, C., The sng2 mutant of *Arabidopsis* is defective in the gene encoding the serine carboxypeptidase-like protein sinapoylglucose : choline sinapoyltransferase, *Plant J.*, 2001, 28, 83-94.
113. RAMOS, C., WINTHER, J.R., KIELLANDBRANDT, M.C., Requirement of the Propeptide for in-Vivo Formation of Active Yeast Carboxypeptidase-Y, *J. Biol. Chem.*, 1994, 269, 7006-7012.
114. RAMOS, C., WINTHER, J.R., Exchange of regions of the carboxypeptidase Y propeptide - Sequence specificity and function in folding in vivo, *Eur. J. Biochem.*, 1996, 242, 29-35.
115. CHEN, J.Y., STREB, J.W., MALTBY, K.M., KITCHEN, C.M., MIANO, J.M., Cloning of a novel retinoid-inducible serine carboxypeptidase from vascular smooth muscle cells, *J. Biol. Chem.*, 2001, 276, 34175-34181.
116. CERCOS, M., URBEZ, C., CARBONELL, J., A serine carboxypeptidase gene (PsCP), expressed in early steps of reproductive and vegetative development in Pisum sativum, is induced by gibberellins, *Plant Molec. Biol.*, 2003, 51, 165-174.
117. HAYASHI, R., MOORE, S., STEIN, W.H., Serine at the active center of yeast carboxypeptidase, *J. Biol. Chem.*, 1973, 248, 8366-8369.

118. HAYASHI, R., BAI, Y., HATA, T., Evidence for an essential histidine in carboxypeptidase Y. Reaction with the chloromethyl ketone derivative of benzyloxycarbonyl-L-phenylalanine, *J. Biol. Chem.*, 1975, 250, 5221-5226.
119. BECH, L.M., BREDDAM, K., Inactivation of Carboxypeptidase-Y by Mutational Removal of the Putative Essential Histidyl Residue, *Carlsberg Res. Comm.*, 1989, 54, 165-171.
120. WAJANT, H., MUNDRY, K.W., PFIZENMAIER, K., Molecular-Cloning of Hydroxynitrile Lyase from Sorghum-Bicolor (L) - Homologies to Serine Carboxypeptidases, *Plant Molec. Biol.*, 1994, 26, 735-746.
121. LI, A.X., STEFFENS, J.C., An acyltransferase catalyzing the formation of diacylglucose is a serine carboxypeptidase-like protein, *Proc. Natl. Acad. Sci. U. S. A.*, 2000, 97, 6902-6907.
122. GOUJON, T., SIBOUT, R., EUDES, A., MACKAY, J., JOULANIN, L., Genes involved in the biosynthesis of lignin precursors in *Arabidopsis thaliana*, *Plant Physiol. Biochem.*, 2003, 41, 677-687.
123. BURBULIS, I.E., WINKEL-SHIRLEY, B., Interactions among enzymes of the *Arabidopsis* flavonoid biosynthetic pathway, *Proc. Natl. Acad. Sci. U. S. A.*, 1999, 96, 12929-12934.
124. WINKEL-SHIRLEY, B., Evidence for enzyme complexes in the phenylpropanoid and flavonoid pathways, *Physiol. Plant.*, 1999, 107, 142-149.
125. RASMUSSEN, S., DIXON, R.A., Transgene-mediated and elicitor-induced perturbation of metabolic channeling at the entry point into the phenylpropanoid pathway, *Plant Cell*, 1999, 11, 1537-1551.
126. DIXON, R.A., CHEN, F., GUO, D.J., PARVATHI, K., The biosynthesis of monolignols: a "metabolic grid", or independent pathways to guaiacyl and syringyl units?, *Phytochemistry*, 2001, 57, 1069-1084.
127. ENDT, D.V., KIJNE, J.W., MEMELINK, J., Transcription factors controlling plant secondary metabolism: what regulates the regulators?, *Phytochemistry*, 2002, 61, 107-114.
128. MOL, J., GROTEWOLD, E., KOES, R., How genes paint flowers and seeds, *Trends Plant Sci.*, 1998, 3, 212-217.
129. SAGASSER, M., LU, G.H., HAHLBROCK, K., WEISSHAAR, B., A-thaliana TRANSPARENT TESTA 1 is involved in seed coat development and defines the WIP subfamily of plant zinc finger proteins, *Genes Dev.*, 2002, 16, 138-149.
130. NESI, N., JOND, C., DEBEAUJON, I., CABOCHE, M., LEPINIEC, L., The *Arabidopsis TT2* gene encodes an R2R3 MYB domain protein that acts as a key determinant for proanthocyanidin accumulation in developing seed, *Plant Cell*, 2001, 13, 2099-2114.
131. NESI, N., DEBEAUJON, D., JOND, C., PELLETIER, G., CABOCHE, M., LEPINIEC, L., The *TT8* Gene encodes a basic helix-loop-helix domain protein required for expression of DFR and BAN genes in *Arabidopsis* siliques, *Plant Cell*, 2000, 12, 1863-1878.
132. BOREVITZ, J.O., XIA, Y.J., BLOUNT, J., DIXON, R.A., LAMB, C., Activation tagging identifies a conserved MYB regulator of phenylpropanoid biosynthesis, *Plant Cell*, 2000, 12, 2383-2393.

133. WALKER, A.R., DAVISON, P.A., BOLOGNESI-WINFIELD, A.C., JAMES, C.M., SRINIVASAN, N., BLUNDELL, T.L., ESCH, J.J., MARKS, M.D., GRAY, J.C., The TRANSPARENT TESTA GLABRA1 locus, which regulates trichome differentiation and anthocyanin biosynthesis in *Arabidopsis*, encodes a WD40 repeat protein, *Plant Cell*, 1999, 11, 1337-1349.
134. JOHNSON, C.S., KOLEVSKI, B., SMYTH, D.R., TRANSPARENT TESTA GLABRA2, a trichome and seed coat development gene of *Arabidopsis*, encodes a WRKY transcription factor, *Plant Cell*, 2002, 14, 1359-1375.
135. TAMAGNONE, L., MERIDA, A., PARR, A., MACKAY, S., CULIANEZ-MACIA, F.A., ROBERTS, K., MARTIN, C., The AmMYB308 and AmMYB330 transcription factors from antirrhinum regulate phenylpropanoid and lignin biosynthesis in transgenic tobacco, *Plant Cell*, 1998, 10, 135-154.
136. JIN, H.L., COMINELLI, E., BAILEY, P., PARR, A., MEHRTENS, F., JONES, J., TONELLI, C., WEISSHAAR, B., MARTIN, C., Transcriptional repression by AtMYB4 controls production of UV-protecting sunscreens in *Arabidopsis*, *EMBO J.*, 2000, 19, 6150-6161.
137. HEMM, M.R., HERRMANN, K.M., CHAPPLE, C., AtMYB4: a transcription factor general in the battle against UV, *Trends Plant Sci.*, 2001, 6, 135-136.
138. STRACKE, R., WERBER, M., WEISSHAAR, B., The R2R3-MYB gene family in *Arabidopsis thaliana*, *Curr. Opin. Plant Biol.*, 2001, 4, 447-456.
139. DIAS, A.P., BRAUN, E.L., MCMULLEN, M.D., GROTEWOLD, E., Recently duplicated maize R2R3 Myb genes provide evidence for distinct mechanisms of evolutionary divergence after duplication, *Plant Physiol.*, 2003, 131, 610-620.
140. COOK, M.E., FRIEDMAN, W.E., Tracheid structure in a primitive extant plant provides an evolutionary link to earliest fossil tracheids, *Int. J. Plant Sci.*, 1998, 159, 881-890.
141. SPERRY, J.S., Evolution of water transport and xylem structure, *Int. J. Plant Sci.*, 2003, 164, S115-S127.

Chapter Four

EVOLUTION OF INDOLE AND BENZOXAZINONE BIOSYNTHESIS IN *Zea mays*

Alfons Gierl,* Sebastian Gruen, Ullrich Genschel, Regina Huettl, and Monika Frey

Lehrstuhl für Genetik
Technische Universität München
Am Hochanger 8
85350 Freising
Germany

* *Author for correspondence, email: gierl@wzw.tum.de*

INTRODUCTION

Plant secondary metabolites constitute a large field of chemical biodiversity. The occurrence of certain metabolites in species sometimes reflects their phylogenetic origin. On the other hand, closely related plant taxa often differ in their spectra of secondary products. The evolution of the synthetic capacity for these substances has accompanied plants from their origin onwards. In *Arabidopsis thaliana*, it is estimated that about 5,000 genes, *i.e.,* about 20% of all genes, are involved in secondary metabolism.[1] This may partly explain the relatively high number of genes present in plant genomes, when compared with genomes of mammals. Primary metabolism represents the platform from which secondary metabolism has evolved. Therefore, many of the "secondary metabolic" genes that encode enzymes or regulatory proteins have probably been recruited from genes encoding primary functions. In order to understand the evolution of secondary metabolism, we have to identify the genes specific for secondary metabolic pathways, determine their function, and try to reconstruct their origins from primary metabolism by sequence and functional comparisons with putative ancestral genes. Ongoing genome projects will be indispensable in this respect.

A secondary metabolic pathway can be defined by the branch point from primary metabolism and the consecutive downstream reactions that lead to specific end products. Obviously, catalysis of the branch reaction is crucial for the establishment of a secondary metabolic pathway. This reaction produces the first intermediate, which can be processed further into "useful" products that may be fixed by natural selection. In this review, indole production and formation of the benzoxazinoid 2,4-dihydroxy-7-methoxy-2*H*-1,4-benzoxazin-3(4*H*)-one (DIMBOA) are used as examples to discuss the evolution of secondary metabolic pathways.

Many plant species respond to herbivore damage by the release of volatile compounds. Herbivore predators and parasitic wasps exploit these chemical signals to locate their prey or hosts. Several such chemically mediated tritrophic interactions have been documented for agrarian systems including lima bean, cotton, and maize.[2,3] Maize seedlings damaged by beet armyworm caterpillars release a specific cocktail of volatile terpenoids and indole that is recognized by parasitic wasps.[4] Volicitin [*N*-(17-hydroxylinolenoyl)-L-glutamine] present in the saliva of beet armyworm caterpillars was identified as the major active elicitor for the formation of volatiles in maize.[5] Recently, three genes, *Igl, stc1,* and *tps1* that are specifically elicited by volicitin have been isolated from maize. *Igl* encodes an indole-3-glycerol phosphate lyase (IGL), *stc1* and *tps1* encode sesquiterpene synthases.[6-8] IGL cleaves indole-3-glycerol phosphate (IGP) to form indole and glycerolaldehyde-3-phosphate (GAP).

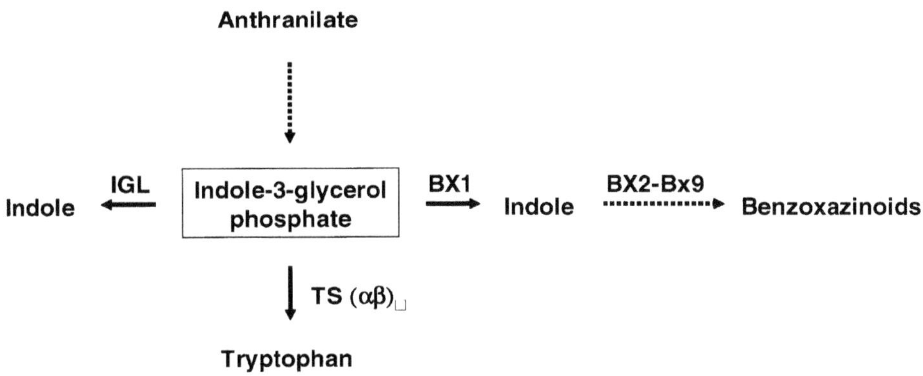

Figure 4.1: Branchpoint from primary metabolism.
Tryptophan synthase (TS) catalyzes the ultimate step in tryptophan biosynthesis (details see Fig. 4.2). Indole and benzoxazinoid secondary metabolite formation branches from this pathway. The two lyases IGL and BX1 cleave indole-3-glycerol phosphate into indole (and glycerolaldehyde-3-phosphate, not shown) and serve as committing enzymes for indole derived secondary metabolites. Indole produced by IGL directly functions as volatile signal. Indole produced by BX1 is converted by other enzymes (BX2-BX9) to benzoxazinoids that have an important function in the chemical defense of grasses.

The benzoxazinoids 2,4-dihydroxy-2*H*-1,4-benzoxazin-3(4*H*)-one (DIBOA) and its methoxy derivative DIMBOA (also called hydroxamic acids) are found predominantly in Gramineae, but also occur sporadically in dicotyledonous plants.[9] Benzoxazinoids are natural pesticides and serve as important factors of host plant resistance against microbial diseases and insects and as allelochemicals.[9] In maize, five genes encode the enzymes to synthesize DIBOA.[10] The first gene in this pathway, *Bx1*, is defined by the benzoxazinless phenotype of the *bx1* mutant, and was identified by transposon tagging.[10] *Bx1* encodes an enzyme function identical to IGL that catalyzes the formation of free indole. Four P450 monooxygenases convert indole to DIBOA, which is further modified by a 2-oxoglutarate-dependent dioxygenase and a O-methyltransferase to DIMBOA.[11] Conversion of IGP to indole is the branch reaction that leads to the production of both secondary metabolites, indole and benzoxazinoids (Fig. 4.1).

Table 4.1: Comparison of kinetic parameters of indole-3-phosphate lyase-type enzymes from *Escherichia coli* and *Zea mays*.

	E. coli TSA		*Z. mays*	
	α	$\alpha_2\beta_2$	Bx1	IGL
K_m^{IGP}	0.5 mM	0.03 mM	0.013 mM	0.1 mM
k^{cat}	0.002 s^{-1}	0.2 s^{-1}	2.8 s^{-1}	2.3 s^{-1}
k^{cat}/K_m^{IGP}	0.004 $mM^{-1}s^{-1}$	7.4 $mM^{-1}s^{-1}$	215 $mM^{-1}s^{-1}$	23 $mM^{-1}s^{-1}$

EVOLUTION OF AN INDOLE-3-GLYCEROL PHOSPHATE LYASE FUNCTION

Tryptophan synthase (TS) catalyzes the conversion of IGP and serine to tryptophan. The well-characterized bacterial TS enzyme consists of α- and β-subunits that join to form two active sites with a hydrophobic tunnel between them. TS is an $\alpha_2\beta_2$heterotetramer linked via the β-subunits.[12] The individual subunits catalyze two independent reactions: IGP is converted by the α-subunit to indole and glyceraldehyd-3-phosphate, and indole and serine are converted by the β-subunit to tryptophan and H_2O. It has been shown for bacterial enzymes that the activity of the isolated subunits is very low in comparison to their activity in the intact TS complex (Table 4.1). Indole is not released from the TS complex but rather travels through the tunnel connecting the active sites of α and β (Fig. 4.2). There is evidence that plant TS, like the bacterial complex, functions as α β heteromers.[13] The α and β subunits are encoded by independent genes (*TSA* and *TSB*) and the interaction of α and β was inferred from complementation experiments.

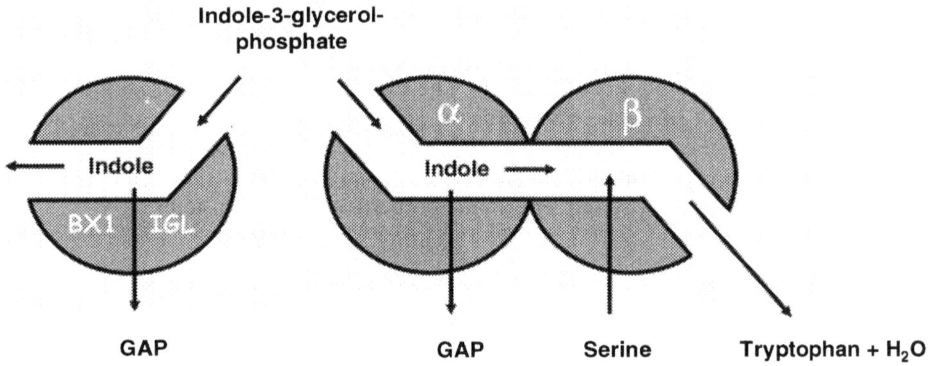

Figure 4.2: Functional comparison of the indole-3-glycerole phosphate lyases IGL and BX1 with the tryptophan synthase complex.
Tryptophan synthase (TS) catalyzes the conversion of indole-3-glycerol phosphate (IGP) and serine to tryptophan. This complex is an $(\alpha\beta)_2$heterotetramer linked via the β-subunits (only one half of the TS complex is shown). The α- and β-subunits catalyze two independent reactions: IGP is converted by the α-subunit to indole and glyceraldehyd-3-phosphate (GAP), and indole and serine are converted by the β-subunit to tryptophan and H_2O. Indole is not released from the TS complex but rather travels through the hydrophobic tunnel connecting the active sites of α and β. BX1 and IGL have homology to α-subunits and catalyze an identical lyase reaction. The difference is, however, that BX1 and IGL are highly active in monomeric form, while α-subunits have substantial activity only in the intact TS complex (Tab. 1).

The BX1 and IGL proteins from maize share an amino acid sequence identity of more than 60 % to plant TSAs. Unlike the isolated TS α subunit, BX1 and IGL can efficiently cleave IGP to form free indole without being activated by a β subunit (Fig. 4.2). Kinetic analysis of purified BX1 and IGL protein expressed in *E. coli* demonstrated that homomeric BX1 and IGL proteins are about 30-fold and 3-fold, respectively, more efficient in catalyzing IGP cleavage than the *E. coli* TS $\alpha_2\beta_2$ heterotetramer (Table 4.1).[10,6]

The genes *Bx1* and *Igl* are evolutionary related to *TSA* genes and were probably generated by gene duplication. *Igl* and TSA_{like}, the maize candidate gene for *TSA*, are separated by only 1.6 kb on chromosome 1 of maize. This close proximity is indicative for a gene duplication event.[6] The exon/intron structure of *Bx1* and *Igl* and the *Arabidopsis thaliana TSA* gene is almost conserved.[6] The amino acid sequence of BX1 and IGL deviates, however, at several positions from the TSA consensus including the domain required for interaction with TSB. These amino acid changes might reflect the different enzymatic properties of these proteins.

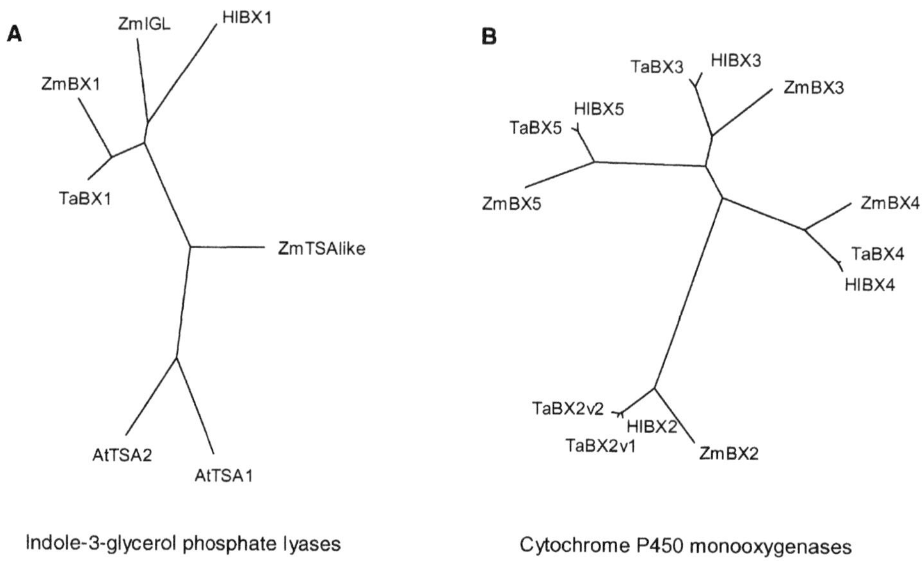

Figure 4.3: Phylogenetic tree of indole-3-glycerole phosphate lyases and cytochrome P450 enzymes involved in benzoxazinoid biosynthesis. Neighbor joining trees were constructed using the ClustalX program.[31] Putative signal sequences and gaps were omitted from the analysis. A) The amino acid sequences of *Bx1* orthologues from *Zea mays, Triticum aestivum*, and *Hordeum lechleri* are compared with *Igl* and *TSA* genes from *Zea mays* and *Arabidopsis thaliana*. B) The comparison of the amino acid sequences of the cytochrome P450 monooxygenases involved in benzoxazinoid biosynthesis included the genes from Zea mays, *Triticum aestivum*, and *Hordeum lechleri*. There are two functional genes for *Bx2* in *Triticum aestivum*. Data were taken from refs. 6, 10, 14, 32. The sequences for the *Hordeum lechleri* proteins have been deposited in GenBank and have the following accession numbers: AY462226 (HIBX1), AY462227 (HIBX2), AY462228 (HIBX3), AY462229 (HIBX4), AY462230 (HIBX5).

Bx1 orthologues have been isolated from wheat and *Hordeum lechleri*, a wild barley variety.[14,15] The comparison of the amino acid sequences of these genes with *Igl*, *TSA*$_{like}$, and the two *TSA* genes from *Arabidopsis thaliana* reveals a close relationship of the *Bx1* orthologous genes and *Igl* relative to the *TSA* genes of maize and *Arabidopsis thaliana* (Fig. 4.3A). However, the *Bx1* genes do not exactly follow the phylogeny of the grass species, *i.e*,. that barley and wheat are more closely related to each other than to maize.[16] Here, *Bx1* from maize and wheat are more closely associated, and *Igl* from maize appears to be more closely related to *Bx1* of *Hordeum lechleri*. This finding suggests that two gene duplication events occurred in the progenitor of maize, wheat, and barley. The duplicates evolved into genes for efficient indole production. In modern maize, both genes, *Bx1* and *Igl*, are active and function in DIMBOA biosynthesis and volatile indole formation, respectively. Wheat has inherited the same *Bx1* gene from the progenitor for benzoxazinoid formation. Nothing is known about an *Igl* gene in this species. The benzoxazinoid pathway is present in several wild barley varieties.[15] In *Hordeum lechleri*, it seems that the first reaction in benzoxazinoid biosynthesis is catalyzed by an enzyme encoded by the other gene duplicate. The original *Bx1* function might have been lost in this lineage and *Igl* recruited to function in DIBOA biosynthesis.

In summary, a gene from primary metabolism (*TSA*) was duplicated twice and subsequently recruited for secondary metabolism. In this process, *Bx1* and *Igl* evolved to obtain their specific functions. Not only did the enzymatic properties have to be modified such that free indole is produced, but the expression pattern also had to be altered in order for the genes to function in secondary metabolism. While *TSA* transcripts are expressed in the whole plant at a relatively low level, *Bx1* is under developmental control in the young seedling and expressed strongly in certain tissues. *Igl* is massively induced at a later developmental stage in leaves in response to herbivore damage.[6,10]

The synthesis of several other plant metabolites, such as auxin, indole glucosinolates, anthranilate-derived alkaloids, and tryptamine derivatives, could depend on indole as an intermediate.[17,18] Indole is also found in the scent of flowers such as lilac and robinia. Therefore, it is possible that the recruitment of an indole-3-glycerol phosphate lyase function from TSA genes might have occurred independently several times during plant evolution.

There are two other examples for the recruitment of genes from primary metabolism. The homospermidine synthase from *Senecio vernalis* is derived from deoxyhypusine synthase, an enzyme required for activation of translation factor 5A,[19] and a serine carboxypeptidase-like protein that functions as an acyltransferase in secondary metabolism has been found in *Arabidopsis thaliana*.[20]

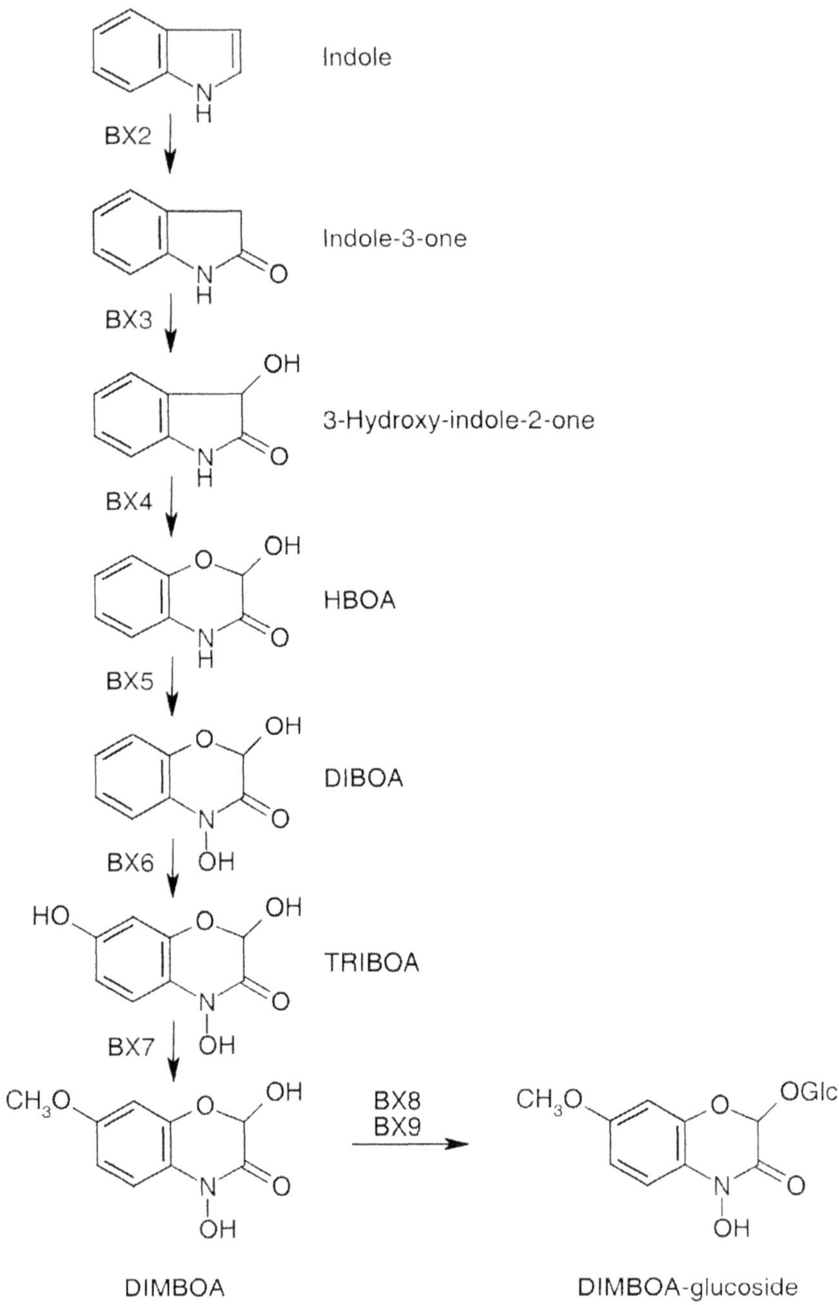
Indole
BX2
Indole-3-one
BX3
3-Hydroxy-indole-2-one
BX4
HBOA
BX5
DIBOA
BX6
TRIBOA
BX7
BX8
BX9
DIMBOA
DIMBOA-glucoside

CONVERSION OF INDOLE TO BENZOXAZINOIDS

The biosynthesis of benzoxazinones commences by conversion of indole to DIBOA. In certain grasses like rye, DIBOA is glycosylated and stored in the vacuole. In other species like maize and wheat, DIBOA is first converted to its 7-methoxy derivative DIMBOA and then glycosylated for vacuolar storage (Fig. 4.4).[9] The introduction of four oxygen atoms into the indole moiety that yields DIBOA is catalyzed by four cytochrome P450-dependent monooxygenases. These enzymes are membrane-bound heme-containing mixed function oxidases. They utilize NADPH or NADH to reductively cleave molecular oxygen to produce functionalized organic products and a molecule of water. In this generalized reaction, reducing equivalents from NADPH are transferred to the P450 enzyme via a flavin-containing NADPH-P450 reductase. In plants, P450 enzymes are involved mainly in hydroxylation or oxidative demethylation reactions of a large variety of primary and secondary metabolites including hormones, phytoalexins, xenobiotics, and pharmaceutically relevant compounds. The plant P450 genes represent a fairly large gene family. In *Arabidopsis thaliana,* 286 P450 genes have been annotated.[1] Even a greater number of P450 genes can be expected in plants containing more secondary metabolites.

The four P450 genes involved in DIBOA biosynthesis have been termed *Bx2-Bx5*.[10] They are members of the *CYP71C* subfamily of plant cytochrome P450 genes and share an overall amino acid identity of 45 to 65%. The stepwise conversion of indole to DIBOA occurs as follows (Fig. 4.4): BX2 catalyzes the formation of indolin-2(1*H*)-one, which is converted to 3-hydroxy-indolin-2(1*H*)-one by BX3. Then, BX4 catalyzes the conversion of 3-hydroxy-indolin-2(1*H*)-one to 2-hydroxy-2*H*-1,4-benzoxazin-3(*4H*)-one (HBOA). This unusual ring expansion was investigated by labeling experiments, and a mechanism for this transformation was proposed.[21] The N-hydroxylation of HBOA to DIBOA is catalyzed by BX5. The presence of the N-hydroxyl in the cyclic hemiacetal is a unique feature of benzoxazinones. From the chemist's point of view, this is the structural source of a certain instability, which is essential to obtain the chemical reactivity required for the

Figure 4.4: Benzoxazinoid biosynthetic pathway in maize.
Indole is synthesized in the plastid by BX1. The cytochrome P450 Enzymes BX2 through BX5 convert indole to DIBOA (2-hydroxy-2*H*-1,4-benzoxazin-3(4*H*)-one). BX6, a 2-oxo-glutarate-dependent dioxygenase, catalyzes the subsequent formation of TRIBOA (2,4,7-trihydroxy-2*H*-1,4-benzoxazin-3(*H*)-one). The O-methyltransferase for conversion of TRIBOA to DIMBOA has not yet been isolated. DIMBOA (2,4-dihydroxy-7-methoxy-2*H*-1,4-benzoxazin-3(4*H*)-one) is converted to the respective D-glucosides by the glucosyltransferases BX8 and BX9.

defense reaction. This reactivity explains the broad resistance against microbes, fungi, and insects that is conferred by benzoxazinones.

The sequence homology, the similar exon/intron structure, and the gene clustering of *Bx2-Bx5* (see below) indicate that these genes have been derived by gene duplications from one precursor.[10] However, each of the four P450 enzymes has evolved a high degree of distinct substrate specificity. Only one intermediate in the pathway is converted by each respective P450 enzyme to a specific product. Each enzyme is specific for the introduction of one specific oxygen atom in the DIBOA molecule. The relatively high specificity of the enzymes seems to support the idea that plant P450s generally have a much greater substrate specificity than their animal homologues. However, there is emerging evidence that plant P450s in addition to their normal physiological function, also can convert certain xenobiotics with varying efficiencies. For example, the artificial substrate *p*-chloro-*N*-methylaniline (pCMA) is efficiently demethylated by BX2 and by several other plant P450 enzymes.[22]

The function of *Bx2-Bx5* was also determined in wheat[14] and *Hordeum lechleri*.[15] These genes are true orthologues. The phylogenetic comparison (Fig. 4.3B) shows that the four P450 genes were already present in the progenitor of maize, wheat, and barley. In the four branches of the phylogenetic tree, the orthologous genes of wheat and barley are always more closely related to each other than to the maize genes and, thus, reflect the expected phylogeny.[16]

In maize, DIBOA is converted to its 7-methoxy derivative DIMBOA via hydroxylation and consecutive methylation (Fig. 4.4). The hydroxylation at C-7 is catalyzed by a 2-oxoglutarate-dependent dioxygenase, which is encoded by *Bx6*.[23] Hence, two functionally different classes of oxygenases are involved in the biosynthesis of DIMBOA. P450 enzymes and 2-oxoglutarate-dependent dioxygenases catalyze (among other reactions) oxidation reactions that lead to the incorporation of oxygen atoms from molecular oxygen.[24,25] Like the P450 genes, the 2-oxoglutarate-dependent dioxygenases represent a fairly large gene family. In *Arabidopsis thaliana*, 54 genes encoding these dioxygenases have been annotated.[1] It has been demonstrated recently that apparent gene duplication and diversification of 2-oxoglutarate-dependent dioxygenases genes have a significant impact on diversity of the secondary metabolism in plants.[26] The conversion of TRIBOA to DIMBOA is catalyzed by an O-methyltransferase encoded by *Bx7*. This gene is defined genetically and remains to be molecularly cloned and investigated *in vitro*.[6]

The maize genes *Bx8* and *Bx9* encode specific benzoxazinoid UDP-glucosyltransferases Their gene products convert DIBOA and DIMBOA to the respective D-glucosides.[11] Both enzymes are specific for DIBOA and DIMBOA as substrates. Glucosylation is probably required for vacuolar storage of benzoxazinoids. According to the sequence similarity and the conserved exon/intron structure, *Bx8* and *Bx9* also represent duplicated genes and are members of a gene family. In *Arabidopsis thaliana*, this divergent gene family comprises 112 members.[27]

CELLULAR COMPARTMENTATION OF THE BENZOXAZINOID BIOSYNTHETIC ENZYMES

Formation of indole by BX1 takes place in the plastid.[10] The conversion of indole to DIBOA by consecutive oxidation is catalyzed by BX2-BX5. These P450 enzymes are localized in the endoplasmatic reticulum. Very likely, conversion of DIBOA to DIMBOA by BX6 and BX7 takes place in the cytoplasm. Biosynthesis commences by glycosylation followed by transport and storage of the glucosides in the vacuole. The β-glycosidases GLU1 and GLU2 required for activation of the glucosides are stored in the plastid.[28] In the case of cell wounding, the two cellular organelles are damaged and the toxic aglucones are produced.

BX GENES ARE CLUSTERED ON ONE CHROMOSOME

A unique feature of the *Bx* genes in maize is that a completed set of the biosynthetic genes is clustered on the short arm of chromosome 4 (Fig. 4.5).[23] Gene clustering is often associated with gene duplication. Therefore, the relative close arrangement of the P450 genes *Bx2-Bx5* within 6 cM is not unexpected. However, the P450 genes are tightly linked to the *Bx1* gene and to *Bx8* encoding the DIBOA/DIMBOA specific glucosyltransferase. *Bx1* and *Bx2* are separated by only 2.5 kb, but the exact position of *Bx8* relative to these two genes remains to be determined. Bx6 and Bx7 are also associated with the cluster. The gene cluster comprises five different enzymatic functions and a complete set of genes for the biosynthesis of DIBOA and DIMBOA glucosides. Only *Bx9*, the duplicate of Bx8, is located outside of the cluster on chromosome 1. At present, there is no other example of plant genes integrated in one biosynthetic pathway that are all arranged in one gene cluster.

The clustering of some *Bx* genes was analyzed in two other cereals that are distantly related to maize.[10] In hexaploid wheat, the *Bx1* and *Bx2* orthologues are present on chromosome 4 of all three genomes (A,B,D), and *Bx3-Bx5* orthologues have been localized on the short arms of chromosomes 5A, 5B, 5D. In rye, homeo loci for *Bx1* and *Bx2* are located on chromosome 7R and those for *Bx3-Bx5* on chromosome 5R.[14] Triticeae chromosomes 5R are synthenic to maize 4S.[29] This suggests that the *Bx* gene cluster is an ancient feature. The maize genome could represent the original gene organization, while in the Triticeae the cluster has been separated into two parts by chromosomal translocation.

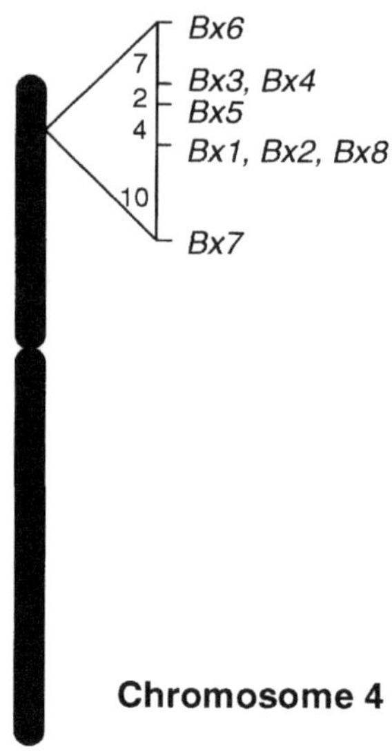

Figure 4.5: The *Bx* gene cluster.
In maize a complete set of the benzoxazinoid biosynthetic genes is clustered on the short arm of chromosome 4. The genetic distances are given in centi Morgan.

It is unclear if gene clustering has any influence on the expression of *Bx* genes. Since the *Bx* genes are genetically linked, they will frequently be transferred to the next generation as one functional unit, encoding all enzymes required for the biosynthesis of DIMBOA. Whether this genetic co-segregation is of any advantage for maize is presently unclear. One could speculate that the loss of one enzyme would interrupt the pathway, which could lead to the formation of a potentially deleterious intermediate.

EVOLUTION OF BENZOXAZINOID BIOSYNTHESIS

Benzoxazinoids are widely distributed in grasses and are found only sporadically in three dicotyledoneous species, the Acanthaceae, Ranunculaceae, and Scrophulariaceae, suggesting that the acquisition of this pathway occurred relatively early in the evolution of Gramineae, probably even before monocots and dicots diverged.[9] However, the orthologous nature of the genes has only been proven thus far for *Bx1-Bx5* in gramineae.[22] It remains to be shown whether *Bx6-Bx9* have also a common evolutionary origin.

SUMMARY AND FUTURE DIRECTIONS

Gene duplications seem to play an important role in the evolution of secondary metabolic pathways. In the examples presented, duplicated *TSA* genes from primary metabolism are recruited for production of free indole. This compound is either used directly for signaling in the tritrophic interaction with insects or converted to a defense chemical. For the latter steps, genes have been duplicated and recruited for benzoxazinoid biosynthesis. All these genes are members of gene families that include cytochrome P450 monooxygenases, 2-oxoglutarate-dependent dioxygenases, and UDPG-glycosyltransferases. All enzymes have evolved such that they exhibit a high degree of substrate specificity. The DIMBOA pathway is a good example to illustrate that redundancy potentially created by gene duplication does not necessarily result in functional or genetic redundancy, because the gene products have evolved towards a defined substrate specificity, and their specific expression patterns generate non-overlapping functions. In the *Arabidopsis thaliana* genome sequence, a fairly high degree of gene duplication was detected.[1] Detailed analysis indicated that these duplications are not due to a single polyploidization event.[30] Rather, they have accompanied the evolution of *Arabidopsis thaliana* for the last 200 million years. The detailed analysis of other plant genomes suggests that a high degree of gene duplications may also be characteristic for their evolution.

The structures of the biosynthetic genes for indole and benzoxazinoid formation have been identified and it has now been shown that these genes are expressed in a tissue-specific manner during early stages of maize development. In the future, the *cis*-elements and *trans*-factors controlling the expression of these genes can be analyzed. The benzoxazioid biosynthesis can also serve as a model for the evolution of the regulatory requirements of other secondary metabolic pathways.

REFERENCES

1. The Arabidopsis genome initiative, Analysis of the genome sequence of the flowering plant *Arabidopsis thaliana*. *Nature*, 2001, **408**, 796-814.
2. DICKE, M., SABELIS, M.W., TAKABAYASHI, J., BRUIN, J., POSTHUMUS, M.A., Plant strategies for manipulating predator-prey interactions through allelochemicals: Prospects for the application in pest-control, *J. Chem. Ecol.*, 1990, **16**, 3091-3118.
3. TURLINGS, T.C.J., TUMLINSON, J.H., LEWIS, W.J., Exploitation of herbivore-induced plant odors by host-seeking parasitic wasps, *Science*, 1990, **250**, 1251-1253.

4. TURLINGS, T.C., TUMLINSON, J.H., HEATH, J.H., PROVEAU, A.T., DOOLITTLE, R.E., Isolation and identification of allelochemicals that attract the larval parasitoid *Cortesia marginiventris* (Cresson) to the microhabitat of one of its hosts, *J. Chem. Ecol.*, 1991, **17**, 2235-2251.
5. ALBORN, H.T., TURLINGS, T.C., JONES, T.H., STENHAGEN, G., LOUGHRIN, J.H., TUMLINSON, J.H., An elicitor of plant volatiles from beet armyworm oral secretion, *Science*, 1997, **276**, 945-949.
6. FREY, M., STETTNER, C., PARE, P.W., SCHMELZ, E.A., TUMLINSON, J.H., GIERL, A., A herbivore elicitor activates the gene for indole emission in maize, *Proc. Natl. Acad. Sci. USA*, 2000, **97**, 14801-14806.
7. SHEN, B., ZHENG, Z., DOONER, H.K., A maize sesquiterpene cyclase gene induced by insect herbivory and volicitin: Characterization of wild-type and mutant alleles, *Proc. Natl. Acad. Sci. USA*, 2000, **97**, 14807-14812.
8. SCHNEE, C., KOLLNER, T.G., GERSHENZON, J., DEGENHARDT, J., The maize gene terpene synthase 1 encodes a sesquiterpene synthase catalyzing the formation of (E)-beta-farnesene, (E)-nerolidol, and (E,E)-farnesol after herbivore damage, *Plant Physiol.*, 2002, **130**, 2049-2060.
9. SICKER, D,, FREY, M., SCHULZ, M., GIERL, A., Role of natural benzoxazinones in the survival strategy of plants, *Int. Rev. Cytol.*, 2000, **198**, 319-346.
10. FREY, M., CHOMET, P., GLAWISCHNIG, E., STETTNER, C., GRUN, S., WINKLMAIR, A., EISENREICH, W., BACHER, A., MEELEY, R.B., BRIGGS, S.P., SIMCOX, K., GIERL, A., Analysis of a chemical plant defense mechanism in grasses, *Science*, 1997, **277**, 696-699.
11. VON RAD, U., HÜTTL, R., LOTTSPEICH, F., GIERL, A., AND FREY, M., Two glucosyltransferases are involved in detoxification of benzoxazinoids in maize, *Plant J.*, 2001, **28**, 633-642.
12. CREIGHTON, T.E., YANOFSKY, C., Association of the alpha and beta-2 subunits of the tryptophan synthetase of *Escherichia coli*, J. Biol. Chem., 1966, **241**, 980-990.
13. RADWANSKI, E.R., LAST, R.L., Tryptophan biosynthesis and metabolism: Biochemical and molecular genetics, *Plant Cell*, 1995, **7**, 921-934.
14. NOMURA, T., ISHIHARA, A., IMAISHI, H., OHKAWA, H., ENDO, T.R., IWAMURA, H., Rearrangement of the genes for the biosynthesis of benzoxazinones in the evolution of Triticeae species. *Planta*, 2003, **217**, 776-782.
15. GRUEN, S., Die Evolution der Benzoxazinoid-Biosynthese in den Gramineae. PhD thesis, 2001, Technische Universität München, Germany.
16. GAUT, B.S., LE THIERRY D'ENNEQUIN, M., PEEK, A.S., SAWKINS, M.C., Maize as a model for the evolution of plant nuclear genomes, *Proc. Natl. Acad. Sci. USA*, 2000, **97**, 7008-7015.
17. RADWANSKI, E.R., ZHAO, J., Last, R.L., *Arabidopsis thaliana* tryptophan synthase alpha: gene cloning, expression, and subunit interaction, *Mol. Gen. Genet.*, 1995, **248**, 657-667
18. KUTCHAN, T.M., Alkaloid biosynthesis-The basis for metabolic engineering of medical plants, *Plant Cell*, 1995, **7**, 1059-1070.

19. OBER, D., HARTMANN, T., Homospermidine synthase, the first pathway-specific enzyme of pyrrolizidine alkaloid biosynthesis, evolved from deoxyhypusine synthase, *Proc. Natl. Acad. Sci. USA*, 1999, **96**, 14777-14782.
20. LEIGHTON, V., NIEMEYER, H.M., JONSSON, L.M.V., Substrate specificity of a glucosyltransferase and a *N*-hydroxylase involved in the biosynthesis of cyclic hydroxamic acids in Gramineae, *Phytochemistry*, 1994, **36**, 887-892.
21. SPITELLER, P., GLAWISCHNIG, E., GIERL, A., STEGLICH, W., Studies on the biosynthesis of 2-hydroxy-1,4-benzoxazin-3-one (HBOA) from 3-hydroxy-indolin-2-one in *Zea mays*, *Phytochemistry*, 2001, **57**, 373-376.
22. GLAWISCHNIG, E., GRUEN, S., FREY, M., GIERL, A., Cytochrome P450 monooxygenases of DIBOA biosynthesis: Specificity and conservation among grasses, *Phytochemistry*, 1999, **50**, 925-930.
23. FREY, M., HUBER, K., PARK, W,J., SICKER, D., LINDBERG, P., MEELEY, R.B., SIMMONS, C.R., YALPANI, N., GIERL, A., A 2-oxoglutarate-dependent dioxygenase is integrated in DIMBOA-biosynthesis, *Phytochemistry*, 2003, **62**, 371-376.
24. HALKIER, B.A., Catalitic reactivities and strukture/fuction relationships of cytochrome P450 enzymes, *Phytochemistry*, 1996, **43**, 1-21.
25. QUE, L.J., HO, R.Y.N., Dioxygen activation by enzymes with mononuclear non-heme iron active sites, *Chem. Rev.*, 1996, **96**, 2607-2624.
26. KLIEBENSTEIN, D.J., LAMBRIX, V.M., REICHELT, M., GERSHENZON, J., MITCHELL-OLDS, T., Gene duplication in the diversification of secondary metabolism: Tandem 2-oxoglutarate-dependent dioxygenases control glucosinolate biosynthesis in *Arabidopsis*, *Plant Cell*, 2001, **13**, 681-693.
27. PAQUETTE, S., MOLLER, B.L., BAK, S., On the origin of family 1 plant glycosyltransferases, *Phytochemistry*, 2003, **62**, 399-413.
28. CICEK,M., ESEN, A., Expression of soluble and catalytically active plant (monocot) beta-glucosidases in *E. coli*, *Biotechnol Bioeng.*, 1999, **63**, 392-400.
29. DEVOS, K.M., GALE, M.D., Comparative genetics in the grasses, *Plant Mol. Biol.*, 1997, **35**, 3-15.
30. VISION, T.J., BROWN, D.G., TANKSLEY, S.D., The origins of genomic duplications in *Arabidopsis*, *Science*, 2000, **290**, 2114-2117.
31. THOMPSON, J.D., GIBSON, T.J., PLEWNIAK, F., JEANMOUGIN. F., HIGGINS, D.G., The CLUSTALX windows interface: Flexible strategies for multiple sequence alignment aided by quality analysis tools, *Nucleic Acids Res.*, 1997, **25**, 4876-4882.
32. NOMURA, T., ISHIHARA, A., IMAISHI, H., ENDO, T.R., OHKAWA, H., IWAMURA, H., Molecular characterization and chromosomal localization of cytochrome P450 genes involved in the biosynthesis of cyclic hydroxamic acids in hexaploid wheat, *Mol. Genet. Genomics*, 2002, **267**, 210-217.

Chapter Five

GENOMICS, GENETICS, AND BIOCHEMISTRY OF MAIZE CAROTENOID BIOSYNTHESIS

Eleanore T. Wurtzel*

Department of Biological Sciences
Lehman College
The City University of New York (CUNY)
250 Bedford Park Boulevard West
Bronx, New York 10468

and The Graduate School and University Center-CUNY
365 Fifth Avenue
New York, New York 10016

**Author for correspondence:* etwlc@cunyvm.cuny.edu

INTRODUCTION

The Poaceae or grass family represents some of the most important food crops world-wide, and includes the related grasses, maize, wheat, barley, sorghum, pearl millet, and rice.[1] The endosperm tissues of these taxonomically related crops serve as major food staples, though they are deficient in adequate levels of nutritionally essential carotenoids. In humans and animals, various carotenoids derived from plant sources act as antioxidants and protect against certain diseases, while other carotenoids are precursors to vitamin A and to retinoid compounds involved in development.[2-4] Endosperms of these food crops are also low in provitamin A (1-10 %) as compared with nonprovitamin A carotenoids.[5,6] The consumption of carotenoid-poor cereal crops is associated with vitamin A deficiency, affecting 250 million children in developing countries.[7] Effects of vitamin A deficiency are manifested as xerophthalmia (visual impairment), blindness, increased mortality due to increased severity of childhood diseases such as measles, diarrhea, and increased maternal transmission of viruses such as HIV.[8] One approach to alleviating worldwide deficiencies associated with consumption of carotenoid poor food sources is to improve the level and composition of carotenoids in the endosperm of maize, wheat, sorghum, pearl millet, and rice, among others. Maize is an excellent model for the grasses, because of its importance as a food staple worldwide and because of its associated foundation of genetic and biochemical knowledge. To develop a comprehensive understanding of how carotenoid accumulation is regulated in cereal endosperm, genetic tools are being integrated with genomic resources for maize and other grasses, along with molecular/biochemical approaches. These various tools are being used for identification and characterization of the structural and regulatory genes affecting the biosynthetic pathway and are leading to elucidation of the underlying mechanisms regulating carotenoid accumulation in endosperm tissue.

What Are Carotenoids?

Carotenoids are a large class (numbering over 600 structures) of yellow, red, and orange pigments derived from isoprenoids, as represented by beta-carotene, which colors carrots orange and lycopene which colors tomatoes red. Carotenoids are synthesized by all photosynthetic organisms, as well as some bacteria and fungi. Carotenogenic bacteria and other nonphotosynthetic organisms synthesize carotenoids to provide protection in high-light, oxygen-containing environments. In plants, the biosynthesis of carotenoids is essential for growth and development; carotenoids function as accessory pigments in photosynthesis, as photoprotectors preventing photooxidative damage, and as precursors to various apocarotenoids including the plant hormone, abscisic acid (ABA).[9-12] The presence of carotenoids in

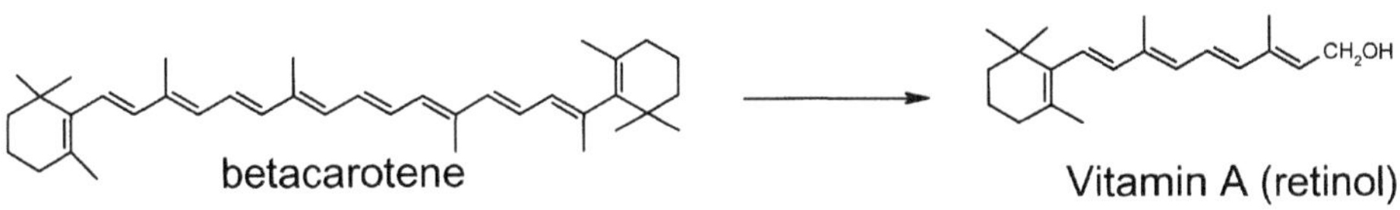

Fig. 5.1: Betacarotene a precursor of vitamin A

plant endosperm tissue adds nutritional value. The symmetrical betacarotene, having beta rings at both ends, can be cleaved into two molecules of vitamin A (Fig. 5.1)[13] and, therefore, has the highest provitamin A activity, compared to other carotenoids such as alphacarotene or betacryptoxanthin that have beta rings at only one end. Nonprovitamin A carotenoids, such as lycopene, lutein, zeaxanthin, and others, also play beneficial roles in human health.[14-16] Geometric isomer states of carotenoids add to a great diversity of structures and influence the biological activities of carotenoids, including intestinal absorption, tissue localization, and biosynthetic metabolic channeling.[17-21] Animals do not have the ability to synthesize carotenoids, but must obtain them typically through dietary plant sources.

THE CAROTENOID BIOSYNTHETIC PATHWAY

Carotenoids are derived from the 20 carbon geranylgeranyl pyrophosphate, GGPP, the first precursor to carotenoids, and to a variety of other isoprenoid-derived pathways, including gibberellins, the phytol chain of chlorophyll, prenylquinones, tocopherols, and other natural products (Fig. 5.2).[22] The carotenoid biosynthetic pathway in maize endosperm requires activity of **PSY** (phytoene synthase), **PDS** (phytoene desaturase), **ZDS** (zetacarotene desaturase) and **ISO** (carotene isomerase), to convert 15-*Z*- phytoene (15-cis-phytoene) to all-*E*-lycopene (*all-trans* lycopene) (Fig. 5.3). With introduction of rings by the cylase enzymes, **LCYB** (lycopene beta cyclase) or **LCYB** in combination with **LCYE** (lycopene epsilon cyclase), the pathway diverges towards two alternate routes to produce betacarotene having two beta rings or alphacarotene containing one beta ring and one epsilon ring, respectively. Hydroxylation of the carotenes to the nonprovitamin A xanthophylls, zeaxanthin and lutein, require activity of hydroxylase enzymes (**HYD**). Therefore, **HYD** and **LCYE** enzymes divert the pathway to compounds that are lower in provitamin A value. The phytohormone, ABA, which plays a role in seed dormancy,is produced from zeaxanthin, though its production does not necessarily have to originate from the endosperm.[23] Enzyme activities required for biosynthesis of carotenoid isoprenoid precursors also control carotenoid pathway flux, which include **DXS** (D-1-deoxyxylulose 5-phosphate synthase or **DXP** synthase), **DXR** (**DXP** reductoisomerase), **IPPI** (isopentenyl pyrophosphate isomerase, IPP isomerase), and **GGPPS** (GGPP synthase), and, therefore, have an "upstream" effect on carotenoid accumulation (Fig. 5.4).

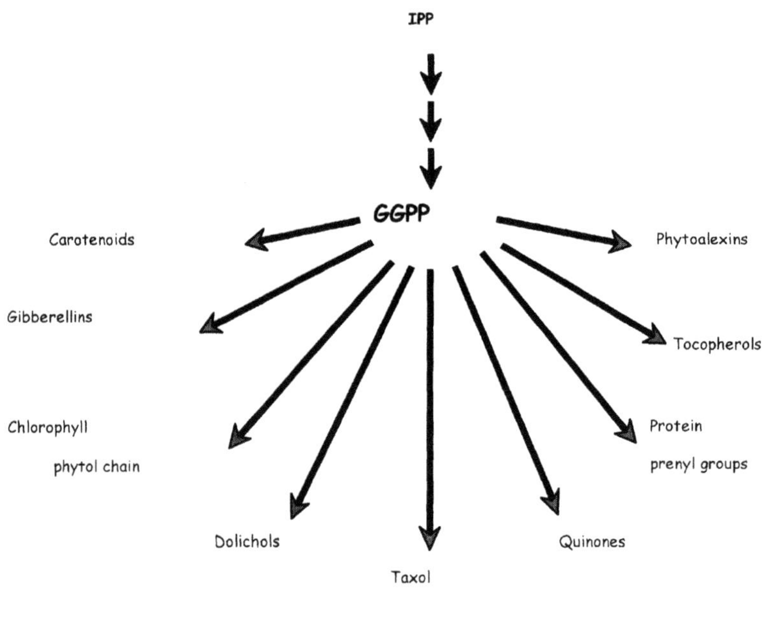

Fig. 5.2: GGPP as a common precursor to multiple terpenoid pathways.

LOCALIZATION

Plastid Localization of Biosynthesis

The biosynthesis of carotenoids occurs on membranes of chloroplasts, chromoplasts, and amyloplasts, genetically identical plastids of very different internal membrane architecture. The enzymes are encoded in the nucleus and targeted to the plastids.[11,24] Therefore, a major question regarding regulation of carotenoid biosynthesis in higher plants is how the pathway is regulated in different plastid types. Carotenoids are found in chloroplasts both on outer envelope membranes and on thylakoid membranes, whereas endosperm amyloplasts possess only envelope membranes. Carotenoid enzymes have been localized to both membrane sites.[25-27] Therefore, the carotenoid pathway should be considered as two pathways that are localized to different membranes, depending on the plastid. It is presently unclear how membrane targeting and metabolon assembly are regulated in plastids of different membrane architecture. Moreover, in the case of single copy genes encoding pathway enzymes, there must be some mechanism to control membrane-specificity of metabolon assembly, and this mechanism is unknown. In chloroplasts, where metabolons may potentially form on two alternate membranes, regulated intraorganellar sorting should facilitate membrane specificity and not depend on a fortuitous process. The possibility of auxiliary factors involved in routing is suggested by *in vitro* chloroplast import experiments; **LCYB** targeting to thylakoid membranes of pea chloroplasts was inhibited by a protease-sensitive thylakoid factor.[25] In addition to these uncharacterized auxiliary factors, there is biochemical evidence of chaperonins, Hsp70 and Cpn60, that facilitate localization of carotenoid enzymes in daffodil flower chromoplasts and whose expression is associated with carotenogenesis.[28,29] In algae, the lipid composition appears to play a role in carotenoid deposition,[30] while in daffodil chromoplasts, galactolipids appear to play a role in the catalytic activity but not membrane anchoring of **PSY**.[31] In some plants, carotenoid binding proteins play a role in carotenoid sequestration.[32]

Accumulation in a Maize Seed

Carotenoids accumulate throughout the maize seed in starch-bearing amyloplasts, primarily in the endosperm and to a lesser extent, in the embryo. Accumulation in developing maize endosperm occurs as early as 10-15 days after pollination (DAP) and reaches a maximum concentration usually around 20-25 DAP, depending on maize variety and environmental condition.[27] The *y1* locus originally

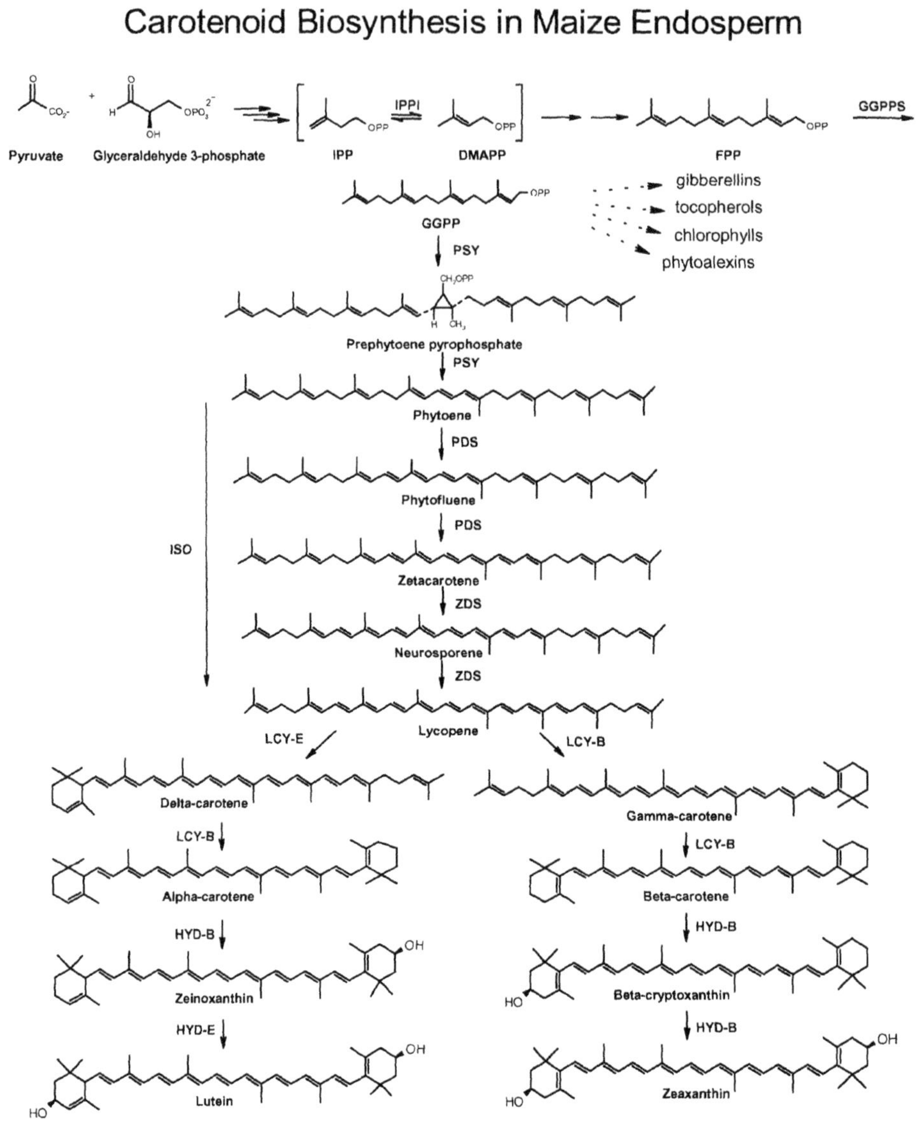
Carotenoid Biosynthesis in Maize Endosperm
Pyruvate
Glyceraldehyde 3-phosphate
IPP
IPPI
DMAPP
FPP
GGPPS
GGPP
gibberellins
tocopherols
chlorophylls
phytoalexins
PSY
Prephytoene pyrophosphate
PSY
Phytoene
PDS
Phytofluene
PDS
ISO
Zetacarotene
ZDS
Neurosporene
ZDS
Lycopene
LCY-E
LCY-B
Delta-carotene
Gamma-carotene
LCY-B
Alpha-carotene
LCY-B
Beta-carotene
HYD-B
Zeinoxanthin
HYD-B
Beta-cryptoxanthin
HYD-E
Lutein
HYD-B
Zeaxanthin

was thought to encode a regulator of maize endosperm carotenoid content as its dosage was correlated with carotenoid level[33,34] and may correspond to a QTL affecting total carotenoid composition in maize kernels.[35] With the isolation and sequencing of the gene by Buckner *et al.*[36,37] and functional testing (Gallagher, Li, and Wurtzel, unpub.), this locus is now recognized as encoding PSY, a rate-controlling pathway enzyme.[11]

GENE REGULATION IN HIGHER PLANT CAROTENOID BIOSYNTHESIS

The biosynthetic pathway is regulated by controlling enzyme activity both within the pathway and upstream of the pathway. From the study of primarily noncereal plants, accumulation of specific carotenoids is commonly regulated by modulating levels of transcripts for the biosynthetic enzymes,[38-40] although this is not the only level of regulation.[29]

Regulation Within the Pathway

Carotenoid accumulation that occurs in the transition of green to red (lycopene-accumulating) tomato fruit chromoplasts is mediated by transcriptional regulation of a gene encoding a fruit-specific **PSY**, and to a lesser degree the gene encoding **PDS**;[38] specific accumulation of lycopene is due to a decrease in transcripts for **LCYB**.[41] Carotenoid accumulation during maize endosperm development is accompanied by increased levels of **PSY** transcripts, whereas **PDS** transcripts are constant.[42,43] In transgenic plant experiments, where the **PSY** transcript level has been increased or decreased, a corresponding change in carotenoids resulted.[44-47] Transcriptional regulation of **PSY** has also been observed in

Fig. 5.3: Carotenoid biosynthesis in maize endosperm. Compounds: IPP, isopentenyl pyrophosphate; FPP, farnesyl pyrophosphate; GGPP, geranylgeranyl pyrophosphate; DMAPP, dimethallyl pyrophosphate. Carotenoid biosynthetic pathway enzymes: PSY, phytoene synthase; PDS, phytoene desaturase; ZDS, zetacarotene desaturase; ISO, carotene isomerase; LCY-B, lycopene beta cyclase; LCY-E, lycopene epsilon cyclase; HYD-B, beta-carotene hydroxylase; HYD-E, alpha-carotene hydroxylase; Isoprenoid biosynthetic pathway enzymes: IPPI (IPP isomerase); GGPPS (GGPP synthase). Structures are not representative of the geometrical isomer substrates (e.g. *Z*-phytoene is a bent structure).

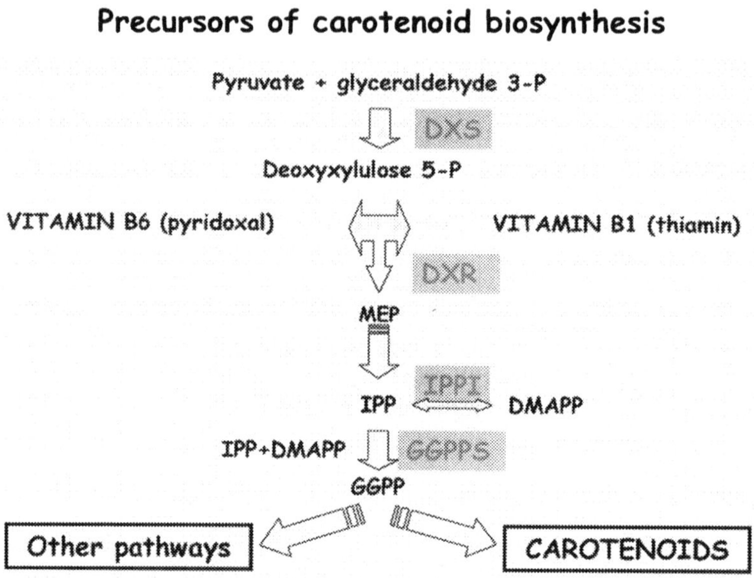

Fig. 5.4: Precursors of carotenoid biosynthesis. Abbreviations for intermediates are: MEP, methylerythritol phosphate; IPP, isopentenyl pyrophosphate; GGPP, geranylgeranyl pyrophosphate; DMAPP, dimethallyl pyrophosphate. Enzymes are shown to the right of the steps catalyzed in plant plastids. DXS (D-1-deoxyxylulose 5-phosphate synthase, DXP synthase); DXR (DXP reductoisomerase); IPPI (IPP isomerase); GGPPS (GGPP synthase).

photomorphogenesis; the potential for inducing carotenoid accumulation associated with photomorphogenesis was regulated at the transcriptional level for **PSY** genes of white mustard and *Arabidopsis thaliana*,[40] however, the accumulation of carotenoids was limited by the photoconversion of protochlorophyllide to chlorophyll.

Regulation Upstream of the Pathway

In pepper, which also has a carotenoid-rich (mainly capsanthin and capsorubin) fruit, the chromoplast, as in daffodil flowers and tomato fruits, is also derived from a chloroplast. Induction of carotenoid accumulation is mediated by transcriptional regulation at a step upstream of the carotenoid biosynthetic pathway;

a dramatic increase in transcripts encoding **GGPPS**, the enzyme responsible for production of the GGPP substrate of **PSY**, is associated with carotenoid biosynthesis and accumulation that accompanies the conversion of fruit chromoplasts from chloroplasts.[48] Furthermore, transgenic plants engineered to over-express enzymes of the carotenoid biosynthetic pathway, without modification of **GGPPS** expression, manifest deficiencies in gibberellins, end-products of a pathway competing for GGPP.[49] This suggests that the pathway can be regulated not only within the pathway, but by modulating the flow of substrates to the pathway, although it is unclear how the **GGPPS** specifically provides GGPP to **PSY** and not to the other competing pathways that also use GGPP as a precursor. Another example of such "upstream regulation" is the light-induced activation of **IPPI** that is associated with the phytochrome-mediated increase of carotenoids.[50]

POTENTIAL FOR IMPROVING MAIZE ENDOSPERM CAROTENOID CONTENT

Compared to other fruits and vegetable, carotenoid accumulation in maize endosperm is orders of magnitude lower.[27,51] The primary compounds accumulating are zeaxanthin and lutein, the ratio of which is highly variable and further accompanied by smaller amounts of the provitamin A compounds, alpha-carotene, betacarotene, and betacryptoxanthin. The earlier pathway intermediates are generally not detected, unless there is a mutation conferring a block in the pathway but which generally causes plant lethality.[52-55] Recent surveys of diverse maize germplasm and F1 hybrids have revealed extensive variation in carotenoid content and composition (T. Rocheford, pers. comm.). Therefore, there is potential for enhancement of carotenoid content and composition in maize endosperm given selection or introduction of the appropriate genes.

TOOLS FOR GENE DISCOVERY AND ENZYME ANALYSIS

Genome Sequence Databases

Genes for most of the pathway enzymes in higher plants have been isolated and it is possible to find many ESTs in GenBank with homology to known genes. Since some of those sequences represent transcripts from different maize cultivars, they are useful in identifying allelic variation. In filling the gaps for the maize genes, it has also been useful to use available sequences to search the fully sequenced rice genome; the rice DNA probes are effective in both isolating the maize genes as well as mapping the genes to chromosomes (Table 5.1). Bacterial artificial chromosome (BAC) libraries representing the maize B73 inbred line are commercially available as

Table 5.1: Carotenoid Biosynthetic Pathway: Integrated mapping of genetic, molecular, and QTL loci

Maize Locus		Chromosome bin	Reference
Genetic loci			
	cl1	3.04	a
	Clm1	8.00-8.09	a
	lty1	??	a
	lw1	1.10	a
	lw2/vp12	5.05	a
	lw3	5.06	a
	lw4	4.06	a
	se1	2.01	a
	vp2	5.04	a
PDS	*vp5*	1.02	a
LCYB	*vp7*	5.04	a
ZDS	*vp9*	7.02	a
	w3 (y11)	2.06	a
	Wc1	9.07-9.08	a
	wlu1	3.07-3.08	a
	wlu2	7.02-7.06	a
	wlu3	8.04-8.09	a
	wlu4	9.03-9.08	a
	wlu5	1.07	a
	wlu7	1.05	a
PSY1	*y1*	6.01	a
	y3-al1	2.00-2.01	a
	y8 (lty2)	7.01	a
	y9 (y12)	10.03-10.04	a
	y10	3.07	a
Molecular marker loci			
	IPPI	7.04	b
	DXS	?	c
	DXR	?	c
Y1	*PSY1*	6.01	[37]
	PSY2	8.07	b
Vp5	*PDS*	1.02	[42]
Vp9	*ZDS*	7.02	[58]
	CRTISO1	2.09	b
	CRTISO2	4.08	b
Vp7	*LCYB*	5.04	[86]
	LCYE	8.05	b
	HYD1	?	c
	HYD2	?	c
	HYD3	7.01	b
QTL loci			
	QTL 1	6.01	[35]
	QTL 2	7.02	[35]
	QTL 3	8.03-8.07	[35]

a Maize Genetics and Genomics Database http://www.maizegdb.org/
b Brutnell and Wurtzel, unpublished.
c Wurtzel *et al.*, unpublished

a consequence of concerted efforts to sequence the maize B73 genome. Where it was once necessary to screen cDNA libraries, it is now routine to use RT-PCR (reverse transcriptase polymerase chain reaction) to directly amplify cDNA from isolated mRNA.[56] Once cDNAs are isolated, it is imperative that the function of the encoded product is confirmed. The assay of hydrophobic enzymes, such as in the carotenoid biosynthetic pathway, would ordinarily present a challenge. However, a convenient heterologous system described below circumvents this problem.

Color Complementation for Functional Testing of Biosynthetic Enzymes

Carotenoid enzymes function in a hydrophobic membrane environment and as a result are difficult to purify in active form. A powerful alternative is the use of the bacterium *Escherichia coli*, which mimics plant plastid biochemistry with required carotenoid isoprenoid precursors; carotenoids will accumulate in *E. coli* if a gene cluster for carotenoid biosynthetic pathway enzymes is introduced.[57] The co-transformation into *E. coli* of higher plant cDNAs and bacterial gene cassettes (from carotenogenic bacteria such as *Erwinia uredovora*) results in pigmented *E. coli* bacteria that can be easily screened by eye; carotenoid pigments and biosynthetic pathway intermediates confer unique (yellow, pink, orange) color to *E. coli* cells that accumulate the pigments. Identification and confirmation of accumulated carotenoids is conducted by HPLC (high pressure liquid chromatography); carotenoids have unique elution profiles on reverse phase columns and unique spectral characteristics that can be monitored with the photodiode array detector. For the maize carotenoid biosynthetic pathway, this heterologous system has been used to demonstrate enzyme function of cDNA gene products,[42,58] to screen for new cDNAs which encode biosynthetic enzymes, [61] to identify novel genes that enhance or inhibit carotenoid accumulation,[59] or to test strategies for engineering plastids for carotenoid accumulation.[57] For example, to test whether maize *PSY* genes encoded functional enzymes, expression constructs were produced and introduced into *E. coli* cells carrying a bacterial gene cluster for the entire pathway except for the PSY bacterial counterpart, *crtB*. Such cells produced zeaxanthin and appeared yellow only when a functional PSY enzyme was present; if the PSY deletion strain was transformed with empty vector, cells remained colorless (Fig. 5.5). This approach, augmented by HPLC analysis, was used to demonstrate the function of two structurally unique maize PSY enzymes, **PSY1** and **PSY2** (Gallagher, Li, and Wurtzel, unpub.).

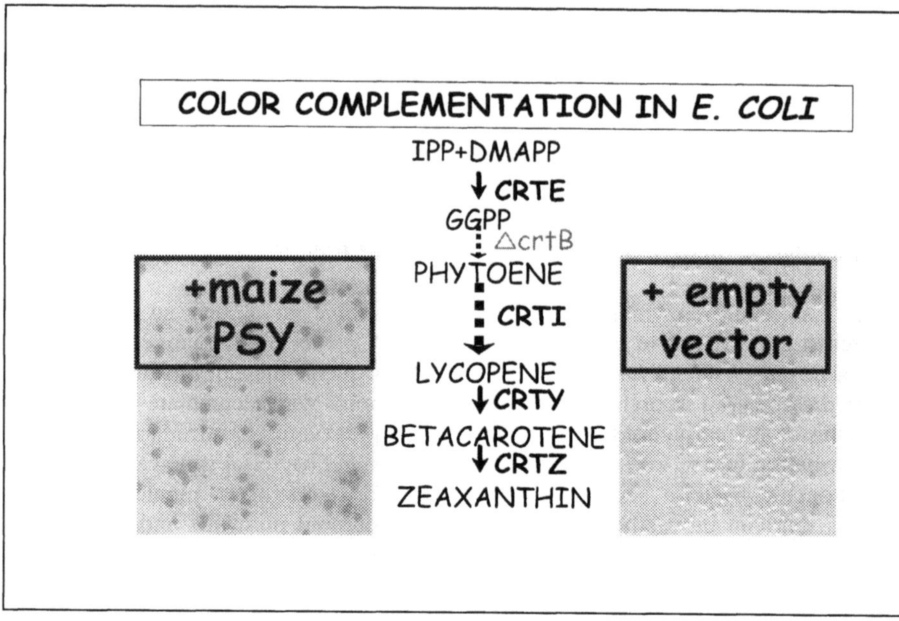

Fig. 5.5: Color complementation in *E. coli*. Bacterial enzymes encoded on a carotenoid gene cluster are denoted by CRT (CRTE=GGPPS; CRTB=PSY; CRTI= phytoene desaturase; CRTY=LCYB; CRTZ=HYDB). The deletion, Δ*crtB*, must be complemented by a gene encoding a plant PSY in order for the yellow-colored zeaxanthin to accumulate in *E. coli* colonies as shown on the left (giving cells a darker appearance). Cells transformed with the carotenoid gene cluster and an empty vector (lacking plant genes) appear lighter as seen on the right.

MAIZE GENETICS AS A TOOL

Identifying Structural and Regulatory Loci

The field of genetics, especially for maize, has provided much information on the genetic loci involved in the carotenoid pathway. Mutations include recessive, dominant, and suppressor alleles; these are putative structural and regulatory loci, some of which have tissue-specific phenotypes and most of which have not been characterized with respect to function. Most of these loci have been mapped as shown in Table 5.1. Genetic loci are associated with particular biosynthetic steps because intermediates accumulate in mutant tissues.[52-55] Some condition a complete absence of carotenoids and/or intermediates (for example, *Wc1*, *lw1*, *lw3*, *lw4*, *cl1*, *Clm*) and may encode transcriptional regulators or factors that function at steps upstream of the pathway. Genetic and biochemical information regarding these loci is useful in identification of putative structural and regulatory genes involved in carotenoid biosynthesis. These genetic resources can be used in combination with transposable elements to isolate unknown genes or to dissect gene function.[37,60] After mapping structural genes to chromosome location using recombinant inbred lines, one can search for linked genes for which alleles are known to alter carotenoid accumulation. Mutants or allelic variants are then tested to compare transcript and/or protein levels for the enzyme, thereby associating a locus with a biosynthetic step or function.[42,58,61] These and other maize genetic stocks are available through the Maize Genetic Stock Center (U. Illinois, Champaign-Urbana) for which further information can be obtained from Maize GDB (Maize Genetics and Genomics Database, http://www.maizegdb.org/).

Quantitative Trait Analysis and Associative Mapping

Regulation of carotenoid accumulation will likely be affected by activity of pathway enzymes, and expression of pathway regulators, or perhaps other yet to be determined factors. While the maize mutants provide one resource to identify key factors required for carotenoid accumulation, analysis of quantitative trait loci (QTL) serves as another approach to identify chromosome regions having significant effects on carotenoid profiles. QTL analysis and associative mapping are two complementary approaches; associative mapping identifies DNA sequence variation of known candidate genes, while QTL analysis scans an entire genome without prior knowledge of candidate genes. Associative mapping was recently applied to study of carotenoid accumulation in maize. This approach exploited allelic variation across a wide germplasm collection to correlate endosperm carotenoids with allelic states of specific nucleotide sequences for maize *PSY1* (*Y1*).[62] Work in the Rocheford lab led to identification of several QTL associated with carotenoid accumulation (see Table 5.1), some of which were linked to *PSY1* (*Y1*).[35,63] These genetic approaches were supported by molecular studies showing endosperm transcript levels of *PSY1* but not *PSY2* correlated with endosperm carotenoid accumulation (Gallagher and Wurtzel, unpub.).

THE MAIZE ENZYMES AND GENES

Some but not all of the maize enzymes are encoded by small gene families. It will be critical to dissect the role of gene family members in contributing to tissue-specific and plastid-specific variation in carotenoid accumulation. Furthermore, the analysis of these genes and gene families in diverse germplasm will provide a broader understanding of the allelic variation that contributes to diversity in maize endosperm carotenoid composition and content. Such information could be exploited to develop advanced breeding methods for improvement of plant metabolism based on carotenoid pathway "marker-assisted selection."

Enzymes and Genes for Carotenoid Precursors

Carotenoids are terpenoids derived from a five-carbon isoprenoid building block, isopentenyl pyrophosphate (IPP), which is common to all terpenoid compounds. All plastids have the ability to manufacture these IPP precursors through a plastid-specific non-mevalonate biosynthetic route that is also found in bacteria,[64] and requires such enzymes as **DXS** (D-1-deoxyxylulose 5-phosphate synthase, DXP synthase) and **DXR** (DXP reductoisomerase) (see Fig. 5.4). In the non-mevalonate route, also referred to as the MEP (methylerythritol phosphate) or DOXP (D-1-deoxyxylulose 5-phosphate) pathway, IPP is derived from deoxyxylulose 5-phosphate (DXP). In *E. coli,* DXP has also been found to be a common precursor to biosynthesis of vitamins B1 (thiamin) and B6 (pyridoxal). **DXS** is responsible for catalyzing the synthesis of DXP from pyruvate and GAP (glyceraldehyde 3-phosphate).[65-67] Wurtzel *et al.*[57] predicted that **DXS**, an enzyme at a metabolic crossroad, would likely be rate-controlling for carotenoid accumulation, and demonstrated this using the color complementation approach described above; the observation was later confirmed to be true in dicot plants.[68] **DXR** catalyzes the next step in the MEP pathway and has also been implicated in controlling pathway flux to carotenoids.[12,69,70] The elevation of both **DXS** and **DXR** transcript levels in maize roots was found to be concomitant with root apocarotenoid accumulation induced by arbuscular mycorrhizal fungi.[12] Maize **DXS** is encoded by a single copy gene, and **DXR** appears to be encoded by at least three genes (Wurtzel *et al.*, unpublished). The maize ***DXR*** gene family contrasts with the single copy gene found in *Arabidopsis.*[70] It is possible that the different maize gene family members may vary in terms of targeting to different organelles.

IPPI and GPPS

Four molecules of IPP, one of which is an isomer produced by IPP isomerase (**IPPI**) are combined to produce the twenty-carbon isoprenoid, GGPP (geranylgeranyl pyrophosphate). This step is mediated by GGPP synthase (**GGPPS**). Light-induced activation of **IPPI** has been associated with the phytochrome-mediated increase of carotenoids in photosynthetic tissue.[50] Similarly, a maize ***IPPI*** was found to positively influence pathway flux for carotenoid accumulation in *E. coli.*[59] Two maize loci for **IPPI** have been detected, one of which has been mapped to chromosome 7 (see Table 5.1). For agronomically important crops such as maize, wheat, and rice, as compared to pepper, *Arabidopsis,*[71] and other dicot plants, there is a paucity of information on **GGPPS** regulation; also, the number and regulation of genes encoding **GGPPS** are unknown. Therefore, the regulation of carotenoid biosynthesis at the key entry point of the pathway is poorly understood in food staples central to human nutrition.

Enzymes and Genes for Carotenoid Biosynthesis

PSY

The biosynthesis of all carotenoids (Fig. 5.3) begins with the combination of two molecules of GGPP to produce the 40-carbon backbone, phytoene, the first compound specific to the carotenoid biosynthetic pathway.[11,24] This step is catalyzed by the enzyme **PSY** (phytoene synthase).[72-75] For maize and throughout the Poaceae, the **PSY** gene appears to be duplicated.[76] The duplicate grass genes are predicted to encode enzymes with variant N- and C-termini, suggesting that the **PSYs** may target to different plastid membranes. Both maize ***PSY1*** and ***PSY2*** encode functional enzymes and maize ***PSY1*** transcripts correlate with endosperm carotenoid accumulation (Gallagher, Li, and Wurtzel, unpub.). It will be of interest to determine for other grass taxa whether there is a correlation between expression of either of the two ***PSY*** paralogs and endosperm carotenoid accumulation.

PDS, ZDS, ISO

Phytoene, a colorless compound, undergoes the addition of double bonds to produce lycopene. In higher plants and cyanobacteria, these steps are catalyzed by two enzymes, **PDS** (phytoene desaturase) and **ZDS** (zeta-carotene desaturase), while in bacteria, such as *Erwinia uredovora*, only one enzyme, **CRTI**, is required to convert phytoene to lycopene. The desaturation steps require oxygen, and as demonstrated in a chromoplast *in vitro* system, are coupled to an electron transport chain, with oxygen being the final acceptor.[77-79] However, progression of metabolites through the carotenoid biosynthetic pathway depends on the geometric isomer states of substrates and products, as carotene desaturases may be stereo-selective, stereo-sensitive, and stereo-specific in their activities.[77,80,81] The bacterial carotene desaturase **CRTI** produces different geometric isomers than do the plant carotene desaturases, which as a consequence require a companion carotene isomerase, **ISO**.[82,83] This stereochemical difference in enzyme specificity is critical when considering genetic engineering efforts using **CRTI** in a plant-based system. Study of the paired maize desaturases provides evidence for such stereochemical specificity in cereals.[58] Since **LCY** requires *E*-lycopene substrates, but **PDS/ZDS** produce the *Z*-

lycopene geometrical isomer, **ISO** may play a rate-controlling role in carotenoid accumulation past lycopene.

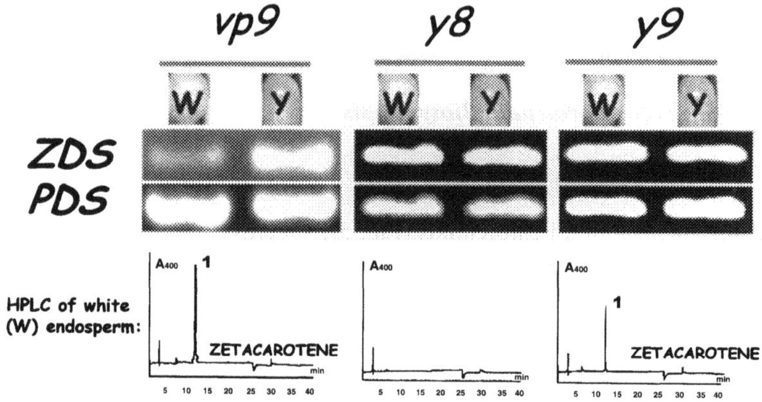

Fig. 5.6: Testing of candidate loci for the *ZDS* structural gene. RT-PCR amplification of transcripts and HPLC analysis of biochemical intermediates in yellow normal (Y) and white mutant (W) maize endosperm.[58]

In maize, **PDS** transcripts are constant throughout endosperm development.[42] Both **PDS** and **ZDS** are encoded by single copy genes and map to single loci (Table 5.1).[42,58] In contrast, **ISO** maps to two loci (Table 5.1). **PDS** and **ZDS** loci have corresponding mutant alleles that confer biochemical lesions expected for blocks in these steps. To identify candidate loci, cDNAs are mapped to chromosome; nearby loci which condition accumulation of predicted intermediates are putative candidates. These can be further tested by gene and transcript analysis of an allelic series to confirm that the locus encodes the enzyme and that mutations cause alterations in gene expression (Table 1). For example, endosperms homozygous for *vp5* accumulate phytoene and the *vp5* locus was established as the structural gene for **PDS**.[42,84] Interestingly, *vp2* and *w3* also condition phytoene accumulation, but mutant alleles have no affect on **PDS** transcript levels and the loci are not **PDS** structural genes; *vp2* is involved in biosynthesis of plastoquinones, which are required for the desaturation steps in carotenoid biosynthesis, but the role of the *w3* product remains unknown.[58] Two candidate genes for **ZDS** were *vp9* and *y9*, mutant alleles which condition zetacarotene accumulation which is predicted for lesions in a gene encoding ZDS, since ZDS catalyzes conversion of zetacarotene to lycopene.[58] Mapping placed ***ZDS*** near *vp9* and *y8*, another carotenoid locus. However, y8 does not condition zetacarotene accumulation, and only *vp9* had reduced ***ZDS*** transcript levels in white mutant endosperm compared to segregating normal yellow endosperm (Fig. 5.6). While **ISO** had been suggested to be encoded by the *y9* locus, this was not supported by mapping of ***ISO*** to chromosomes 2 and 4, while *y9* was mapped to chromosome 10 (Table 5.1). Therefore, the role of *y9* in maize carotenoid biosynthesis is yet to be determined.

LCYB and LCYE

Rings added by the enzyme **LCYB** (lycopene beta cyclase) to both ends of the all-*E*-lycopene molecule result in the most active provitamin A carotenoid, beta-carotene, having two "beta" rings.[85] Alternatively, **LCYE** (lycopene epsilon cyclase), in combination with **LCYB**, catalyzes the biosynthesis of alpha-carotene, with one "epsilon" ring and one "beta" ring.[41] In humans and animals, the central cleavage of beta-carotene results in two molecules of vitamin A; cleavage of alpha-carotene results in only one molecule of vitamin A, which is derived from that half of alpha-carotene having the "beta" ring. Because of the "epsilon" ring, alpha-carotene has only *half* the provitamin A activity compared to that of beta-carotene. Therefore, it is after lycopene formation that the pathway diverges and, produces either more or less provitamin A active carotenoid, depending on relative levels of the two cyclase enzymes **LCYE** and **LCYB**. In maize, both **LCYB** and **LCYE** are encoded by single copy genes (Table 5.1). ***LCYB*** was isolated by transposon mutagenesis and corresponds to the maize *vp7* locus on chromosome 5.[86] ***LCYB*** is the only known pathway locus not to have any introns.[86] ***LCYE*** was identified by a combination of GenBank database mining and use of rice genome sequence and maps to chromosome 8 (Wurtzel *et al.*, unpub.).

HYD

The **HYD** enzymes are responsible for converting provitamin A carotenes to nonprovitamin A xanthophylls. After ring addition, both beta-carotene and alpha-carotene undergo addition of oxygen by **HYD** (hydroxylase) enzymes, giving rise to xanthophylls (oxygenated carotenoids) such as lutein (derived from alpha-carotene) or zeaxanthin (derived from beta-carotene).[87,88] However, addition of oxygen further diminishes provitamin A activity. Intermediates with a single hydroxylation, such as betacryptoxanthin, have some provitamin A activity. Because hydroxylation further diminishes the provitamin A activity of alpha-carotene and beta-carotene, the level of the **HYD** enzyme activities is critical in regulating the level of provitamin A carotenoids; if the **HYD** enzyme levels are high, the level of provitamin A carotenoids will be low. Hydroxylation of the beta rings in betacarotene to produce zeaxanthin is mediated by **HYDB**, a nonheme diiron monooxygenase; the enzyme may also act on the beta ring of alphacarotene. A separate hydroxylase specific for the alpha-carotene epsilon ring, **HYDE**, has recently been identified in *Arabidopsis* as a cytochrome P450-type monooxygenase.[88] The P450-type *Arabidopsis* enzyme has no homology to the previously identified nonheme diiron monooxygenase **HYDB** enzymes, but is structurally related to a second (putative) *Arabidopsis* P450 hydroxylase specific for beta rings, both of which appear to be related to nonplant cytochrome P450s. A P450 specific for beta rings has been recently described in a bacterium[89] but such P450-type enzymes are yet to be discovered in maize or other monocots. An intriguing possibility is that the P450 type enzymes, specific for alpha and beta rings, may function as a heterodimer to convert alpha carotene to lutein, while the nonheme diiron monooxygenase enzymes may utilize betacarotene as the substrate to produce zeaxanthin. This scenario would further suggest that the **LCYE** required for alpha carotene synthesis might form a complex with the P450 type enzymes to channel substrate towards lutein.

Sequence analysis of isolated maize **HYDB** BAC clones revealed that maize B73, which contains endosperm carotenoids, has three ***HYD*** genes, two highly conserved but nonfunctional, and a third functional gene (Gallagher and Wurtzel, unpub.). These genes all encode enzymes predicted to be nonheme diiron monooxygenases and catalyze formation of zeaxanthin from betacarotene, through the mono-hydroxylated intermediate, betacryptoxanthin. Further analysis revealed that other maize cultivars carry variations in the number of functional and nonfunctional copies (Gallagher & Wurtzel, unpublished). These observations raise several questions: 1) if there is only a single functional ***HYDB*** gene in B73, what is the mechanism for targeting the encoded **HYDB** to two potential plastid membrane locations? 2) If maize inbreds possess variant numbers of functional ***HYD*** genes, how does genotype impact endosperm levels of the **HYD** enzyme substrate, beta-carotene? Elucidation of the role of these family members is critical for breeding enhanced levels of provitamin A compounds.

SUMMARY AND FUTURE DIRECTIONS

Metabolic engineering of carotenoid content and composition requires an understanding of how the biosynthetic pathway is regulated in terms of gene expression, localization of enzyme activities, and substrate flow. Preliminary success with metabolic engineering of the pathway in rice, tomato, tobacco, and canola points to the potential of this approach.[90-93] However, unexpected products in transgenic plants indicate that the technology is limited by current deficiencies in understanding of endogenous gene expression. Furthermore, integration of the pathway in local varieties will also entail pyramiding of multiple traits, yet little is known about pathway interactions and competition for common substrates. The discovery that many of the maize enzymes are encoded by small gene families underscores the importance of evaluating the role of these genes in contributing to endosperm carotenoid accumulation. Furthermore, the possibility that the pathway can be assembled on different plastid membranes suggests that future attempts at metabolic engineering or breeding of enhanced carotenoids will require further exploration of this issue. The genetic, genomic, and germplasm resources available for maize will provide a means to develop rational strategies for metabolic engineering and marker-assisted breeding to improve carotenoid content and composition in maize and other grasses of agronomic value.

ACKNOWLEDGMENTS

Current and former members of the Wurtzel lab are acknowledged for their contributions to the research described here. Dr. Cynthia Gallagher is thanked for providing the carotenoid biosynthetic pathway Figure.. Research in the Wurtzel lab has been funded by NIH (S06-GM08225), The American Cancer Society, The McKnight Foundation, The Rockefeller Foundation International Rice Biotechnology Program, New York State and CUNY.

REFERENCES

1. GRASS PHYLOGENY WORKING GROUP, Phylogeny and subfamilial classification of the grasses (Poaceae). *Ann. Mo. Bot. Gard.* 2001, **88**, 373-457.
2. LEE, C., MCCOON, P., and LEBOWITZ, J., Vitamin A value of sweet corn, *J Agric Food Chem.* 1981, **29**, 1294-1295.
3. VAN DEN BERG, H., FAULKS, R., GRANADO, H. F., HIRSCHBERG, J., OLMEDILLA, B., SANDMANN, G., SOUTHON, S., STAHL, W., The potential for the improvement of carotenoid levels in foods and the likely systemic effects, *J. Sci. Food & Agri.* 2000, **80**, 880-912.
4. KIEFER, C., HESSEL, S., LAMPERT, J. M., VOGT, K., LEDERER, M. O., BREITHAUPT, D. E., VON LINTIG, J., Identification and characterization of a mammalian enzyme catalyzing the asymmetric oxidative cleavage of provitamin A, *J. Biol. Chem.* 2001, **276**, 14110-6.
5. GRAHAM, R., Wheat: Research at Waite Agricultural Research Institute in Australia, *CGIAR Micronutrients Project* 1997, Update No. 2.
6. BENDICH, A. and OLSON, J., Biological actions of carotenoids, *FASEB J* 1989, **3**, 1927-1932.
7. UNDERWOOD, B. A., ARTHUR, P., The contribution of vitamin A to public health, *FASEB J.* 1996, **10 (9)**, 1040-1049.
8. SEMBA, R. D., MIOTTI, P. G., CHIPHANGWI, J. D., SAAH, A. J., CANNER, J. K., G.A., D., HOOVER, D. R., Maternal vitamin A deficiency and mother-to-child transmission of HIV-1, *The Lancet* 1994, **343**, 1593-1597.
9. CHERNYS, J. T., ZEEVAART, J. A., Characterization of the 9-*cis*-epoxycarotenoid dioxygenase gene family and the regulation of abscisic acid biosynthesis in avocado., *Plant Physiol.* 2000, **124**, 343-354.

10. ROCK, C. and ZEEVAART, J., The *aba* mutant of *Arabidopsis thaliana* is impaired in epoxy-carotenoid biosynthesis., *PNAS* 1991, **88**, 7496-7499.
11. HIRSCHBERG, J., Carotenoid biosynthesis in flowering plants, *Curr. Opin. Plant Biol.* 2001, **4**, 210-8.
12. WALTER, M. H., FESTER, T., STRACK, D., Arbuscular mycorrhizal fungi induce the non-mevalonate methylerythritol phosphate pathway of isoprenoid biosynthesis correlated with accumulation of the 'yellow pigment' and other apocarotenoids, *Plant J.* 2000, **21**, 571-8.
13. VON LINTIG, J., VOGT, K., Vitamin A formation in animals: molecular identification and functional characterization of carotene cleaving enzymes., *J. Nutr.* 2004, **134**, 251S-256.
14. GIOVANNUCCI, E., ASCHERIO, A., RIMM, E. B., STAMPFER, M. J., COLDITZ, G. A., WILLETT, W. C., Intake of carotenoids and retinol in relation to risk of prostate cancer, *J. Natl. Cancer Inst.* 1995, **87**, 1767-76.
15. SOMMERBURG, O., KEUNEN, J. E., BIRD, A. C., and VAN KUIJK, F. J., Fruits and vegetables that are sources for lutein and zeaxanthin: the macular pigment in human eyes, *Br. J. Ophthalmol.* 1998, **82**, 907-10.
16. KOHLMEIER, L., KARK, J. D., GOMEZ-GRACIA, E., MARTIN, B. C., STECK, S. E., KARDINAAL, A. F., RINGSTAD, J., THAMM, M., MASAEV, V., RIEMERSMA, R., MARTIN-MORENO, J. M., HUTTUNEN, J. K., KOK, F. J., Lycopene and myocardial infarction risk in the EURAMIC Study, *Am. J. Epidemiol.* 1997, **146**, 618-26.
17. KRINSKY, N., RUSSETT, M. D., HANDELMAN, G. J., SNODDERLY, D. M., Structural and geometrical isomers of carotenoids in human plasma., *J. Nutr.* 1990, **120**, 1654-1662.
18. PATRICK, L., Beta-carotene: The controversy continues, *Altern. Med. Rev.* 2000, **5**, 530-45.
19. BJERKENG, B., BERGE, G. M., Apparent digestibility coefficients and accumulation of astaxanthin E/Z isomers in Atlantic salmon (*Salmo salar* L.) and Atlantic halibut (*Hippoglossus hippoglossus* L.), *Comp. Biochem. Physiol. B. Biochem. Mol. Biol.* 2000, **127**, 423-32.
20. HOLLOWAY, D. E., YANG, M., PAGANGA, G., RICE-EVANS, C. A., BRAMLEY, P. M., Isomerization of dietary lycopene during assimilation and transport in plasma, *Free Radic. Res.* 2000, **32**, 93-102.
21. OSTERLIE, M., BJERKENG, B., and LIAAEN-JENSEN, S., Accumulation of astaxanthin all-E, 9Z and 13Z geometrical isomers and 3 and 3' RS optical isomers in rainbow trout (*Oncorhynchus mykiss*) is selective., *J. Nutr.* 1999, **129**, 391-8.
22. CHAPPELL, J., Biochemistry and molecular biology of the isoprenoid biosynthetic pathway in plants., *Annu. Rev. Plant Physiol. Plant Mol. Biol.* 1995, **46**, 521-547.
23. ROBERTSON, D. S., The genotype of the endosperm and embryo as it influences vivipary in maize., *Proc. Natl. Acad. Sci. USA* 1952, **38**, 580-583.
24. CUNNINGHAM, F. X., GANTT, E., Genes and enzymes of carotenoid biosynthesis in plants, *Annu. Rev. Plant Physiol. Plant Mol. Biol.* 1998, **49**, 557-583.
25. BONK, M., HOFFMANN, B., VON LINTIG, J., SCHLEDZ, M., AL-BABILI, S., HOBEIKA, E., KLEINIG, H., BEYER, P., Chloroplast import of four carotenoid

biosynthetic enzymes *in vitro* reveals differential fates prior to membrane binding and oligomeric assembly, *Eur. J. Biochem.* 1997, **247**, 942-50.
26. LINDEN, H., LUCAS, M. M., ROSARIO DE FELIPE, M., and SANDMANN, G., Immunogold localization of phytoene desaturase in higher plant chloroplasts, *Physiol. Plant.* 1993, **88**, 229-236.
27. YU, J., Localization and expression of carotenoid biosynthetic enzymes in endosperms of *Zea mays* and *Oryza sativa*, *Ph.D. Thesis* 1999.
28. AL-BABILI, S., VON LINTIG, J., HAUBRUCK, H., BEYER, P., A novel, soluble form of phytoene desaturase from *Narcissus pseudonarcissus* chromoplasts is Hsp70-complexed and competent for flavinylation, membrane association and enzymatic activation., *Plant J.* 1996, **9**, 601-612.
29. BONK, M., TADROS, M., VANDEKERCKHOVE, J., AL-BABILI, S., BEYER, P., Purification and characterization of chaperonin 60 and heat-shock protein 70 from chromoplast of *Narcissus pseudonarcissus*. Involvement of heat-shock protein 70 in a soluble protein complex containing phytoene desaturase., *Plant Physiol.* 1996, **111**, 931-939.
30. RABBANI, S., BEYER, P., LINTIG, J., HUGUENEY, P., and KLEINIG, H., Induced beta-carotene synthesis driven by triacylglycerol deposition in the unicellular alga *Dunaliella bardawil.*, *Plant Physiol* 1998, **116**, 1239-48.
31. SCHLEDZ, M., AL-BABILI, S., VON LINTIG, J., HAUBRUCK, H., RABBANI, S., KLEINIG, H., BEYER, P., Phytoene synthase from *Narcissus pseudonarcissus*: Functional expression, galactolipid requirement, topological distribution in chromoplasts and induction during flowering., *Plant J.* 1996, **10**, 781-792.
32. CERVANTES-CERVANTES, M., HADJEB, N., NEWMAN, L. A., PRICE, C. A., ChrA is a carotenoid-binding protein in chromoplasts of *Capsicum annuum.*, *Plant Physiol.* 1990, **92**, 1241-1243.
33. MANGELSDORF, P. C., FRAPS, G. S., A direct quantitative relationship between vitamin A in corn and number of genes for yellow pigmentation, *Science* 1931, **73**, 241-242.
34. RANDOLPH, L. F., HAND, D. B., Relation between carotenoid content and number of genes per cell in diploid and tetraploid corn, *J. Agr. Res.* 1940, **60**, 51-64.
35. WONG, J. C., LAMBERT, R. J., WURTZEL, E. T., ROCHEFORD, T. R., QTL and candidate genes phytoene synthase and zetacarotene desaturase associated with the accumulation of carotenoids in maize., *Theor. & Appl. Genet.* 2004, **108**, 349-359.
36. BUCKNER, B., SAN MIGUEL, P., BENNETZEN, J. L., The *y1* gene of maize codes for phytoene synthase, *Genetics* 1996, **143**, 479-488.
37. BUCKNER, B., KELSON, T. L., ROBERTSON, D. S., Cloning of the *y1* locus of maize, a gene involved in the biosynthesis of carotenoids., *Plant Cell* 1990, **2**, 867-876.
38. GIULIANO, G., BARTLEY, G. E., SCOLNIK, P. A., Regulation of carotenoid biosynthesis during tomato development, *Plant Cell* 1993, **5**, 379-387.
39. CORONA, V., ARACRI, B., KOSTURKOVA, G., BARTLEY, G. E., PITTO, L., GIORGETTI, L., SCOLNIK, P. A., GIULIANO, G., Regulation of carotenoid biosynthesis gene promoter during plant development, *Plant J.* 1996, **9**, 505-512.

40. VON LINTIG, J., WELSCH, R., BONK, M., GIULIANO, G., BATSCHAUER, A., KLEINIG, H., Light-dependent regulation of carotenoid biosynthesis occurs at the level of phytoene synthase expression and is mediated by phytochrome in *Sinapis alba* and *Arabidopsis thaliana* seedlings., *Plant J.* 1997, **12**, 625-634.
41. PECKER, I., GABBAY, R., CUNNINGHAM JR., F. X., HIRSCHBERG, J., Cloning and characterization of the cDNA for lycopene β-cyclase from tomato reveals decrease in its expression during fruit ripening., *Plant Mol. Bio.* 1996, **30**, 807-819.
42. LI, Z. H., MATTHEWS, P. D., BURR, B., WURTZEL, E. T., Cloning and characterization of a maize cDNA encoding phytoene desaturase, an enzyme of the carotenoid biosynthetic pathway., *Plant Mol. Biol.* 1996, **30**, 269-279.
43. LI, Z. H., Molecular cloning and characterization of phytoene desaturase cDNA and leucine-rich repeat protein kinase cDNA from maize., Ph.D. dissertation, The Graduate School and University Center of the City University of New York, 1998.
44. BRAMLEY, P., TEULIERES, C., BLAIN, I., BIRD, C., SCHUCH, W., Biochemical characterization of transgenic tomato plants in which carotenoid synthesis has been inhibited through the expression of antisense RNA to pTOM5, *Plant J.* 1992, **2**, 343-349.
45. KUMAGAI, M. H., DONSON, J., DELLA-CIOPPA, G., HARVEY, D., HANLEY, K., GRILL, L. K., Cytoplasmic inhibition of carotenoid biosynthesis with virus-derived RNA, *Proc. Natl. Acad. Sci. U S A* 1995, **92**, 1679-83.
46. FRAY, R. G., GRIERSON, D., Identification and genetic analysis of normal and mutant phytoene synthase genes of tomato by sequencing, complementation and co-suppression, *Plant Mol. Biol.* 1993, **22**, 589-602.
47. BIRD, C. R., RAY, J. A., FLETCHER, J. D., BONIWELL, J. M., BIRD, A. S., TEULIERES, C., BLAIN, I., BRAMLEY, P. M., SCHUCH, W., Using antisense RNA to study gene function: Inhibition of carotenoid biosynthesis in transgenic tomatoes., *Biotechnology* 1991, **9**, 635-639.
48. KUNTZ, M., RÖMER, S., SUIRE, C., HUGUENEY, P., WEIL, J. H., SCHANTZ, R., CAMARA, B., Identification of a cDNA for the plastid-located geranylgeranyl pyrophosphate synthase from *Capsicum annuum*: correlative increase in enzyme activity and transcript level during fruit ripening., *Plant J.* 1992, **2**, 25-34.
49. FRAY, R., WALLACE, A., FRASER, P., VALERO, D., HEDDEN, P., BRAMLEY, P., GRIERSON, D., Constitutive expression of a fruit phytoene synthase gene in transgenic tomatoes causes dwarfism by redirecting metabolites from gibberellin pathway., *Plant J.* 1995, **8**, 693-701.
50. ALBRECHT, M., SANDMANN, G., Light-stimulated carotenoid biosynthesis during transformation of maize etioplasts is regulated by increased activity of isopentenyl pyrophosphate isomerase., *Plant Physiol.* 1994, **105**, 529-534.
51. KURILICH, A., JUVIK, J., Quantification of carotenoid and tocopherol antioxidants in *Zea mays*., *J. Agric. Food Chem.* 1999, **47**, 1948 -1955.
52. ROBERTSON, D., BACHMANN, M., ANDERSON, I., Role of carotenoids in protecting chlorophyll from photodestruction-II. Studies on the effect of four modifiers of the albino *cl1* mutant of maize, *Photochemistry & Photobiology* 1966, **5**, 797-805.

53. TREHARNE, K. J., MERCER, E. I., GOODWIN, T. W., Carotenoid biosynthesis in some maize mutants, *Phytochemistry* 1966, **5**, 581-587.
54. ROBERTSON, D. S., Survey of the albino and white-endosperm mutants of maize., *J. of Hered.* 1975, **66**, 67-74.
55. NEILL, S. J., HORGAN, R., PARRY, A. D., The carotenoid and abscisic acid content of viviparous kernels and seedlings of *Zea mays* L., *Planta* 1986, **169**, 87-96.
56. LI, Z.-H., WURTZEL, E., The *ltk* gene family encodes novel receptor-like kinases with temporal expression in developing maize endosperm, *Plant Mol. Biol.*1998, **37**, 749-761.
57. MATTHEWS, P. D., WURTZEL, E. T., Metabolic engineering of carotenoid accumulation in *Escherichia coli* by modulation of the isoprenoid precursor pool with expression of deoxyxylulose phosphate synthase., *Appl. Microbiol. Biotechnol.* 2000, **53**, 396-400.
58. MATTHEWS, P. D., LUO, R., WURTZEL, E. T., Maize phytoene desaturase and zetacarotene desaturase catalyze a poly-Z desaturation pathway: implications for genetic engineering of carotenoid content among cereal crops., *J. Exp. Botany* 2003, **54**, 2215-2230.
59. GALLAGHER, C. E., CERVANTES-CERVANTES, M., and WURTZEL, E. T., Surrogate biochemistry: use of *Escherichia coli* to identify plant cDNAs that impact metabolic engineering of carotenoid accumulation., *App. Microbiol & Biotech.*2003, **60**, 713-719.
60. WURTZEL, E. T., Use of a *Ds* chromosome breaking element to examine maize *Vp5* expression, *J. Hered.* 1992, **83**, 109-113.
61. LUO, R., Molecular and genetic studies related to zeta-carotene desaturation and carotenoid biosynthesis in maize and rice, Ph.D. Dissertation, City University of New York, 2000.
62. PALAISA, K. A., MORGANTE, M., WILLIAMS, M., RAFALSKI, A., Contrasting effects of selection on sequence diversity and linkage disequilibrium at two phytoene synthase loci, *Plant Cell* 2003, **15,** 1795-806.
63. WONG, J. C., LAMBERT, R. J., ROCHEFORD, T. R., Comparing QTL and candidate genes for carotenoids and tocopherols in two maize populations., pp. 145 - 170 *in Proc. 38th Annu. Illinois Corn Breeders School* 2002.
64. LICHTENTHALER, H. K., The 1-deoxy-d-xylulose-5-phosphate pathway of isoprenoid biosynthesis in plants, *Annu. Rev. Plant Physiol. Plant Mol. Biol.* 1999, **50**, 47-65.
65. SPRENGER, G. A., SCHORKEN, U., WIEGERT, T., GROLLE, S., DE GRAAF, A. A., TAYLOR, S. V., BEGLEY, T. P., BRINGER-MEYER, S., SAHM, H., Identification of a thiamin-dependent synthase in Escherichia coli required for the formation of the 1-deoxy-D-xylulose 5-phosphate precursor to isoprenoids, thiamin, and pyridoxol, *Proc. Natl. Acad. Sci. U S A* 1997, **94**, 12857-62.
66. LANGE, B. M., WILDUNG, M. R., MCCASKILL, D., CROTEAU, R., A family of transketolases that directs isoprenoid biosynthesis via a mevalonate-independent pathway, *Proc. Natl. Acad. Sci. U S A* 1998, **95**, 2100-4.

67. LOIS, L. M., CAMPOS, N., PUTRA, S. R., DANIELSEN, K., ROHMER, M., BORONAT, A., Cloning and characterization of a gene from *Escherichia coli* encoding a transketolase-like enzyme that catalyzes the synthesis of D-1-deoxyxylulose 5-phosphate, a common precursor for isoprenoid, thiamin, and pyridoxol biosynthesis, *Proc. Natl. Acad. Sci. U S A* 1998, **95**, 2105-10.
68. ESTEVEZ, J. M., CANTERO, A., ROMERO, C., KAWAIDE, H., JIMENEZ, L. F., KUZUYAMA, T., SETO, H., KAMIYA, Y., LEON, P., Analysis of the expression of *CLA1*, a gene that encodes the 1- deoxyxylulose 5-phosphate synthase of the 2-C-methyl-D-erythritol-4- phosphate pathway in *Arabidopsis*., *Plant Physiol.* 2000, **124**, 95-104.
69. KIM, S. W., KEASLING, J. D., Metabolic engineering of the nonmevalonate isopentenyl diphosphate synthesis pathway in *Escherichia coli* enhances lycopene production., *Biotechnol. Bioeng.* 2001, **72**, 408-15.
70. CARRETERO-PAULET, L., AHUMADA, I., CUNILLERA, N., RODRIGUEZ-CONCEPCION, M., FERRER, A., BORONAT, A., CAMPOS, N., Expression and molecular analysis of the *Arabidopsis* DXR Gene encoding 1-deoxy-D-xylulose 5-phosphate reductoisomerase, the first committed enzyme of the 2-C-methyl-D-erythritol 4-phosphate pathway., *Plant Physiol.* 2002, **129**, 1581-1591.
71. ZHU, X., SUZUKI, K., SAITO, T., OKADA, K., TANAKA, K., NAKAGAWA, T., MATSUDA, H., and KAWAMUKAI, M., Geranylgeranyl pyrophosphate synthase encoded by the newly isolated gene *GGPS6* from *Arabidopsis thaliana* is localized in mitochondria, *Plant Mol. Biol.* 1997, **35**, 331-341.
72. KREUZ, K., BEYER, P., KLEINIG, H., The site of carotenogenic enzymes in chromoplasts from *Narcissus pseudonarcissus* L., *Planta* 1982, **154**, 66-69.
73. LÜTKE-BRINKHAUS, F., LIEDVOGEL, B., KREUZ, K., KLEINIG, H., Phytoene synthase and phytoene dehydrogenase associated with envelope membranes from spinach chloroplasts, *Planta* 1982, **156**, 176-180.
74. BEYER, P., WEISS, G., KLEINIG, H., Solubilization and reconstitution of the membrane bound carotenogenic enzymes from daffodil chromoplasts, *Eur. J. Biochem.* 1985, **153**, 341-346.
75. MAYFIELD, S. P., NELSON, T., TAYLOR, W. C., MALKIN, R., Carotenoid synthesis and pleiotropic effects in carotenoid-deficient seedlings of maize, *Planta* 1986, **169**, 23-32.
76. MATTHEWS, P. D., Carotenogenesis in Maize and Rice., Graduate School and University Center, The City University of New York, 2001.
77. BEYER, P., MAYER, M., KLEINIG, K., Molecular oxygen and the state of geometric isomerism of intermediates are essential in the carotene desaturation and cyclization reactions in daffodil chromoplasts., *Eur. J. Biochem.* 1989, **184**, 141-150.
78. MAYER, M. P., BEYER , P., KLEINIG, K., Quinone compounds are able to replace molecular oxygen as terminal electron acceptor in phytoene desaturation in chromoplasts of *Narcissus pseudonarcissus* L., *Eur. J. Biochem.* 1990, **191**, 359-363.
79. MAYER, M. P., NIEVELSTEIN, V., BEYER, P., Purification and characterization of a NADPH dependent oxidoreductase from chromoplasts of *Narcissus*

pseudonarcissus: a redox mediator possibly involved in carotene desaturation., *Plant Physiol. & Biochem.* 1992, **30**, 389-398.

80. BREITENBACH, J., KUNTZ, M., TAKAICHI, S., SANDMANN, G., Catalytic properties of an expressed and purified higher plant type zeta-carotene desaturase from *Capsicum annuum.*, *Eur. J. Biochem.* 1999, **265**, 376-83.
81. BARTLEY, G. E., SCOLNIK, P. A., BEYER, P., Two *Arabidopsis thaliana* carotene desaturases, phytoene desaturase and zeta-carotene desaturase, expressed in *Escherichia coli*, catalyze a poly-cis pathway to yield pro-lycopene., *Eur. J. Biochem.* 1999, **259**, 396-403.
82. PARK, H., KREUNEN, S. S., CUTTRISS, A. J., DELLAPENNA, D., POGSON, B., Identification of the carotenoid isomerase provides insight into carotenoid biosynthesis, prolamellar body formation, and photomorphogenesis., *Plant Cell* 2002, **14**, 321-332.
83. ISAACSON, T., RONEN, G., ZAMIR, D., HIRSCHBERG, J., Cloning of *tangerine* from tomato reveals a carotenoid isomerase essential for the production of β–carotene and xanthophylls in plants., *Plant Cell* 2002, **14**, 333-342.
84. HABLE, W. E., OISHI, K. K., SCHUMAKER, K. S., *Viviparous-5* encodes phytoene desaturase, an enzyme essential for abscisic acid (ABA) accumulation and seed development in maize., *Mol. Gen. Genet.* 1998, **257**, 167-76.
85. CUNNINGHAM JR., F. X., POGSON, B., SUN, Z., MCDONALD, K. A., DELLAPENNA, D., GANTT, E., Functional analysis of the β and ε lycopene cyclase enzymes of *Arabidopsis* reveals a mechanism for control of cyclic carotenoid formation, *Plant Cell* 1996, **8**, 1613-1626.
86. SINGH, M., LEWIS, P. E., HARDEMAN, K., BAI, L., ROSE, J. K. C., MAZOUREK, M., CHOMET, P., BRUTNELL, T. P., *Activator* mutagenesis of the *pink scutellum1/viviparous7* locus of maize., *Plant Cell* 2003, **15 (4)**, 874-884.
87. SUN, Z., GANTT, E., CUNNINGHAM, J., F. X., Cloning and functional analysis of the β-carotene hydroxylase of *Arabidopsis thaliana*, *J. of Biol. Chem.* 1996, **271**, 24349-24352.
88. TIAN, L., MUSETTI, V., KIM, J., MAGALLANES-LUNDBACK, M., DELLAPENNA, D., The Arabidopsis *LUT1* locus encodes a member of the cytochrome P450 family that is required for carotenoid ε-ring hydroxylation activity., *Proc. Natl. Acad. Sci.: USA* 2004, **101**, 402-407.
89. BLASCO, F., KAUFFMANN, I., SCHMID, R., CYP175A1 from Thermus thermophilus HB27, the first beta-carotene hydroxylase of the P450 superfamily., *Appl. Microbiol. Biotechnol.* 2004.
90. YE, X., AL-BABILI, S., KLOTI, A., ZHANG, J., LUCCA, P., BEYER, P., POTRYKUS, I., Engineering the provitamin A (beta-carotene) biosynthetic pathway into (carotenoid-free) rice endosperm., *Science* 2000, **287**, 303-5.
91. SHEWMAKER, C. K., SHEEHY, J. A., DALEY, M., COLBURN, S., KE, D. Y., Seed-specific overexpression of phytoene synthase: increase in carotenoids and other metabolic effects., *Plant J.* 1999, **20**, 401-412.
92. MANN, V., HARKER, M., PECKER, I., HIRSCHBERG, J., Metabolic engineering of astaxanthin production in tobacco flowers., *Nat. Biotechnol.* 2000, **18**, 888-92.

93. ROSATI, C., AQUILANI, R., DHARMAPURI, S., PALLARA, P., MARUSIC, C., TAVAZZA, R., BOUVIER, F., CAMARA, B., GIULIANO, G., Metabolic engineering of beta-carotene and lycopene content in tomato fruit, *Plant J.* 2000, **24**, 413-9.

Chapter Six

GENOMIC SURVEY OF METABOLIC PATHWAYS IN RICE

Bernd Markus Lange[1]* and Gernot Presting[2]

Torrey Mesa Research Institute
Syngenta Research & Technology
3115 Merryfield Row,
San Diego, CA 92121

[1]*Institute of Biological Chemistry*
Washington State University
PO Box 646340
Pullman, WA 99164-6340

[2]*Department of Molecular Biosciences & Bioengineering*
College of Tropical Agriculture and Human Resources
University of Hawaii
Honolulu, HI 96822

**Author for correspondence, e-mail: lange-m@wsu.edu*

INTRODUCTION

Functional genomics, the science of deciphering DNA sequence structure, variation, and function, is expected to become the engine driving the discovery of traits and to help solve intractable problems in crop production. The recent completion of rice (*Oryza sativa*) draft genome sequences represents an enormous pool of information for rice improvement through marker-aided selection or genetic engineering.[1,2] Yet, a full exploitation of this wealth of information will not be possible until we understand the biological functions encoded by the sequenced DNA. A genome-wide experimental approach will be instrumental in dissecting metabolic pathways important for increasing rice productivity and nutritional content. In this article, we focus on progress toward the elucidation of specific metabolic pathways linked to key quality traits in rice.

THE RICE GENOME - AN INVALUABLE RESOURCE FOR FUNCTIONAL GENOMICS

Rice, wheat, and maize account for approximately half of the world's food production. Over the last 30 years, world rice production has doubled as the result of the introduction of new varieties and improved technology. However, the annual rate of rice production has slowed to the point that it is no longer keeping pace with the growth in the number of consumers. Thus, there will be great demands on biotechnology to improve rice production. The genome of the dicot weed *Arabidopsis thaliana,* the first plant genome to become available, has fostered rapid progress in understanding metabolism in this model species.[3] The rice genome represents the first genome of a commercially important crop to be sequenced and will be highly valuable as a monocot model. The rice genome is roughly three times the size of the *Arabidopsis thaliana* genome and, with a predicted gene density of one gene every 15 kb, ranks as the smallest genome of the major cereals. Rice should be an important model because genes are highly conserved among the cereal species, which also include maize, wheat, barley, sorghum, millet, and oat. Therefore, linking important traits, such as disease resistance, yield, and nutritional content, to genes in rice, could be translated to other crops.

RICE METABOLISM – CURRENT KNOWLEDGE AND FUTURE CHALLENGES

Based on sequence homology to genes of known function, roughly 25% of rice genes are involved in metabolism.[1] To evaluate the metabolic capabilities of rice computationally, BLASTP searches were conducted in which the peptide

Table 6.1: Rice gene products with a demonstrated role in metabolic pathways.

Encoded enzyme	Reference	Encoded enzyme	Reference
S-Adenosyl-L-methionine synthetase	40	Isopentenyl diphosphate isomerase	83
ADP-glucose pyrophosphorylase	41	Lipoxygenase	84
Aldehyde dehydrogenase	42, 43	Methionine synthase	85
α-Amylase	44, 45, 46	Methionyl-tRNA synthetase	86
1-Aminocyclopropane-1-carboxylate oxidase	47	Monosaccharide transporter	87
1-Aminocyclopropane-1-carboxylate synthase	48, 49	Nitrate transporter	88
Ammonium transporter	50	Nitrite reductase	89
Anthranilate synthase	51	12-Oxophytodienoic acid reductase	90
Arginine decarboxylase	52	Pantothenate synthetase	91
Aspartate aminotransferase	53	Peroxidase	92
Brassinosteroid C6-oxidase	54, 55	Phenylalanine ammonia-lyase	93
Catalase	56, 57	Phosphate transporter	94
Chalcone isomerase	58	Phosphatidylinositol 4-kinase	95
Chalcone synthase	59	Phosphoglucose isomerase	96
Cellulose synthase	60	Phospholipase D	97
Chitinase (class I)	61, 62	Phospholipid hydroperoxide glutathione peroxidase	98
Chitinase (class III)	63	Proline transporter	99
Cytokinin oxidase/dehydrogenase	64	Δ1-Pyrroline-5-carboxylate synthetase	100
Dehydroascorbate reductase	65	Pyruvate decarboxylase	101, 102
Farnesyl diphosphate synthase	66	Pyruvate orthophosphate dikinase	103
Formaldehyde dehydrogenase	67	Ribulose-1,5-bisphosphate carboxylase/oxygenase	104
β-D-Fructofuranosidase	68	Ribulose-1,5-bisphosphate carboxylase/oxygenase activase	105, 106
Fructokinase	69	D-Ribulose-5-phosphate 3-epimerase	107
Fructose-1,6-bisphosphate aldolase	70	Squalene synthase	108
Gibberellin 2-oxidase	71	Starch branching enzyme	109, 110
Gibberellin 3 β-hydroxylase	72	Starch debranching enzyme	111
Gibberellin 20-oxidase	73, 74	Starch synthase	112, 113, 114
β-1,3-Glucanase	75	Sucrose-6(F)-phosphate phosphohydrolase	115
Glucose 6-phosphate/phosphate translocator	76	Sucrose-phosphate synthase	116
Glutamate synthase	77	Sucrose transporter	117, 118
Glutamine synthetase	78	Sulphite oxidase	119
Glutathione reductase	79	Superoxide dismutase	120, 121
3-Hydroxy-3-methylglutaryl coenzyme A reductase	80, 81	UDP-glucose pyrophosphorylase	122
myo-Inositol 1-phosphate synthase	82	UDP-glucuronic acid decarboxylase	123

sequences of enzymes known to be involved in metabolic pathways were used to query rice predicted protein databases (http://www.tigr.org/tdb/e2k1/osa1/; http://portal.tmri.org/rice/RicePublicAccess.html; http://rgp.dna.affrc.go.jp/IRGSP/index.html). Our search results indicated that pathways involved in all central metabolic processes (glycolysis; citric acid cycle; pentose phosphate pathway; photosynthesis and respiration; synthesis and degradation of amino acids, nucleotides, fatty acids and lipids, cofactors, carbohydrates, and cell wall materials) and nutrient exchange (assimilation of carbon, nitrogen, sulfur, and phosphorus; absorption of minerals) are present in rice (Fig. 6.1). However, the functions of only few rice genes involved in metabolic pathways have been characterized in detail thus far (Table 6.1), which is in contrast to the structurally diverse natural products isolated from rice tissues (Fig. 6.2). For example, rice synthesizes sakuranetin (a flavanone), the momilactones A and B (diterpenes), the oryzalexins A-F and S (diterpenes), and the phytocassanes A-D (diterpenes) as antifungal phytoalexins.[4-10] Constitutive antifungal compounds

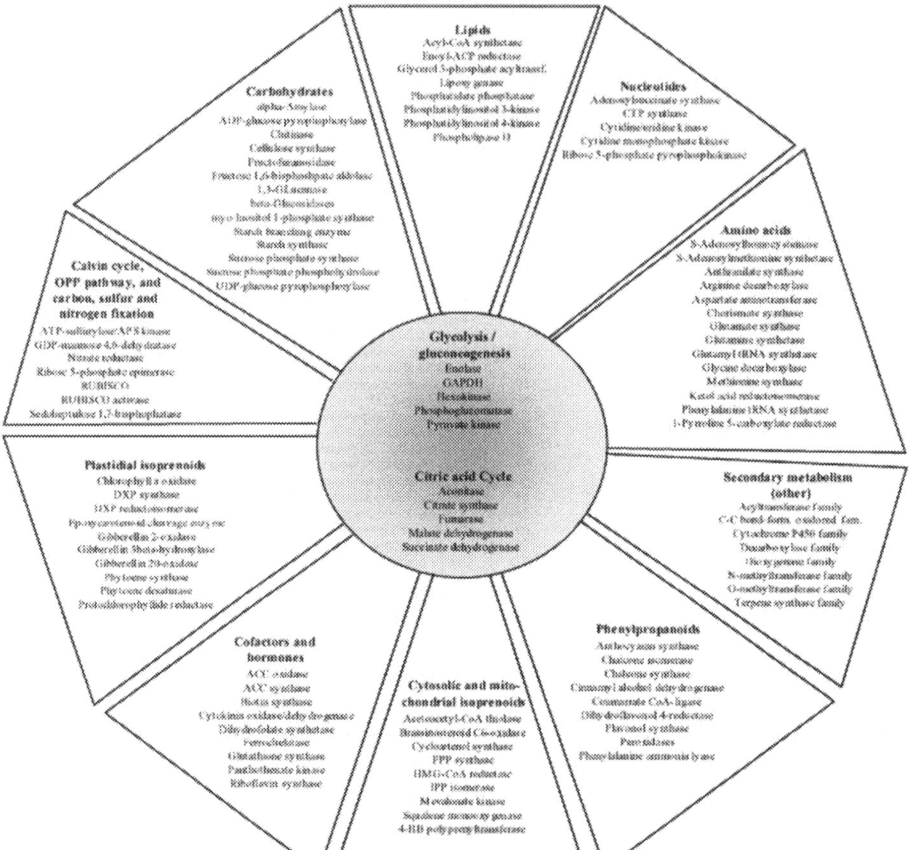

Figure 6.1: Survey of metabolic pathways in rice. BLASTP searches were conducted in which the peptide sequences of enzymes known to be involved in metabolic pathways were used to query rice predicted protein databases (http://portal.tmri.org/rice/RicePublicAccess.html; http://rgp.dna.affrc.go.jp/IRGSP/index.html; http://www.tigr.org/tdb/e2k1/osa1/). Rice hits were sorted based on metabolic pathways and were visualized in a pie chart. The center of the chart is occupied by central metabolic pathways that provide the precursors for all branch pathways, which are depicte as pie slices.

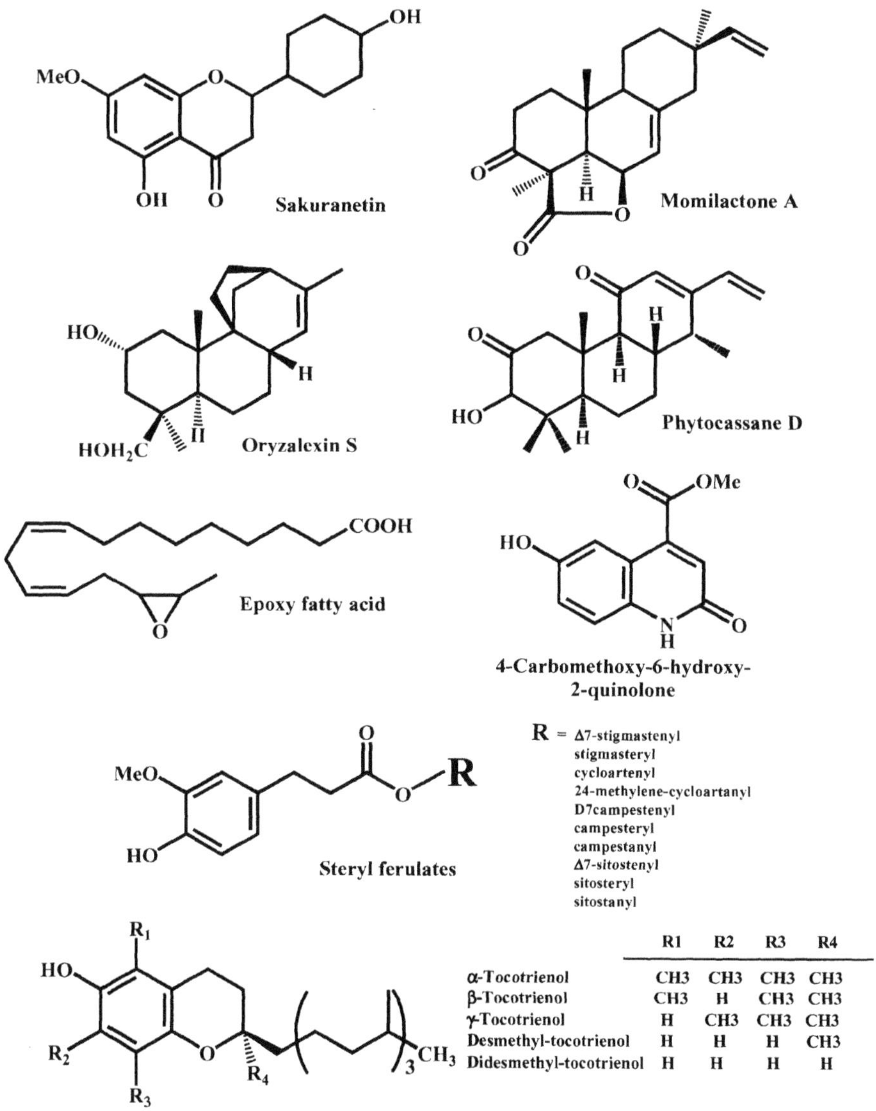

Figure 6.2: Structures of secondary metabolites isolated from different rice tissues. For details see text.

include hydroxy and epoxy fatty acids derived from linolenic acid.[11] Rice bran, a byproduct of the rice milling process, constitutes about 10 % (w : w) of rough rice grain. The hypocholesterolemic activity of rice bran has been attributed to the presence of γ-oryzanols (ferulate esters of triterpene alcohols) and tocotrienols.[12,13] The aleurone layer of anthocyanin-pigmented rice was shown to contain a quinolone alkaloid that was identified as part of a screen for grain antioxidants.[14] Taken together, these examples illustrate that, besides expressing ubiquitous primary metabolic pathways, rice is capable of producing certain classes of secondary metabolites, with implications for plant disease resistance and human health. It will be a challenge for the years to come to elucidate the biochemical pathways involved.

In rice, as in *Arabidopsis*, extensive gene redundancy exists across all metabolic pathways. It has been hypothesized that multiple-copy genes may facilitate the tightly regulated expression of specific isozymes in specialized tissues, at certain developmental stages, or in response to environmental challenges.[15,16] In rice, large gene families for a number of enzymes putatively involved in the biosynthesis of secondary metabolites have been detected.[1] In general, these structurally diverse compounds are generated by only a few types of reactions, which are catalyzed by (i) enzymes forming core structures (*e.g.*, chalcone/stilbene synthases, (+)-pinoresinol-forming dirigent proteins, terpene synthases, strictosidine synthases, berberine bridge enzymes), (ii) redox enzymes (*e.g.*, cytochrome P450-dependent oxidoreductases, oxoglutarate-dependent dioxygenases, phenol oxidases, desaturases, dehydratases, dehydrogenases, reductases), and (iii) substitution enzymes (*e.g.*, aminotransferases, methyltransferases, glycosyltransferases, acyltransferases).[17] Furthermore, metabolic diversity in plants is facilitated by the occurrence of multifunctional enzymes. For example, certain terpene synthases are known for their ability to synthesize multiple products from a single substrate and 2-oxoglutarate-dependent dioxygenases can typically accept multiple substrates and produce multiple products.[18,19]

Interestingly, the genomes of rice and *A. thaliana* contain gene families putatively involved in several pathways of alkaloid biosynthesis that are not known to operate in these organisms (Table 6.2). Pichersky and Gang have recently discussed a model in which repeated evolution - a process that leads to orthologous or paralogous genes with modified biochemical functions – would play a major role in secondary metabolism.[20] Hence, enzymes encoded by such gene families should be regarded as representatives of enzyme classes with common catalytic mechanisms (*e.g.*, berberine-bridge enzyme is a C-C bond-forming oxidoreductase), the functions of which need to be determined biochemically and can as yet not be assigned solely on the basis of sequence similarity. In many cases, distant homologs of the genes putatively encoding plant enymzes involved in secondary metabolism occur in the eubacterial and animal kingdoms (Figs. 6.3 and 6.4). Based on these observations, it can be speculated that the capabilities of rice and *A. thaliana* to produce secondary metabolites may have been vastly underestimated and/or that the members of these

Table 6.2: Rice genes putatively involved in alkaloid biosynthesis.

Gene product	Species	Functional classification	Phylogenetic distribution*
Berbamunine synthase	*Berberis stolonifera*	cytochrome P450-dependent monooxygenase	AODSE
Berberine-bridge enzyme	*Papaver somniferum*	C-C bond-forming oxidoreductase	AO--E
Caffeine synthase	*Camellia sinensis*	S-adenosylmethionine-dependent N-methyltransferase	AO---
Codeinone reductase	*Papaver somniferum*	aldo/keto reductase	AODSE
Deacetylvindoline acetyltransferase	*Catharanthus roseus*	acetyl-CoA-dependent acetyltransferase	AO---
Desacetoxyvindoline 4-hydroxylase	*Catharanthus roseus*	2-oxoglutarate-dependent dioxygenase	AODS-
Hyoscyamine 6β-hydroxylase	*Hyoscyamus niger*	2-oxoglutarate-dependent dioxygenase	AOD-E
3'-Hydroxy-N-methylcoclaurine-4'-O-methyltransferase	*Coptis japonica*	S-adenosylmethionine-dependent O-methyltransferase	AO--E
N-Methylcoclaurine 3'-hydroxylase	*Berberis stolonifera*	cytochrome P450-dependent monooxygenase	AODSE
Norcoclaurine 6-O-methyltransferase	*Coptis japonica*	S-adenosylmethionine-dependent O-methyltransferase	AO--E
Putrescine N-methyltransferase	*Atropa belladonna*	S-adenosylmethionine-dependent N-methyltransferase	AODSE
Scoulerine 9-O-methyltransferase	*Coptis japonica*	S-adenosylmethionine-dependent O-methyltransferase	AO--E
Strictosidine-β-D-glucosidase	*Catharanthus roseus*	membrane-associated glucosidase	AOD-E
Strictosidine synthase	*Catharanthus roseus*	vacuolar glycoprotein	AOD-E
Tropinone reductase I	*Datura stramonium*	short-chain alcohol dehydrogenase	AODSE
Tryptophan decarboxylase	*Catharanthus roseus*	pyridoxal-5'-phosphate decarboxylase	AOD-E
Tyrosine decarboxylase	*Papaver somniferum*	pyridoxal-5'-phosphate decarboxylase	AOD-E

* A, *Arabidopsis thaliana*; O, *Oryza sativa*; D, *Drosophila melanogaster*; S, *Saccharomyces cerevisiae*; E, *Escherichia coli*; -, not present.

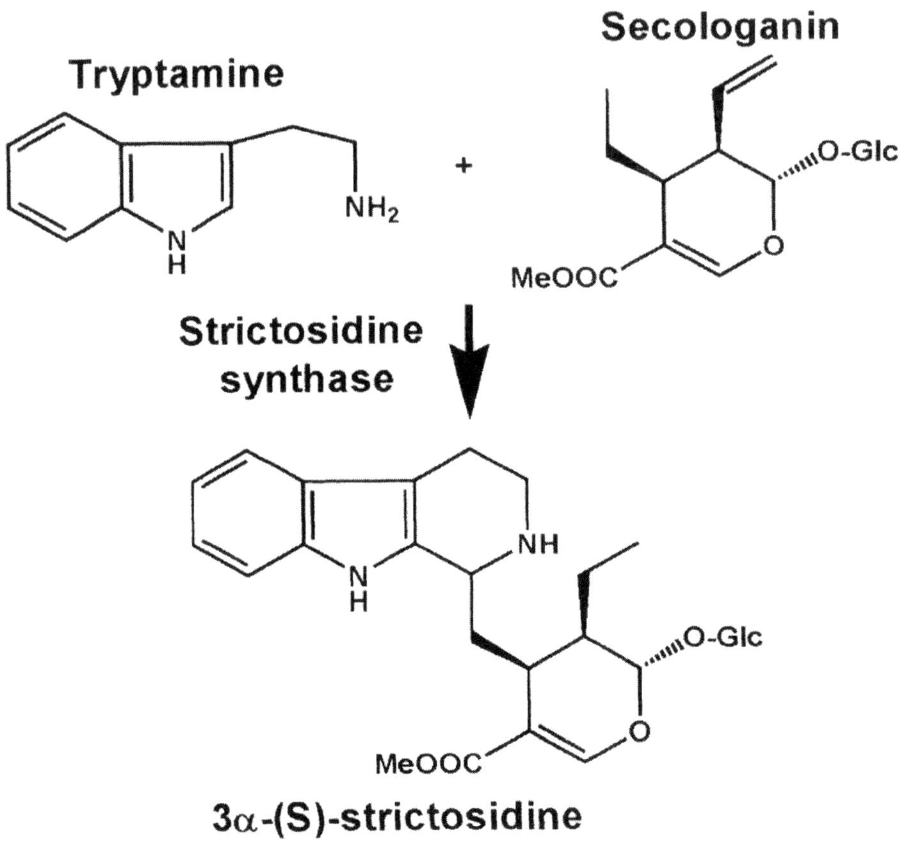

Figure 6.3: Fifteen *A. thaliana* and 45 rice proteins with homology (maximum expectation value of 1e-05) to strictosidine synthase (S22464) were identified in the current protein databases for the two genomes. These plant proteins were then used to identify homologs in a set of 75 completely sequenced organisms. All proteins were aligned using CLUSTALW, and the resulting tree was displayed using the TreeView software.[124,125] (See facing page). Accession numbers for protein sequences derived from differrent species are are symbolized as follows: rice, **R**; *A. thaliana,* **A**; other plants, **O**; animals, **An**; bacteria, **B**.

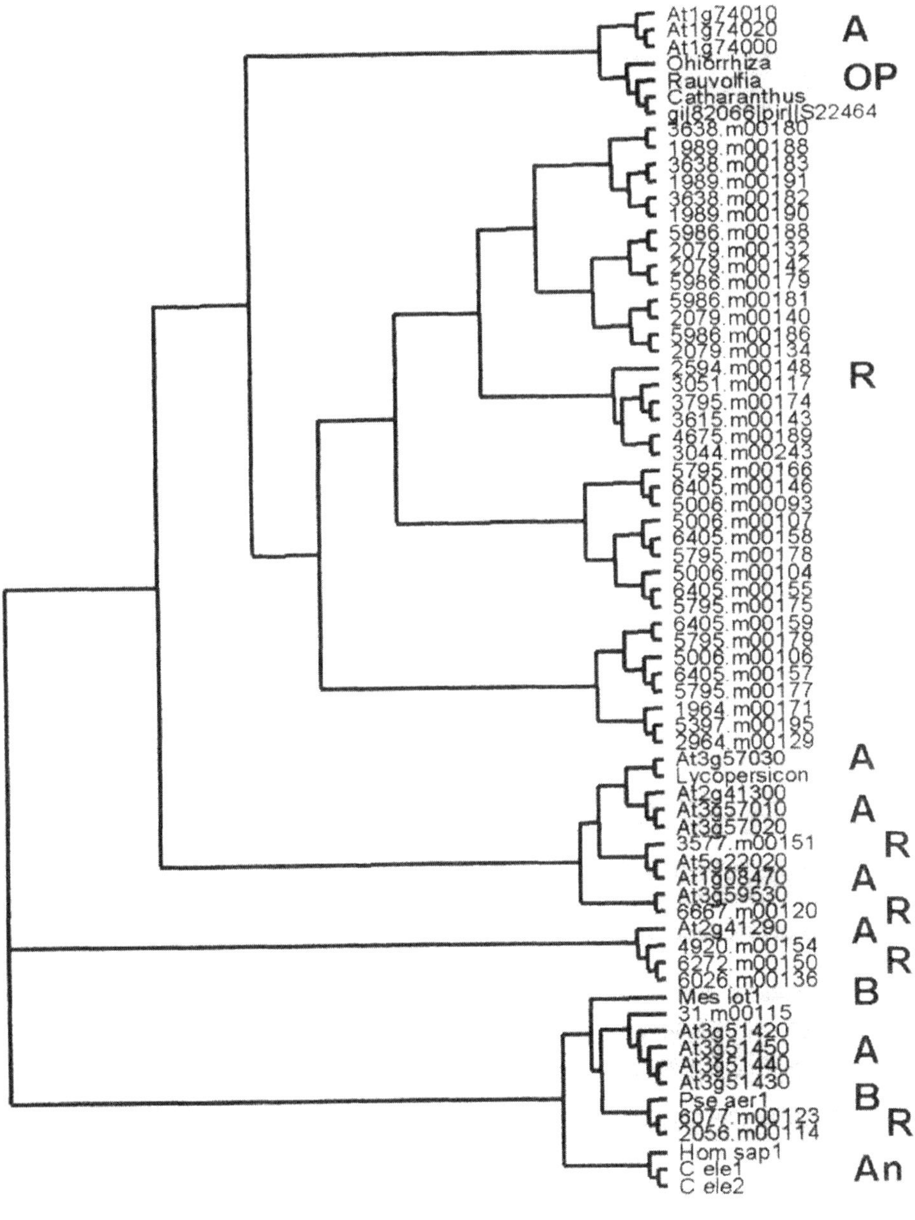
At1g74010
At1g74020
At1g74000
Ohiorrhiza
Rauvolfia
Catharanthus
gi|82066|pir||S22464
3638.m00180
1989.m00188
3638.m00183
1989.m00191
3638.m00182
1989.m00190
5986.m00188
2079.m00132
2079.m00142
5986.m00179
5986.m00181
2079.m00140
5986.m00186
2079.m00134
2594.m00148
3051.m00117
3795.m00174
3615.m00143
4675.m00189
3044.m00243
5795.m00166
6405.m00146
5006.m00093
5006.m00107
6405.m00158
5795.m00178
5006.m00104
6405.m00155
5795.m00175
6405.m00159
5795.m00179
5006.m00106
6405.m00157
5795.m00177
1964.m00171
5397.m00195
2964.m00129
At3g57030
Lycopersicon
At2g41300
At3g57010
At3g57020
3577.m00151
At5g22020
At1g08470
At3g59530
6667.m00120
At2g41290
4920.m00154
6272.m00150
6026.m00136
Mes lot1
31.m00115
At3g51420
At3g51450
At3g51440
At3g51430
Pse aer1
6077.m00123
2056.m00114
Hom sap1
C ele1
C ele2
A
OP
R
A
A
R
A
R
A
R
B
A
B
R
An

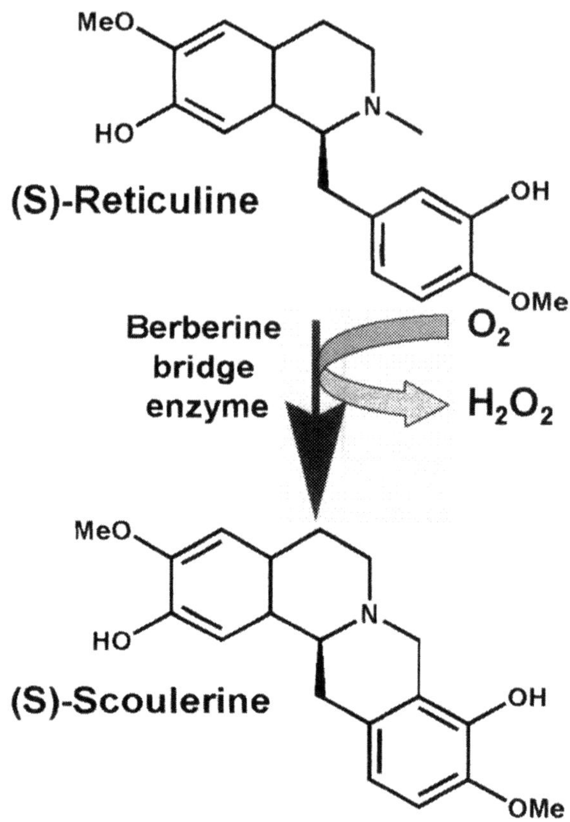

Figure 6.4: Twentyseven *A. thaliana* and 28 rice proteins with homology (maximum expectation value of 1e-05) to the berberine bridge forming enzyme (P93479) were identified in the current protein databases for the two genomes. These plant proteins were then used to identify homologs in a set of 75 completely sequenced organisms. All proteins were aligned using CLUSTALW and the resulting tree was displayed using the TreeView software.[124,125] (See facing page). Accession numbers for protein sequences derived from differrent species are are symbolized as follows: rice, **R**; *A. thaliana,* **A**; other plants, **O**; bacteria, **B**. Note the gene family expansion in plants as opposed to other higher eukaryotes. Furthermore, where these enzymes do occur in bacteria, the sequences tend to be very similar to their plant homologs.

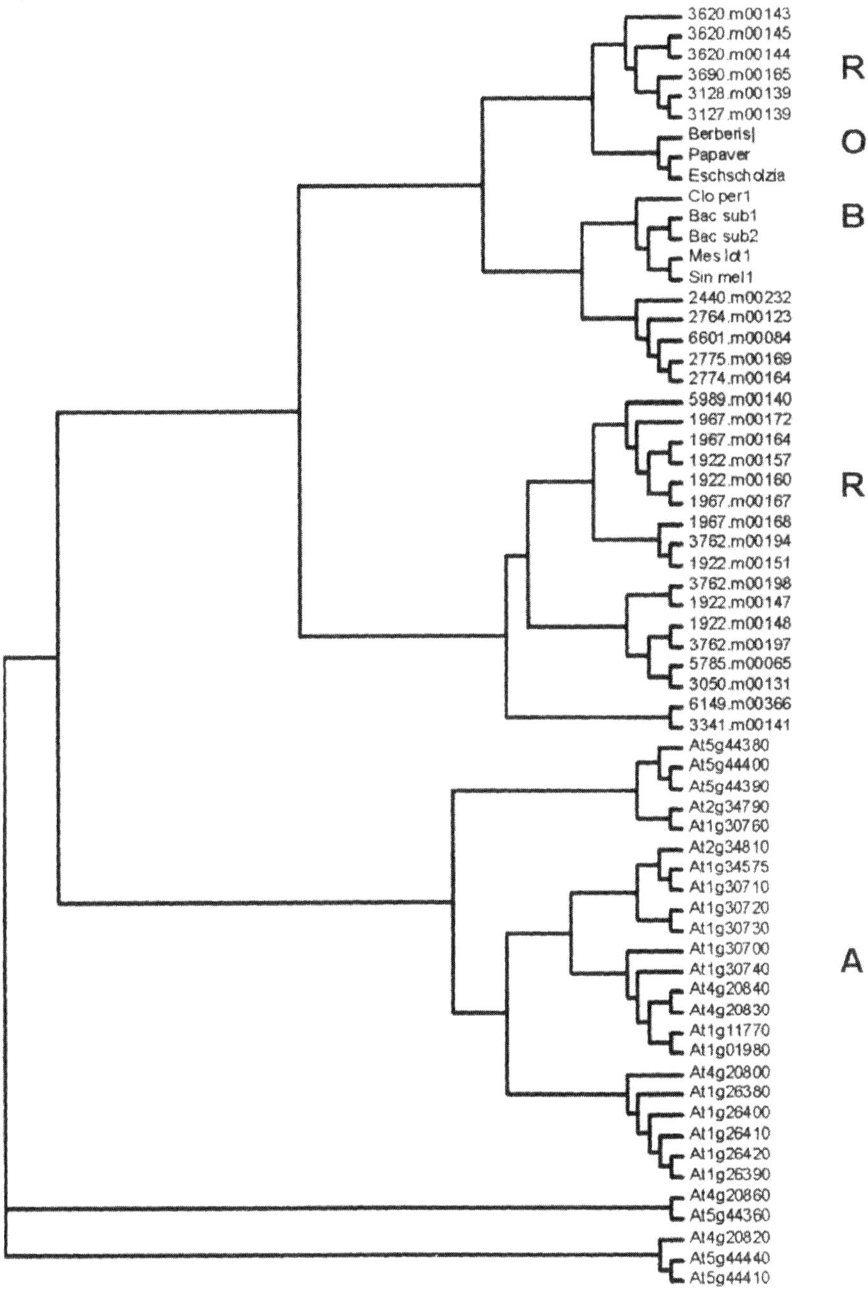
3620.m00143
3620.m00145
3620.m00144
3690.m00165
3128.m00139
3127.m00139
Berberis|
Papaver
Eschscholzia
Clo per1
Bac sub1
Bac sub2
Mes lot1
Sin mel1
2440.m00232
2764.m00123
6601.m00084
2775.m00169
2774.m00164
5989.m00140
1967.m00172
1967.m00164
1922.m00157
1922.m00160
1967.m00167
1967.m00168
3762.m00194
1922.m00151
3762.m00198
1922.m00147
1922.m00148
3762.m00197
5785.m00065
3050.m00131
6149.m00366
3341.m00141
At5g44380
At5g44400
At5g44390
At2g34790
At1g30760
At2g34810
At1g34575
At1g30710
At1g30720
At1g30730
At1g30700
At1g30740
At4g20840
At4g20830
At1g11770
At1g01980
At4g20800
At1g26380
At1g26400
At1g26410
At1g26420
At1g26390
At4g20860
At5g44360
At4g20820
At5g44440
At5g44410
R
O
B
R
A

gene families putatively related to secondary metabolism encode enzymes with novel functions in primary pathways.

RICE AROMA - MAPPING THE FRAGRANCE GENE

The genetic map constructed at the Japanese Rice Genome Project has been the basis for the international public effort and the industrial rice genome sequencing projects of Syngenta and Monsanto.[21] A genetic map contains markers that are ordered based on meiotic recombination events. These markers are linked to phenotypic traits that may be of commercial interest. Many commercially important traits (*e.g.*, grain size, number of grains) have values that are continuously distributed and are conferred by the interaction of many different genes at different map locations. These traits are termed quantitative traits, and each gene or locus that affects such a trait is known as a quantitative trait locus or QTL. Most such QTLs are mapped to fairly large intervals (10-20 cM) of the rice genome using any of a number of available molecular marker sets (*e.g.*, Restriction Fragment Length Polymorphisms, RFLP; Amplification Fragment Length Polymorphisms, AFLP; Simple Sequence Repeats, SSR; Cleaved Amplified Polymorphic Sequences, CAPS; Random Amplified Polymorphic DNA, RAPD). With the advent of genomics, physical maps have been constructed of the rice genome, primarily by ordering overlapping large insert clones (Bacterial Artificial Chromosomes or BACs) based on their restriction fingerprints. This yields BAC contigs (contiguous DNA sequences assembled using overlapping DNA sequences) of up to several megabases in length, which can be anchored to the rice genetic map using the genetic markers mapped and provided by the Japanese group. Anchored BAC contigs can then be used to generate a minimum tiling path of overlapping BACs for sequencing. Several rough drafts of the rice genome have been generated in this way, and a complete final high quality sequence of the rice genome is expected to be completed within 2004. The completed sequence can be overlaid with the molecular markers and QTL data to clearly define a sequence stretch (often covering many megabases of DNA) that contains one or more candidate genes for an agronomically important trait. Here, we illustrate this concept using the rice fragrance gene.

Certain rice varieties, most prominently the Basmati- and jasmine-style fragrant rice lines, are very popular owing to their characteristic aroma and flavor. Buttery et al. and Lorieux et al. established 2-acetyl-1-pyrroline (AP) as the key aroma component of aromatic rice varieties.[22,23] Despite the commercial importance of AP, relatively little progress has been made toward elucidating its biosynthesis in rice. Suprasanna et al. found that L-proline supplementation yielded an increase in aroma production in cell cultures of Basmati rice.[24] Recently, the role of proline as a precursor of the N-heterocycle moiety of AP in the Thai rice variety Khao Dawk Mali 105 was confirmed based on tracer feeding studies.[25] According to this study, the acetyl group of AP was not derived from praline, but the authors did not provide

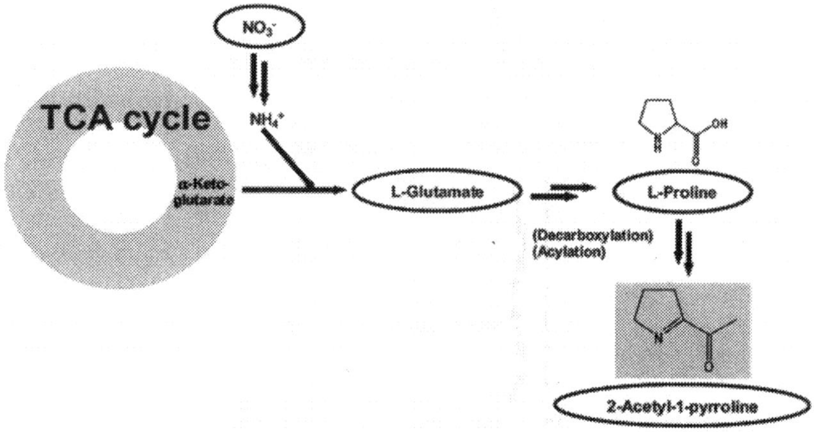

Figure 6.5: Hypothetical pathway for the biosynthesis of 2-acetyl-1-pyrroline (AP) in rice.[22-25]

evidence for its biosynthetic origin.[25] It can be speculated, however, that AP biosynthesis from proline, proceeds via decarboxylation and subsequent acyl transfer (Fig. 6.5).

Because of the great demand for aromatic rice, breeders have made efforts to improve productivity and yield while retaining aroma and texture. Australian breeding programs have relied on sensory detection of fragrances, but a number of technical challenges have been reported.[26] The chemical detection of AP, which has been used as a chemical marker in several rice improvement projects, is time-consuming and requires relatively large amounts of sample.[27] These issues highlight the utility of molecular breeding approaches to screen germplasm resources. A major advance was reported by Garland et al., who identified a small mononucleotide repeat that was polymorphic between a pair of fragrant and non-fragrant cultivars, and holds promise to be developed into a co-dominant PCR-based marker.[28] Attempts to map the loci related to fragrance in rice established that a single recessive fragrance gene (*fgr*) was responsible, which linked to the RFLP clone RG28 on chromosome 8, at a genetic distance of 4.5 cM.[23,29] RFLP marker RG1 flanks the fragrance gene on the other side. Using the existing physical map, all BACs spanning this region of the rice genome between 67.3 cM and 82.8 cM can be identified. Twentyfour BACs make up the minimum tiling path across this region; as of July 2003 nine of them were in the finishing stage, ten in the annotation stage, and five were completed. Sequence analysis of these BACs and subsequent gene prediction revealed all genes encoded in this region (Fig. 6.6). A total of 645 genes,

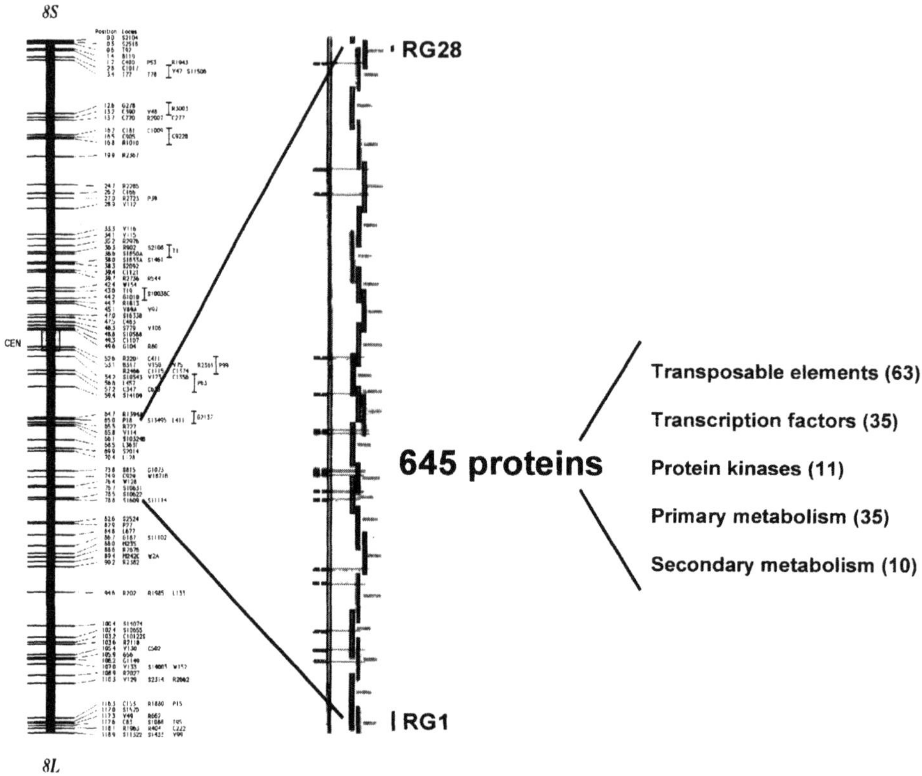

Figure 6.6: Genetic map (left) of rice chromosome 8 and physical map (right) in the form of overlapping BAC clones making up a minimum tile of the rice genome region containing the fragrance gene fgr. Both maps were obtained from the Japanese Rice Genome Sequencing website at http://rgp.dna.affrc.go.jp/. The flanking markers RG1 and RG28 were mapped to the rice genome using BLASTN. Gene sequences from this region were obtained from www.tigr.org and categorized based on function.

among them 35 transcription factors, are predicted from the sequence between the markers, making them potential candidates for the fragrance gene. Genes that are differentially expressed between fragrant and non-fragrant rice varieties, as well as those whose products are likely involved in the AP biosynthesis pathway would be particularly strong candidates. A functional confirmation could be obtained by generating transgenic rice lines that over-express these candidate genes, and a subsequent evaluation of correlation between transgene expression patterns and AP production.

RICE NUTRITION – PROTEOMIC APPROACHES TO EXPLORE STARCH METABOLISM

At the Torrey Mesa Research Institute (TMRI), we were interested in investigating the utility of genomic technologies to survey the tissue-specific expression of metabolic pathways in plant and animal model systems. As part of a systematic analysis of rice leaf, root, and seed tissues, two independent proteomic technologies (two-dimensional gel electrophoresis followed by tandem mass spectrometry and multidimensional protein identification technology) were employed to identify 2,528 unique proteins.[30] The expression patterns of proteins identified and classified as being involved in metabolic pathways were visualized on an interactive map to illustrate the contribution of these enzymes to tissue-specific metabolic pathways. In this article, we discuss the significance of our findings in understanding the compartmentation of enzymes involved in the starch metabolic pathway.

Starch is composed of two D-glucose homopolymers, amylose (linear polymer of α-1,4-linked glucosyl monomers) and amylopectin (branched polymer of α-1,4- and α-1,6-linked glucosyl monomers). In leaves, starch is synthesized during the day directly from photosynthetically fixed carbon dioxide in the stroma of chloroplasts, where it serves as a short-term carbohydrate reserve termed transitory starch. During the night, this pool of starch is degraded to provide a carbon supply for sucrose synthesis and export, and for respiration.[31] In seeds, starch accumulates in the endosperm, where it serves as an energy reserve, and plays an important role as the primary carbohydrate component in the diets of humans and livestock.[32]

The starch biosynthetic pathway starts with the conversion of glucose 1-phosphate into ADP-glucose, a key regulatory step catalyzed by ADP-glucose pyrophosphorylase (AGPase). The enzyme is now known to be largely extra-plastidial (*i.e.*, 85–95 % cytosolic) in cereal endosperm but plastidial in other cereal tissues and in all tissues of non-cereal plants.[33,34] AGPase occurs as a heterotetramer, composed of two small and two large subunits.[35] The small subunits are mainly responsible for the catalytic properties, whereas the large subunits appear to be of regulatory importance.[36] The two isoforms of the small AGPase subunit, for

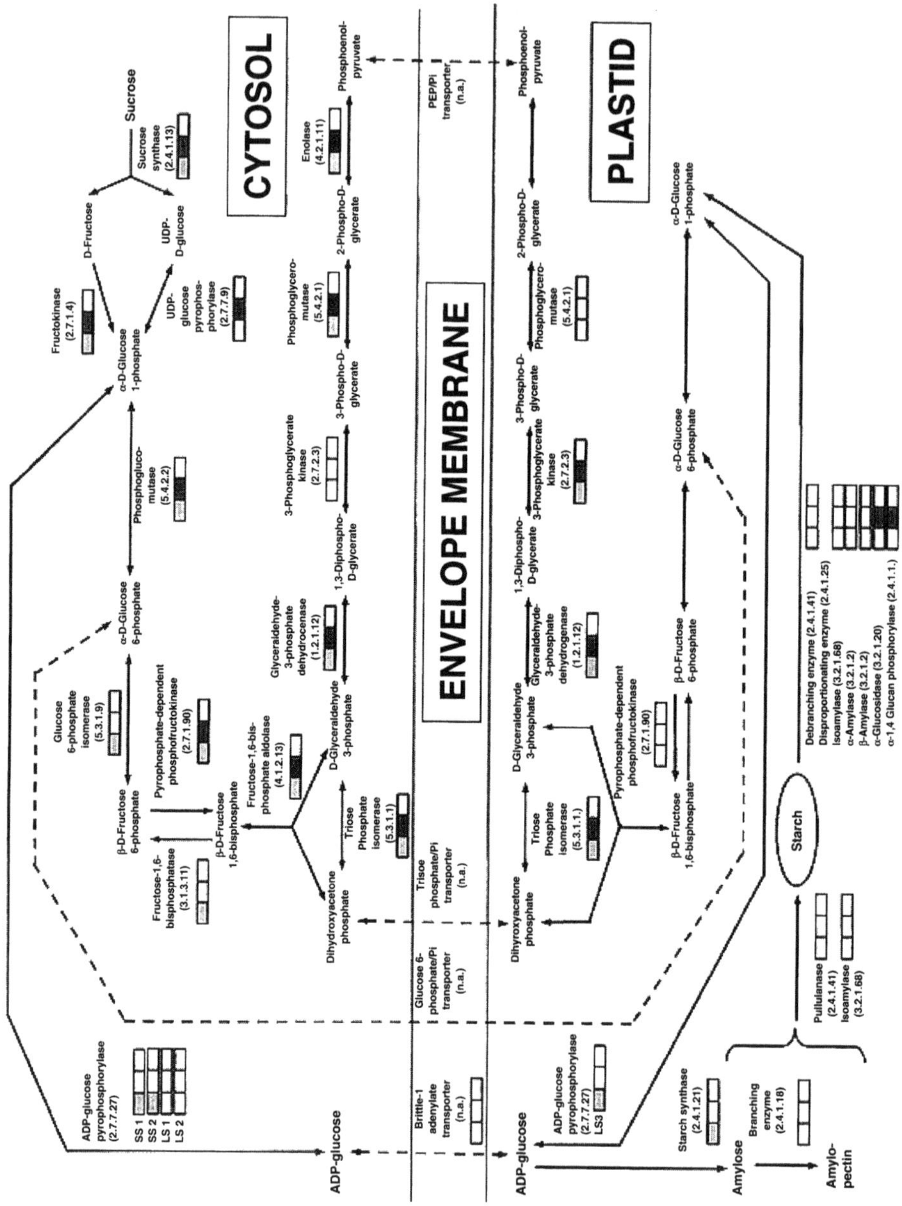
CYTOSOL
ENVELOPE MEMBRANE
PLASTID
Sucrose
Sucrose synthase (2.4.1.13)
D-Fructose
UDP-D-glucose
Fructokinase (2.7.1.4)
UDP-glucose pyrophosphorylase (2.7.7.9)
α-D-Glucose 1-phosphate
Phosphoglucomutase (5.4.2.2)
α-D-Glucose 6-phosphate
Glucose 6-phosphate isomerase (5.3.1.9)
β-D-Fructose 6-phosphate
Pyrophosphate-dependent phosphofructokinase (2.7.1.90)
β-D-Fructose 1,6-bisphosphate
Fructose-1,6-bisphosphatase (3.1.3.11)
Fructose-1,6-bisphosphate aldolase (4.1.2.13)
D-Glyceraldehyde 3-phosphate
Dihydroxyacetone phosphate
Triose Phosphate isomerase (5.3.1.1)
Glyceraldehyde-3-phosphate dehydrogenase (1.2.1.12)
1,3-Diphospho-D-glycerate
3-Phosphoglycerate kinase (2.7.2.3)
3-Phospho-D-glycerate
Phosphoglyceromutase (5.4.2.1)
2-Phospho-D-glycerate
Enolase (4.2.1.11)
Phosphoenolpyruvate
PEP/Pi transporter (n.a.)
Triose phosphate/Pi transporter (n.a.)
Glucose 6-phosphate/Pi transporter (n.a.)
Brittle-1 adenylate transporter (n.a.)
ADP-glucose pyrophosphorylase (2.7.7.27)
SS 1
SS 2
LS 1
LS 2
ADP-glucose
LS3
Starch synthase (2.4.1.21)
Amylose
Branching enzyme (2.4.1.18)
Amylopectin
Pullulanase (2.4.1.41)
Isoamylase (3.2.1.68)
Starch
Debranching enzyme (2.4.1.41)
Disproportionating enzyme (2.4.1.25)
Isoamylase (3.2.1.68)
α-Amylase (3.2.1.2)
β-Amylase (3.2.1.2)
α-Glucosidase (3.2.1.20)
α-1,4 Glucan phosphorylase (2.4.1.1.)

which peptides identical to published rice sequences were identified in our proteomics study, were detected in both leaves and seeds (Fig. 6.7).[30] Two isoforms of the large AGPase subunit were detected only in seeds, whereas the third isoform was detected exclusively in leaves.[30] Interestingly, based on an analysis of the subcellular compartmentation by ChloroP (http://www.cbs.dtu.dk/services/ChloroP/), the two seed-specific isozymes of the large subunit are devoid of a plastidial targeting sequence, which is in agreement with previously published reports indicating that AGPase activity is mainly localized to the cytosol of the graminaceous endosperm.[30,33] It has been speculated that the cytosolic AGPase may facilitate the partitioning of carbon from sucrose into starch when there is a sufficient supply of sucrose in the endosperm, which would require the import of cytosolic ADP-glucose into the plastids.[34] This import has been proposed to be accomplished through the action of the brittle-1 protein, an adenylate translocator with a common ADP-glucose-binding, for which we detected a seed-specific expression, thus supporting the evidence of cytosolic AGPase pools in seed tissues.[30,37] In plants that express exclusively plastidial AGPase, the sucrose-to-starch pathway involves plastid import of hexose phosphates that can also be used in pathways other than starch synthesis. In cereals, however, carbon entering the plastid as ADP-glucose is committed to starch synthesis and cannot be diverted into other metabolic pathways within the plastid.[32]

In the next phase of starch biosynthesis, starch synthases utilize ADP-glucose to generate linear chains by the formation of α-1,4 linked glucose building blocks. Cereal endosperms contain at least five starch synthase isoforms that are categorized according to conserved sequence relationships. Four isoforms are believed to have unique functions in amylopectin synthesis, although their precise roles have not been identified.[38] In our proteomic study, we identified peptide sequences corresponding to isoforms that occur preferentially in leaves, and other isoforms that are present mainly in seeds (Fig. 6.7).[30] Following the formation of α-1,4 linked glucose chains by starch synthases, branching enzymes generate α-1,6-linkages by cleaving internal α-1,4-bonds and transferring the released reducing ends to C6 hydroxyls. The

Figure 6.7: Tissue localization of proteins involved in the starch biosynthesis and degradation pathways as indicated by proteomic analysis. Enzymes and corresponding EC numbers involved in each pathway are shown. The three boxes below each enyzme name display in which tissue this enzyme has been found; gray color in the left box indicates its presence in leaves, a black box in the middle box its presence in roots, and light gray shading in the right box its presence in seeds.

temporal and spatial patterns of expression vary between branching enzyme isoforms. In our proteomic survey, we identified seed-specific branching enzyme isoforms, whereas peptides released from leaf-specific isoforms were not detected.[30] Mutations in many species indicate that starch synthesis involves debranching enzymes in addition to starch synthases and branching enzymes. Two debranching enzyme families exist in plants, the isoamylase-type and the pullulanase-type.[30] Both types hydrolyze α-1,6-linkages, but they differ in substrate specificity. The pullulanase-type enzyme appears to provide a function that overlaps with that of the isoamylase-type enzyme during biosynthesis. In addition, the pullulanase-type enzyme, termed ZPU1 in maize, participates in starch degradation in the endosperm.[39] Based on our proteomics results, isoamylases and pullulanases were both expressed preferentially in seed tissue (Fig. 6.7).[30] Enzymes for the starch degradation pathway, which comprise debranching enzyme, disproportionating enzyme, α-amylase, β-amylase, α-glucosidase, and starch phosphorylase, were exclusively detected in seeds (Fig. 6.7).[30] The absence of the starch catabolic enzymes in the leaf sample, as indicated by our proteomics results, might be explained by the fact that the leaves were picked about 4 h after dawn and were, thus, synthesizing transitory starch with very little starch degradation activity.[30]

Future research in this area will identify direct interactions among starch biosynthetic enzymes, as well as modifying factors that regulate enzyme activity. In addition, comparative genomics will help to distinguish components of the starch-synthesizing machinery that are conserved among all plant species from those that are unique to cereals, and to identify those components that differ among the various cereal species. Furthermore, in-depth phenotypic analyses of mutants will allow us to decipher the role of specific isoforms of enzymes involved in starch metabolism.

REFERENCES

1. GOFF, S. A., RICKE, D., LAN, T. H., PRESTING, G., WANG, R., DUNN, M., GLAZEBROOK, J., SESSIONS, A., OELLER, P., VARMA, H., *et al.*, A draft sequence of the rice genome (*Oryza sativa* L. ssp. *japonica*), *Science*, 2002, **296**, 92–100.
2. YU, J., HU, S., WANG, J., WONG, G.K., LI, S., LIU, B., DENG, Y., DAI, L., ZHOU, Y., ZHANG, X., et al., A draft sequence of the rice genome (*Oryza sativa* L. ssp. Indica), *Science*, 2002, **296,** 79-92.
3. *ARABIDOPSIS* GENOME INITIATIVE, Analysis of the genome sequence of the flowering plant *Arabidopsis thaliana*, *Nature*, 2000, **408**, 796-815.
4. KODAMA, O., MIYAKAWA, J., AKATSUKA, T., KIYOSAWA, S., Sakuranetin, a flavanone phytoalexin from ultraviolet-irradiated rice leaves, *Phytochemistry*, 1992, **31,** 3807-3813.
5. CARTWRIGHT, D.W., LANGCAKE, P., PRYCE, R.J., LEWORTHY, D.P., RIDE J.P., Isolation and characterization of two phytoalexins from rice as momilactones A and B, *Phytochemistry*, 1981, **20,** 535-537.

6. AKATSUKA, T., KODAMA, O., SEKIDO, H., KONO, Y., TAKEUCHI, S., Novel phytoalexins (oryzalexins A, B and C) isolated from rice blast leaves infected with *Pyricularia oryzae*. Part I: isolation, characterization and biological activities of oryzalexins, *Agric. Biol. Chem.*, 1985, **49,** 1689-1694.
7. SEKIDO, H., ENDO, T., SUGA, R., KODAMA, O., AKATSUKA, T., KONO, Y., TAKEUCHI, S., Oryzalexin D (3,7-dihydroxy-(+)-sandaracopimaradiene), a new phytoalexin isolated from blast-infected rice leaves, *J. Pest. Sci.*, **11,** 369-372.
8. KODAMA, O., LI, W.X., TAMOGAMI, S., AKATSUKA, T., Oryzalexin S, a novel stemarane-type diterpene rice phytoalexin, *Biosci. Biotechnol. Biochem.*, 1992, **56,** 1002-1003.
9. KATO, H., KODAMA, O., AKATSURA, T., Oryzalexin E, a diterpene phytoalexin from UV-irradiated rice leaves, *Phytochemistry*, 1993, **33,** 79-81.
10. KOGA, J., SHIMURA, M., OSHIMA, K., OGAWA, N., YAMAUCHI, T., OGASAWARA, N., Phytocassanes A, B, C and D, novel diterpene phytoalexins from rice, *Tetrahedron*, 1995, **51,** 7907-7918.
11. KATO, T., YAMAGUCHI, Y., NANAI, T., HIRUKAWA, T., Oxygenated fatty acids with anti-rice blast fungus activity in rice plants, *Biosci., Biotechnol. Biochem.*, 1993, **57,** 283-287.
12. NICOLOSI, R.J., AUSMAN, L.M., HEGSTED, D.M., Rice bran oil lowers serum total and low-density lipoprotein cholesterol and apoB levels in nonhuman primates, *Atherosclerosis*, 1991, **88,** 133-142.
13. QURESHI, A.A., HUANBIAO, M., PACKER, L., PETERSON, D. Isolation and identification of novel tocotrienols from rice bran with hypocholesteremic, antioxidant and antitumor properties, *J. Agric. Food Chem.*, 2000, **48,** 3130-3140.
14. CHUNG, H.S., WOO, W.S., A quinolone alkaloid with antioxidant activity from the aleurone layer of anthocyanin-pigmented rice, *J. Nat. Prod.*, 2001, **64,** 1579-1580.
15. LANGE, B.M., RUJAN, T., MARTIN, W., CROTEAU, R., Isoprenoid biosynthesis: The evolution of two ancient and distinct pathways across genomes, *Proc. Natl. Acad. Sci. U.S.A.*, 2000, **97**, 13172-13177.
16. DIXON, R.A., Natural products and plant disease resistance, *Nature*, 2001, **411**, 843-847.
17. FACCHINI, P.J., Plant secondary metabolism: out of the evolutionary abyss, *Trends Plant Sci.*, 1999, **4**, 382.
18. BOHLMANN, J., MEYER-GAUEN, G., CROTEAU, R., Plant terpenoid synthases: Molecular biology and phylogenetic analysis, *Proc. Natl. Acad. Sci. U.S.A.* **95**, 1998, 4126-4133.
19. PRESCOTT, A.G., JOHN, P., Dioxygenases: Molecular structure and role in plant metabolism, *Annu. Rev. Plant Physiol. Plant Mol. Biol.*, 1996, **47**, 245-271.
20. PICHERSKY, E., GANG, D.R., Genetics and biochemistry of secondary metabolites in plants: An evolutionary perspective, *Trends Plant Sci.*, 2000, **5**, 439-445.
21. HARUSHIMA, Y. , YANO, M. , SHOMURA, A. , SATO, M. , SHIMANO, T. , KUBOKI, Y. , YAMAMOTO, T. , LIN, S. Y. , ANTONIO, B. A. , PARCO, A. *et*

al., A high-density rice genetic linkage map with 2275 markers using a single F_2 population, *Genetics*, 1998, **148,** 479-494.
22. BUTTERY, R.G., LING, L.C., JULIANO, B.O., Cooked rice aroma and 2-acetyl-1-pyrroline, *J. Agric. Food. Chem.*, 1993, **31,** 823-826.
23. LORIEUX, M., PETROV, M., HUANG, N., GUIDERDONI, E., GHESQUIÈRE, A., Aroma in rice: Genetic analysis of a quantitative trait, *Theor. Appl. Genet.*, 1996, **93,** 1145-1151.
24. SUPRASANNA, P., GANAPATHI, T.R., RAMASWAMY, N.K., SURENDRANATHAN, K.K., RAO, P.S., Aroma synthesis in cell and callus cultures of rice, *Rice Genet. Newsl.*, 1998, **15,** 123-125.
25. YOSHIHASHI, T., HUONG, M.T.T., INATOMI, H., Precursors of 2-acetyl-1-pyrroline, a potent flavor compound of an aromatic rice variety, *J. Agric. Food Chem.*, 2002, **50,** 2001-2004.
26. GARLAND, S., HENRY, R., Application of Molecular Markers to Rice Breeding in Australia. Rural Industries Research and Development Corp., 2001, 21p.
27. WIDJAJA, R., CRASKE, J.D., WOOTTON, M., Comparative studies on volatile components of non-fragrant and fragrant rices, *J. Sci. Food Agric.*, 1996, **70,** 151-161.
28. GARLAND, S., LEWIN, L., BLAKENEY, A., REINKE, R., HENRY, R., PCR-based molecular markers for the fragrance gene in rice (*Oryza sativa* L.), *Theor. Appl. Genet.*, 2000, **101,** 364-371.
29. AHN, S.N., BOLLICH, C.N., TANKSLEY, S.D., RFLP tagging of a gene for aroma in rice, *Theor. Appl. Genet.*, 1992, **84,** 825-828.
30. KOLLER, A., WASHBURN, M.P., LANGE, M., ANDON, N. L., DECIU, C., HAYNES, P.A., HAYS, L., SCHIELTZ, D., ULASZEK, R., WEI, J., WOLTERS, D., YATES, J.R., Proteomic survey of metabolic pathways in rice, *Proc. Natl. Acad. Sci. U.S.A.,* 2002, **99,** 11969-11974.
31. ZEEMAN, S.C., NORTHROP, F., SMITH, A.M., REES, T., A starch-accumulating mutant of *Arabidopsis thaliana* deficient in a chloroplastic starch-hydrolysing enzyme, *Plant J.*, 1998, **15,** 357-365.
32. JAMES, M.G., DENYER, K., MYERS, A.M., Starch biosynthesis in the cereal endosperm, *Curr. Op. Plant Biol.*, 2003, **6,** 215-222.
33. DENYER, K., DUNLAP, F., THORBRJØRNSEN, T., KEELING, P., SMITH, A.M., The major form of ADP-glucoase pyrophosphorylase in maize nedosperm is extra-plastidial, *Plant Physiol.*, 1996, **112,** 779-785.
34. BECKLES, D.M., SMTH, A.M., RESS, T., A cytosolic ADP-glucose pyrophosphorylase is a feature of graminaceaous endosperms, but not of other starch-storing organs, *Plant Physiol.*, 2001, **125,** 818-827.
35. SMITH-WHITE, B.J., PREISS, J., Comparison of proteins of ADP-glucose pyrophosphorylase from diverse sources, J. Mol. Evol., 1992, **34,** 449-464.
36. FU, Y., BALLICORA, M.A., PREISS, J., Mutagenesis of the glucose-1-phosphate-binding site of potato tuber ADP-glucose pyrophosphorylase, *Plant Physiol.*, 1998, **117,** 989-996.

37. SHANNON, J.C., PIEN, F.M., CAO, H., LIU, K.C., Brittle-1, an adenylate translocator, facilitates transfer of extraplastidial synthesized ADP-glucose into amyloplasts of maize endosperm, *Plant Physiol.*, 1998, **117,** 1235-1252.
38. CAO, H., JAMES, M.G., MYERS, A.M., Purification and characterization of soluble starch synthases from maize endosperm, *Arch. Biochem. Biophys.*, 2000, **373,** 135-146.
39. DINGES, J.R., COLLEONI, C., MYERS, A.M., JAMES, M.G., Molecular structure of three mutations at the maize sugary1 locus and their allele-specific phenotypic effects, *Plant Physiol.*, 2001, **125,** 1406-1418.
40. LEE, J.H., CHAE, H.S., LEE, J.H., HWANG, B., HAHN, K.W., KANG, B.G., KIM, W.T., Structure and expression of two cDNAs encoding S-adenosyl-L-methionine synthetase of rice (*Oryza sativa* L.), *Biochim. Biophys. Acta*, 1997, **1354,** 13-18.
41. SMIDANSKY, E.D., MARTIN, J.M., HANNAH, L.C., FISCHER, A.M., GIROUX, M.J., Seed yield and plant biomass increases in rice are conferred by deregulation of endosperm ADP-glucose pyrophosphorylase, *Planta*, 2003, **216,** 656-664.
42. TSUJI, H., TSUTSUMI, N., SASAKI, T., HIRAI, A., NAKAZONO, M., Organ-specific expressions and chromosomal locations of two mitochondrial aldehyde dehydrogenase genes from rice (*Oryza sativa* L.), ALDH2a and ALDH2b, *Gene*, 2003, **305,** 195-204.
43. LI, Y., NAKAZONO, M., TSUTSUMI, N., HIRAI, A., Molecular and cellular characterizations of a cDNA clone encoding a novel isozyme of aldehyde dehydrogenase from rice, *Gene*, 2000, **249,** 67-74.
44. ABE, R., YOSHIDA, K., AOYAGI, M., KASAHARA, S., ICHISHIMA, E., NAKAJIMA, T., Characterization of chimeric enzymes constructed between two distinct alpha-amylase cDNAs from cultured rice cells, *Biosci. Biotechnol. Biochem.*, 1999, **63,** 1329-35.
45. MITSUI, T., YAMAGUCHI, J., AKAZAWA, T., Physicochemical and serological characterization of rice alpha-amylase isoforms and identification of their corresponding genes, *Plant Physiol.*, 1996, **110,** 1395-1404.
46. HUANG, N., SUTLIFF, T.D., LITTS, J.C., RODRIGUEZ, R.L., Classification and characterization of the rice alpha-amylase multigene family, *Plant Mol. Biol.*, 1990, **14,** 655-668.
47. CHAE, H.S., CHO, Y.G., PARK, M.Y., LEE, M.C., EUN, M.Y., KANG, B.G., KIM, W.T., Hormonal cross-talk between auxin and ethylene differentially regulates the expression of two members of the 1-aminocyclopropane-1-carboxylate oxidase gene family in rice (*Oryza sativa* L.), *Plant Cell Physiol.*, 2000, **41,** 354-62.
48. ZHOU, Z., ALMEIDA, D.E., ENGLER, J., ROUAN, D., MICHIELS, F., VAN MONTAGU, M., VAN DER STRAETEN, D., Tissue localization of a submergence-induced 1-aminocyclopropane-1-carboxylic acid synthase in rice, *Plant Physiol.*, 2002, **129,** 72-84.

49. ZAREMBINSKI, T.I., THEOLOGIS, A., Anaerobiosis and plant growth hormones induce two genes encoding 1-aminocyclopropane-1-carboxylate synthase in rice (*Oryza sativa* L.), *Mol. Biol. Cell*, 1993, **4,** 363-373.
50. SONODA, Y., IKEDA, A., SAIKI, S., VON WIREN, N., YAMAYA, T., YAMAGUCHI, J., Distinct expression and function of three ammonium transporter genes (OsAMT1;1-1;3) in rice, *Plant Cell Physiol.*, 2003, **44,** 726-734.
51. TOZAWA, Y., HASEGAWA, H., TERAKAWA, T., WAKASA, K., Characterization of rice anthranilate synthase alpha-subunit genes OASA1 and OASA2. Tryptophan accumulation in transgenic rice expressing a feedback-insensitive mutant of OASA1, *Plant Physiol.*, 2001, **126,** 1493-1506.
52. CHATTOPADHYAY, M.K., GUPTA, S., SENGUPTA, D.N., GHOSH, B., Expression of arginine decarboxylase in seedlings of indica rice (*Oryza sativa* L.) cultivars as affected by salinity stress, *Plant Mol. Biol.*, 1997, **34,** 477-83.
53. SONG, J., YAMAMOTO, K., SHOMURA, A., YANO, M., MINOBE, Y., SASAKI, T., Characterization and mapping of cDNA encoding aspartate aminotransferase in rice, *Oryza sativa* L., *DNA Res.*, 1996, **3,** 303-310.
54. HONG, Z., UEGUCHI-TANAKA, M., SHIMIZU-SATO, S., INUKAI, Y., FUJIOKA, S., SHIMADA, Y., TAKATSUTO, S., AGETSUMA, M., YOSHIDA, S., WATANABE, Y., UOZU, S., KITANO, H., ASHIKARI, M., MATSUOKA, M., Loss-of-function of a rice brassinosteroid biosynthetic enzyme, C-6 oxidase, prevents the organized arrangement and polar elongation of cells in the leaves and stem, *Plant J.*, 2002, **32,** 495-508.
55. MORI, M., NOMURA, T., OOKA, H., ISHIZAKA, M., YOKOTA, T., SUGIMOTO, K., OKABE, K., KAJIWARA, H., SATOH, K., YAMAMOTO, K., HIROCHIKA, H., KIKUCHI, S., Isolation and characterization of a rice dwarf mutant with a defect in brassinosteroid biosynthesis, *Plant Physiol.*, 2002, **130,** 1152-1161.
56. HIGO, K., HIGO, H., Cloning and characterization of the rice CatA catalase gene, a homologue of the maize Cat3 gene, *Plant Mol. Biol.*, 1996, **30,** 505-521.
57. MORITA, S., TASAKA, M., FUJISAWA, H., USHIMARU, T., TSUJI, H., A cDNA clone encoding a rice catalase isozyme, *Plant Physiol.*, 1994, **105,** 1015-1016.
58. DRUKA, A., KUDRNA, D., ROSTOKS, N., BRUEGGEMAN, R., VON WETTSTEIN, D., KLEINHOFS, A., Chalcone isomerase gene from rice (*Oryza sativa*) and barley (*Hordeum vulgare*): physical, genetic and mutation mapping, *Gene*, 2003, **302,** 171-178.
59. REDDY, A.R., SCHEFFLER, B., MADHURI, G., SRIVASTAVA, M.N., KUMAR, A., SATHYANARAYANAN, P.V., NAIR, S., MOHAN, M., Chalcone synthase in rice (*Oryza sativa* L.): detection of the CHS protein in seedlings and molecular mapping of the chs locus, *Plant Mol. Biol.*, 1996, **32,** 735-743.
60. HAZEN, S.P., SCOTT-CRAIG, J.S., WALTON, J.D., Cellulose synthase-like genes of rice, *Plant Physiol.*, 2002, **128,** 336-340.
61. PAN, C.H., RHIM, S.L., KIM, S.I., Expression of two cDNAs encoding class I chitinases of rice in *Escherichia coli*, *Biosci. Biotechnol. Biochem.*, 1996, **60,** 1346-1348.

62. KIM, Y.K., BAEK, J.M., PARK, H.Y., CHOI, Y.D., KIM, S.I., Isolation and characterization of cDNA clones encoding class I chitinase in suspension cultures of rice, *Biosci. Biotechnol. Biochem.*, 1994, **58,** 1164-1166.
63. PARK, H.Y., PAN, C.H., SO, M.Y., AH, J.H., JO, D.H., KIM, S., Purification, characterization, and cDNA cloning of rice class III chitinase, *Mol. Cells*, 2002, **13,** 69-76.
64. SCHMULLING, T., WERNER, T., RIEFLER, M., KRUPKOVA, E., BARTRINA Y MANNS, I., Structure and function of cytokinin oxidase/dehydrogenase genes of maize, rice, *Arabidopsis* and other species, *J. Plant Res.*, 2003, **116,** 241-252.
65. URANO, J., NAKAGAWA, T., MAKI, Y., MASUMURA, T., TANAKA, K., MURATA, N., USHIMARU, T., Molecular cloning and characterization of a rice dehydroascorbate reductase, *FEBS Lett.*, 2000, **466,** 107-111.
66. SANMIYA, K., IWASAKI, T., MATSUOKA, M., MIYAO, M., YAMAMOTO, N., Cloning of a cDNA that encodes farnesyl diphosphate synthase and the blue-light-induced expression of the corresponding gene in the leaves of rice plants, *Biochim. Biophys. Acta*, 1997, **1350,** 240-246.
67. DOLFERUS, R., OSTERMAN, J.C., PEACOCK, W.J., DENNIS, E.S., Cloning of the *Arabidopsis* and rice formaldehyde dehydrogenase genes: Implications for the origin of plant ADH enzymes, *Genetics*, 1997, **146,** 1131-1141.
68. FU, R.H., WANG, Y.L., SUNG, H.Y., Cloning, characterization and functional expression of a new beta-D-fructofuranosidase (Os beta fruct2) cDNA from *Oryza sativa*, *Biotechnol. Lett.*, 2003, **25,** 455-459.
69. JIANG, H., DIAN, W., LIU, F., WU, P., Isolation and characterization of two fructokinase cDNA clones from rice, *Phytochemistry*, 2003, **62,** 47-52.
70. CHEN, H.B., YANG, C.S., Cloning and high expression in *E. coli* of a chimeric gene coding for rice fructose-l,6-bisphosphate aldolase, *Sheng Wu Hua Xue Yu Sheng Wu Wu Li Xue Bao (Shanghai)*, 1997, **29,** 613-616.
71. SAKAI, M., SAKAMOTO, T., SAITO, T., MATSUOKA, M., TANAKA, H., KOBAYASHI, M., Expression of novel rice gibberellin 2-oxidase gene is under homeostatic regulation by biologically active gibberellins, *J. Plant Res.*, 2003, **116,** 161-164.
72. ITOH, H., UEGUCHI-TANAKA, M., SENTOKU, N., KITANO, H., MATSUOKA, M., KOBAYASHI, M., Cloning and functional analysis of two gibberellin 3 beta-hydroxylase genes that are differently expressed during the growth of rice, *Proc. Natl. Acad. Sci. U.S.A.*, 2001, **98,** 8909-8914.
73. MONNA, L., KITAZAWA, N., YOSHINO, R., SUZUKI, J., MASUDA, H., MAEHARA, Y., TANJI, M., SATO, M., NASU, S., MINOBE, Y., Positional cloning of rice semidwarfing gene, sd-1: rice "green revolution gene" encodes a mutant enzyme involved in gibberellin synthesis, *DNA Res.*, 2002, **9,** 11-17.
74. SPIELMEYER, W., ELLIS, M.H., CHANDLER, P.M., Semidwarf (sd-1), "green revolution" rice, contains a defective gibberellin 20-oxidase gene, *Proc. Natl. Acad. Sci. U.S.A.*, 2002, **99,** 9043-9048.
75. YAMAGUCHI, T., NAKAYAMA, K., HAYASHI, T., TANAKA, Y., KOIKE, S., Molecular cloning and characterization of a novel beta-1,3-glucanase gene from rice, *Biosci. Biotechnol. Biochem.*, 2002, **66,** 1403-1406.

76. JIANG, H.W., DIAN, W.M., LIU, F.Y., WU, P., Cloning and characterization of a glucose 6-phosphate/phosphate translocator from *Oryza sativa*, *J. Zhejiang Univ. Sci.*, 2003, **4,** 331-335.
77. GOTO, S., AKAGAWA, T., KOJIMA, S., HAYAKAWA, T., YAMAYA, T., Organization and structure of NADH-dependent glutamate synthase gene from rice plants, *Biochim. Biophys. Acta*, 1998, **1387,** 298-308.
78. HOSHIDA, H., TANAKA, Y., HIBINO, T., HAYASHI, Y., TANAKA, A., TAKABE, T., TAKABE, T., Enhanced tolerance to salt stress in transgenic rice that overexpresses chloroplast glutamine synthetase, *Plant Mol. Biol.*, 2000, **43,** 103-111.
79. KAMINAKA, H., MORITA, S., NAKAJIMA, M., MASUMURA, T., TANAKA, K., Gene cloning and expression of cytosolic glutathione reductase in rice (*Oryza sativa* L.), *Plant Cell Physiol.*, 1998, **39,** 1269-1280.
80. HA, S.H., LEE, S.W., KIM, Y.M., HWANG, Y.S., Molecular characterization of Hmg2 gene encoding a 3-hydroxy-methylglutaryl-CoA reductase in rice, *Mol. Cells*, 2001, **11,** 295-302.
81. NELSON, A.J., DOERNER, P.W., ZHU, Q., LAMB, C.J., Isolation of a monocot 3-hydroxy-3-methylglutaryl coenzyme A reductase gene that is elicitor-inducible, *Plant Mol. Biol.*, 1994, **25,** 401-412.
82. YOSHIDA, K.T., WADA, T., KOYAMA, H., MIZOBUCHI-FUKUOKA, R., NAITO, S., Temporal and spatial patterns of accumulation of the transcript of *myo*-inositol-1-phosphate synthase and phytin-containing particles during seed development in rice, *Plant Physiol.*, 1999, **119,** 65-72.
83. CUNNINGHAM, F.X., GANTT, E., Identification of multi-gene families encoding isopentenyl diphosphate isomerase in plants by heterologous complementation in *Escherichia coli*, *Plant Cell Physiol.*, 2000, **41,** 119-23.
84. OHTA, H., SHIRANO, Y., TANAKA, K., MORITA, Y., SHIBATA, D., cDNA cloning of rice lipoxygenase L-2 and characterization using an active enzyme expressed from the cDNA in *Escherichia coli*, *Eur. J. Biochem.*, 1992, **206,** 331-336.
85. XIE, G.S., LIU, S.K., TAKANO, T., YOU, Z.B., ZHANG, D.P., Cloning and expression of VB12-independent methionine synthase gene responsive to alkaline stress in rice, *Yi Chuan Xue Bao*, 2002, **29,** 1078-1084.
86. DENIZIAK, M., MIRANDE, M., BARCISZEWSKI, J., Cloning and sequencing of cDNA encoding the rice methionyl-tRNA synthetase, *Acta Biochim. Pol.*, 1998, **45,** 669-76.
87. NGAMPANYA, B., SOBOLEWSKA, A., TAKEDA, T., TOYOFUKU, K., NARANGAJAVANA, J., IKEDA, A., YAMAGUCHI, J., Characterization of rice functional monosaccharide transporter, OsMST5, *Biosci. Biotechnol. Biochem.*, 2003, **67,** 556-562.
88. LIN, C.M., KOH, S., STACEY, G., YU, S.M., LIN, T.Y., TSAY, Y.F., Cloning and functional characterization of a constitutively expressed nitrate transporter gene, *OsNRT1*, from rice, *Plant Physiol.*, 2000, **122,** 379-388.

89. TERADA, Y., AOKI, H., TANAKA, T., MORIKAWA, H., IDA, S., Cloning and nucleotide sequence of a leaf ferredoxin-nitrite reductase cDNA of rice, *Biosci. Biotechnol. Biochem.*, 1995, **59,** 2183-2185.
90. SOBAJIMA, H., TAKEDA, M., SUGIMORI, M., KOBASHI, N., KIRIBUCHI, K., CHO, E.M., AKIMOTO, C., YAMAGUCHI, T., MINAMI, E., SHIBUYA, N., SCHALLER, F., WEILER, E.W., YOSHIHARA, T., NISHIDA, H., NOJIRI, H., OMORI, T., NISHIYAMA, M., YAMANE, H., Cloning and characterization of a jasmonic acid-responsive gene encoding 12-oxophytodienoic acid reductase in suspension-cultured rice cells, *Planta*, 2003, **216,** 692-698.
91. GENSCHEL, U., POWELL, C.A., ABELL, C., SMITH, A.G., The final step of pantothenate biosynthesis in higher plants: Cloning and characterization of pantothenate synthetase from *Lotus japonicus* and *Oryza sativum* (rice), *Biochem. J.*, 1999, **341,** 669-678.
92. HIRAGA,. S., YAMAMOTO, K., ITO, H., SASAKI, K., MATSUI, H., HONMA, M., NAGAMURA, Y., SASAKI, T., OHASHI, Y., Diverse expression profiles of 21 rice peroxidase genes. *FEBS Lett.*, 2000, **471,** 245-250.
93. ZHU, Q., DABI, T., BEECHE, A., YAMAMOTO, R., LAWTON, M.A., LAMB, C., Cloning and properties of a rice gene encoding phenylalanine ammonia-lyase, *Plant Mol. Biol.*, 1995, **29,** 535-550.
94. TAKABATAKE, R., HATA, S., TANIGUCHI, M., KOUCHI, H., SUGIYAMA, T., IZUI, K., Isolation and characterization of cDNAs encoding mitochondrial phosphate transporters in soybean, maize, rice, and *Arabidopsis*, *Plant Mol. Biol.*, 1999, **40,** 479-486.
95. KONG, X.F., XU, Z.H., XUE, H.W., Isolation and functional characterization of the C-terminus of rice phosphatidylinositol 4-kinase *in vitro*, *Cell Res.*, 2003, **13,** 131-139.
96. NOZUE, F., UMEDA, M., NAGAMURA, Y., MINOBE, Y., UCHIMIYA, H., Characterization of cDNA coding for phosphoglucose isomerase of rice (*Oryza sativa* L.), *DNA Seq.*, 1996, **6,** 127-135.
97. UEKI, J., MORIOKA, S., KOMARI, T., KUMASHIRO, T., Purification and characterization of phospholipase D (PLD) from rice (*Oryza sativa* L.) and cloning of cDNA for PLD from rice and maize (*Zea mays* L.), *Plant Cell Physiol.*, 1995, **36,** 903-14.
98. LI, W.J., FENG, H., FAN, J.H., ZHANG, R.Q., ZHAO, N.M., LIU, J.Y., Molecular cloning and expression of a phospholipid hydroperoxide glutathione peroxidase homolog in *Oryza sativa*, *Biochim. Biophys. Acta*, 2000, **1493,** 225-230.
99. IGARASHI, Y., YOSHIBA, Y., TAKESHITA, T., NOMURA, S., OTOMO, J., YAMAGUCHI-SHINOZAKI, K., SHINOZAKI, K., Molecular cloning and characterization of a cDNA encoding proline transporter in rice, *Plant Cell Physiol.*, 2000, **41,** 750-756.
100. IGARASHI, Y., YOSHIBA, Y., SANADA, Y., YAMAGUCHI-SHINOZAKI, K., WADA, K., SHINOZAKI, K., Characterization of the gene for delta1-pyrroline-5-carboxylate synthetase and correlation between the expression of the gene and salt tolerance in *Oryza sativa* L., *Plant Mol. Biol.*, 1997, **33,** 857-865.

101. RIVOAL, J., THIND, S., PRADET, A., RICARD, B., Differential induction of pyruvate decarboxylase subunits and transcripts in anoxic rice seedlings, *Plant Physiol.*, 1997, **114,** 1021-1029.
102. HOSSAIN, M.A., HUQ, E., GROVER, A., DENNIS, E.S., PEACOCK, W.J., HODGES, T.K., Characterization of pyruvate decarboxylase genes from rice, *Plant Mol. Biol.*, 1996, **31,** 761-770.
103. MOONS, A., VALCKE, R., VAN MONTAGU, M., Low-oxygen stress and water deficit induce cytosolic pyruvate orthophosphate dikinase (PPDK) expression in roots of rice, a C3 plant, *Plant J.*, 1998, **15,** 89-98.
104. GESCH, R.W., BOOTE, K.J., VU, J.C., ALLEN, L.H., BOWES, G., Changes in growth CO_2 result in rapid adjustments of ribulose-1, 5-bisphosphate carboxylase/oxygenase small subunit gene expression in expanding and mature leaves of rice, *Plant Physiol.*, 1998, **118,** 521-529.
105. ZHANG, Z., KOMATSU, S., Molecular cloning and characterization of cDNAs encoding two isoforms of ribulose-1,5-bisphosphate carboxylase/oxygenase activase in rice (*Oryza sativa* L.), *J. Biochem. (Tokyo)*, 2000, **128,** 383-389.
106. TO, K.Y., SUEN, D.F., CHEN, S.C., Molecular characterization of ribulose-1,5-bisphosphate carboxylase/oxygenase activase in rice leaves, *Planta*, 1999, **209,** 66-76.
107. KOPRIVA, S., KOPRIVOVA, A., SUSS, K.H., Identification, cloning, and properties of cytosolic D-ribulose-5-phosphate 3-epimerase from higher plants, *J. Biol. Chem.*, 2000, **275,** 1294-1299.
108. HATA, S., SANMIYA, K., KOUCHI, H., MATSUOKA, M., YAMAMOTO, N., IZUI, K., cDNA cloning of squalene synthase genes from mono- and dicotyledonous plants, and expression of the gene in rice, *Plant Cell Physiol.*, 1997, **38,** 1409-1413.
109. KAWASAKI, T., MIZUNO, K., BABA, T., SHIMADA, H., Molecular analysis of the gene encoding a rice starch branching enzyme, *Mol. Gen. Genet.*, 1993, **237,** 10-16.
110. MIZUNO, K., KOBAYASHI, E., TACHIBANA, M., KAWASAKI, T., FUJIMURA, T., FUNANE, K., KOBAYASHI, M., BABA, T., Characterization of an isoform of rice starch branching enzyme, RBE4, in developing seeds, *Plant Cell Physiol.*, 2001, **42,** 349-357.
111. FRANCISCO, P.B., ZHANG, Y., PARK, S.Y., OGATA, N., YAMANOUCHI, H., NAKAMURA, Y., Genomic DNA sequence of a rice gene coding for a pullulanase-type of starch debranching enzyme, *Biochim. Biophys. Acta*, 1998, **1387,** 469-477.
112. DIAN, W., JIANG, H., CHEN, Q., LIU, F., WU, P., Cloning and characterization of the granule-bound starch synthase II gene in rice: Gene expression is regulated by the nitrogen level, sugar and circadian rhythm, *Planta*, 2003, [published on website ahead of print on Aug 30].
113. TANAKA, K., OHNISHI, S., KISHIMOTO, N., KAWASAKI, T., BABA, T., Structure, organization, and chromosomal location of the gene encoding a form of rice soluble starch synthase, *Plant Physiol.*, 1995, **108,** 677-683.

114. BABA, T., NISHIHARA, M., MIZUNO, K., KAWASAKI, T., SHIMADA, H., KOBAYASHI, E., OHNISHI, S., TANAKA, K., ARAI, Y., Identification, cDNA cloning, and gene expression of soluble starch synthase in rice (*Oryza sativa* L.) immature seeds, *Plant Physiol.*, 1993, **103,** 565-73.
115. LUNN, J.E., ASHTON, A.R., HATCH, M.D., HELDT, H.W., Purification, molecular cloning, and sequence analysis of sucrose-6F-phosphate phosphohydrolase from plants, *Proc. Natl. Acad. Sci. U.S.A.*, 2000, **97,** 12914-12919.
116. VALDEZ-ALARCON, J.J., FERRANDO, M., SALERNO, G., JIMENEZ-MORAILA, B., HERRERA-ESTRELLA, L., Characterization of a rice sucrose-phosphate synthase-encoding gene, *Gene*, 1996, **170,** 217-222.
117. AOKI, N., HIROSE, T., SCOFIELD, G.N., WHITFELD, P.R., FURBANK, R.T., The sucrose transporter gene family in rice, *Plant Cell Physiol.*, 2003, **44,** 223-32.
118. HIROSE, T., IMAIZUMI, N., SCOFIELD, G.N., FURBANK, R.T., OHSUGI, R., cDNA cloning and tissue specific expression of a gene for sucrose transporter from rice (*Oryza sativa* L.). *Plant Cell Physiol.*, 1997, **38,** 1389-1396.
119. NAKAMURA, T., MEYER, C., SANO, H., Molecular cloning and characterization of plant genes encoding novel peroxisomal molybdoenzymes of the sulphite oxidase family, *J. Exp. Bot.*, 2002, **53,** 1833-1836.
120. KAMINAKA, H., MORITA, S., YOKOI, H., MASUMURA, T., TANAKA, K., Molecular cloning and characterization of a cDNA for plastidic copper/zinc-superoxide dismutase in rice (*Oryza sativa* L.), *Plant Cell Physiol.*, 1997, **38,** 65-69.
121. SAKAMOTO, A., OKUMURA, T., KAMINAKA, H., TANAKA, K., Molecular cloning of the gene (*SodCc1*) that encodes a cytosolic copper/zinc-superoxide dismutase from rice (*Oryza sativa* L.), *Plant Physiol.*, 1995, **107,** 651-652.
122. ABE, T., NIIYAMA, H., SASAHARA, T., Cloning of cDNA for UDP-glucose pyrophosphorylase and the expression of mRNA in rice endosperm, *Theor. Appl. Genet.*, 2002, **105,** 216-221.
123. SUZUKI, K., SUZUKI, Y., KITAMURA, S., Cloning and expression of a UDP-glucuronic acid decarboxylase gene in rice, *J. Exp. Bot.*, 2003, **54,** 1997-1999.
124. THOMPSON, J.D., HIGGINS, D.G. AND GIBSON, T.J., CLUSTAL W: improving the sensitivity of progressive multiple sequence alignment through sequence weighting, position specific gap penalties and weight matrix choice. *Nucl. Acids Res.*, 1994, **22,** 4673-4680.
125. EISEN, M.B., SPELLMAN, P.T., BROWN, P.O., BOTSTEIN, D., Cluster analysis and display of genome-wide expression patterns, *Proc Natl Acad Sci U.S.A.*, 1998, **95,** 14863-14868.

Chapter Seven

INTEGRATING GENOME AND METABOLOME TOWARD WHOLE CELL MODELING WITH THE E-CELL SYSTEM

Emily Wang, Yoichi Nakayama, Masaru Tomita*

Institute for Advanced Biosciences
Keio University
Fujisawa, Japan 252-8520

**Author for correspondence, e-mail: mt@sfc.keio.ac.jp*

INTRODUCTION

Advances in experimental techniques and tools applicable to computational biology have spurred great expectations in the field of cell simulation and have allowed scientists to probe deeper into ever more complex systems. The development of research strategies capable of integrating the various data types currently in use and of producing cohesive information is essential for providing further insights into such systems. Systems biology attempts to integrate individual components and analyze a system, *e.g.,,* cells and organisms as a whole, in order to understand and predict properties globally. Over the past decade, applications in systems biology have mainly focused on bacteria such as *Escherichia coli* because of their simplicity and the availability of abundant data, and on human cells because of their importance for medical applications (myocardial cells, erythrocytes). As methods in computational biology become established, attention has also turned towards eukaryotic systems including plants. However, the multi-dimensionality of eukaryotic organisms, their sheer size, and their complexity present difficulties in plant cell modeling and simulation. The application of systems biology to plants in order to automate and efficiently analyze available data will facilitate the generation of predictions, and aid in the engineering of plants and microorganisms.

As technologies for collecting transcriptome, proteome, metabolome, even systeome data improve, an ever-increasing amount of information will become available to define cellular processes. Our research aims at 1) developing and investigating methods for the modeling of a large-scale plant system and 2) at implementing the model by using computer simulations. Our current method mainly utilizes genomic and metabolomic data for whole cell metabolic components. These components are integrated with data from public databases and quantitative data from the literature, applying novel algorithms to compose a model for simulation using the E-Cell system.

The E-Cell project (http://www.e-cell.org) was launched in 1996 with the goal of simulating a whole cell and developing a software environment for building integrative models to facilitate large-scale simulations.[1] Members of the project are currently working on a wide range of cell types and processes, including the human erythrocyte, *E.coli*, plant cells, myocardial cells, and nerve growth cone signal transduction. The E-Cell Simulation Environment (Fig.. 7.1) allows users to develop simulations using a combination of calculation algorithms in different timescales and compartments.[2] Such features are essential as cellular functions vary, requiring corresponding computational methods for reliable simulations. The flexibility of the system also allows the definition of object classes such as flux distribution analysis, the S-system, and definitions of other non-kinetic reactions in cells. The latest versions of E-Cell Simulation Environment can be downloaded free from http://www.e-cell.org/software/.

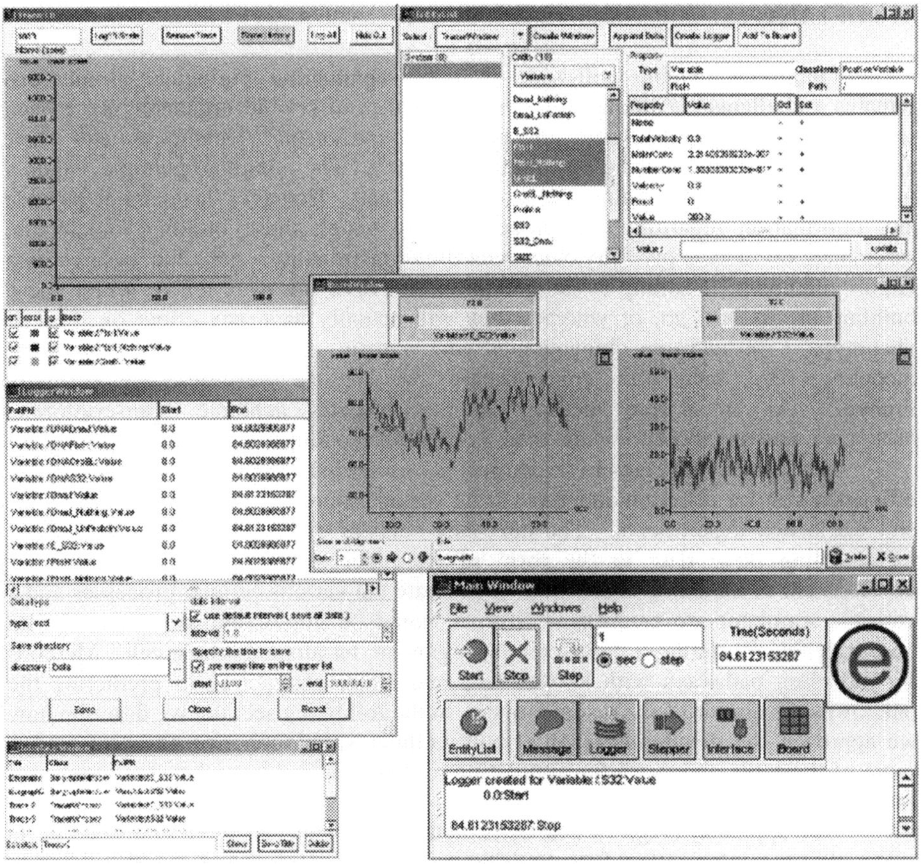

Figure 7.1: Screenshot of E-Cell Simulation Environment Version3 running a heat-shock model for *E.coli* using both deterministic and stochastic modeling methods.

INTEGRATIVE SYSTEMS BIOLOGY FOR LARGE-SCALE MODELING

Obtaining reliable cell-wide data for conducting biological simulations remains a challenge. Although the entire genomes of several organisms have been sequenced, the link between advances in experimental techniques and their application in simulation biology remains limited. The complete genome yields a picture of all possible building blocks within a cell. However, it does not provide any information regarding the process(es) by which these building blocks are assembled, nor does it give any clues regarding functionality – gene functions remain largely unknown. Looking at the genome alone does not even tell us where these building blocks will go, or whether they will actually have any effect on the cell phenotype. In addition, biochemical and kinetic data necessary for detailed modeling, *e.g.,*, metabolite concentration and enzyme activities, are currently limited. An integrative method that incorporates genomic, transcriptomic, proteomic, and metabolomic data, thus, is ideal for modeling biological systems.

One of our first tasks in the project is to establish modeling methods that can efficiently convert different data types from various sources, such as public databases and wet bench experiments. The next step consists of effectively integrating and reassembling these data in the form of computer simulations. Therefore, the development of modeling algorithms to simulate the various cellular processes and a versatile simulator are essential.[3] Still, it would be impossible to gather all data necessary for simulating a multi-cellular organism, let alone a whole cell. Methods for modeling pathways with known data and either abstracting or predicting the unknowns are also crucial for simulations. In the following section, we describe how we approach the development of a multi-cellular simulation; we also summarize some of the tools necessary for simulating a plant cell and how they are integrated for simulation using the E-Cell system.

Our approach (Fig. 7.2) to constructing the plant cell model consists of 1) using genomic information for acquiring a comprehensive data set and predicting existing proteins in the entire cell, 2) using metabolome concentrations and time-series dynamics of the major metabolites, 3) applying traditional kinetic modeling of dynamic reactions based on mathematical modeling, and 4) integrating the data from (1), (2), and (3) using new algorithms for modeling cellular pathways.

The following sections present an overview of the tools and methods currently being developed towards large-scale pathway modeling. These include 1) the Genome to E-Cell System (GEM System) for modeling pathways from genomic information, 2) the integration of rice metabolome data using capillary electrophoresis mass spectrometry (CE-MS), 3) Atomic Reconstruction of Metabolism (ARM), and 4) the hybrid static/dynamic algorithm for integrating static models based on pathway stoichiometry and dynamic models using kinetics.

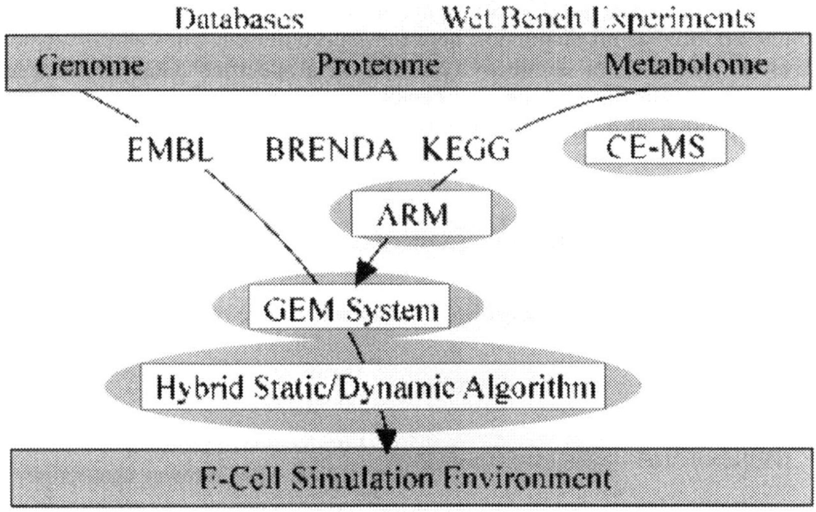

Figure 7.2: Approach to integrative large-scale modeling. Seamless integration of data from both genome and metabolome for simulation using the E-Cell system with modeling algorithms and tools.

THE GENOME-BASED E-CELL MODELING (GEM) SYSTEM

The Genome-based E-Cell Modeling System (GEM System), developed by Arakawa and members of the G-language Project (http://www.g-language.org), allows automatic generation of pathway models using only the genome sequence as input data.[4] The output results in a cell-wide metabolic pathway viewer, a list of metabolites, and a simulation model ready to run in the E-Cell system. A test run of the GEM system for *E.coli*, based on the EMBL Nucleotide Sequence Database (http://www.ebi.ac.uk/embl/), resulted in 1580 metabolites with 2273 reactions.

The GEM System retrieves genomes from public databases such as GenBank, Fasta, and EMBL. The next step involves searching for open reading frames (ORFs) to decipher candidate genes; subsequently, these sequences are matched with meta-databases to look for probable enzymes. Gene sequences are matched to enzymes through a combined method of reference through annotation, homology, and orthology.[5] Results are then stored in an internal GEM System relational database with consistent data, nomenclature, and reactions. Proteins are localized using an algorithm similar to PSORTII.[6] Since the stoichiometric reaction list may still be incomplete and makes no distinction between heteromer enzymes and isozymes, all pathways are checked for connectivity based on a specified database such as the Kyoto Encyclopedia of Genes and Genomes (KEGG, http://www.genome.ad.jp/kegg/) (Fig. 7.3).

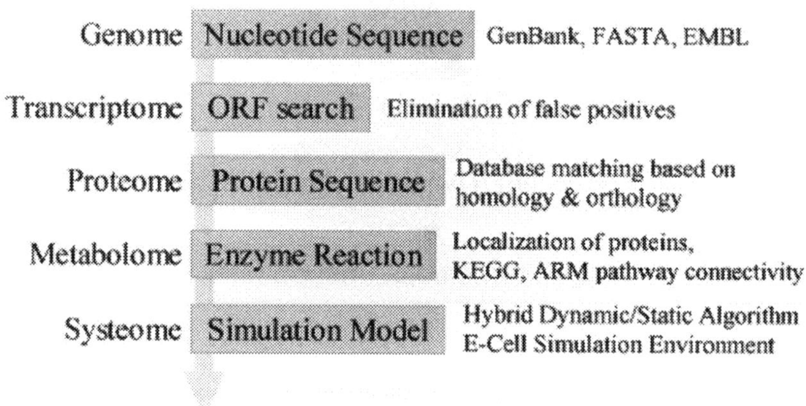

Figure 7.3: System flow of Genome-based E-Cell Modeling (GEM).

For each reaction in the generated pathway, users can manually select the reactions to be represented by dynamic equations. Initially, the GEM system would automatically search for static reactions based on monomer enzymes found in Brenda and Swiss-PROT databases. Based on the search results, the static part of the model is then generated using the hybrid dynamic/static simulation algorithm detailed below. Finally, the generated pathway model can be used for simulations in E-Cell System. Users can then specify dynamic equations by selecting an appropriate reaction mechanism and input reaction parameters, or they can program their own set of reaction process description files.

INTEGRATING METABOLOME DATA

Genome- and enzyme data from public databases provide a large picture into a sequenced plant. However, these data are insufficient for modeling, and current methods include the prediction of gene function and localization. Sequence homology provides insight into probable function whose identification would be based on the quality of the matched sequence rather than direct analysis of the target gene. In addition, mRNA levels may not correlate with protein levels,[7] nor are all proteins necessarily enzymatically active. Cellular processes such as alternative reading frames, gene fusion, alternative splicing, and post-translational modifications complicate the process of assigning functions to specific sequences and genes.[8, 9]

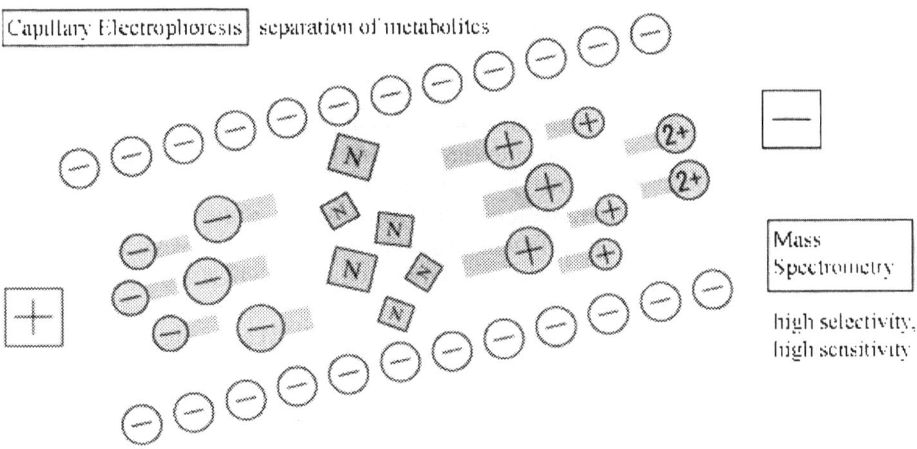

Figure 7.4: Capillary Electrophoresis Mass Spectrometry (CE-MS). Metabolites are roughly separated based on their charge and size using CE and measured by MS.

An alternative approach to investigating functional aspects of an organism is to integrate information and predictions from genome data with metabolome data. As the latter directly apply to the functional aspects of a cell and its biochemical status, cellular function and active reactions can be deduced on a large scale.[10] Metabolome research into the dynamics of rice is currently being conducted by Sato et al. at the Institute for Advanced Biosciences, Keio University.[11] The goal of that group is to establish metabolome analysis methods in rice and to identify pathway networks. Using capillary electrophoresis coupled to mass spectrometry (CE-MS)-based metabolomics,[12-14] the intracellular concentrations of 88 main metabolites involved in energy metabolism and photosynthesis have been analyzed in the third leaf of rice (*Oryza sativa* L. ssp. *Japonica*, Haenuki). The CE-MS method roughly separates metabolites based on their charge and size using CE, where all cations migrate towards the cathode and anions towards the anode (Fig.7.4). The separated metabolites are then detected by MS for direct measurements. This method allows highly sensitive and selective analysis of various ionic metabolites without derivatization. In addition, analysis of one sample is efficient, requiring less than 30 minutes. Biochemical networks of particular plant cells in distinct locations may be predicted using the CE-MS method for obtaining a wide range of metabolite data at various time and cell phases.

ATOMIC RECONSTRUCTION OF METABOLISM (ARM)

Automated analysis that can generate hypotheses regarding unknown pathways is required, especially because higher organisms synthesize secondary (*i.e.*, not directly necessary for the maintenance of life) metabolites, many of which remain to be classified. Since the type of secundciy varies depending on the growth phase and location, the determination of pathways poses a difficult challenge. [15] Metabolome data alone are insufficient to predict candidate pathways for the vast number of secondary metabolites, and a combination of analytical and mathematical methods is necessary.[16]

The Atomic Reconstruction of Metabolism (ARM) software, developed by Arita at the University of Tokyo and Institute for Advanced Biosciences, realized the computer simulation of radioisotope tracer experiments.[17] Each molecule was represented at the atomic scale to describe its structural features (mappings) within a database. Data from the ENZYME database (http://us.expasy.org/enzyme) were curated and precomiled with all reactions in KEGG, EcoCyc (http://www.ecocyc.org), and reactions in the basic metabolism from the Roche Biochemical Pathways chart.[18] Using graph representation, the software detected structural correspondences between substrates and products for each enzymatic reaction to generate all logically possible pathways between any two given metabolites in the ARM database. Each reaction in the newly computed pathway was validated and checked for atomic balance. The computed pathways were then visualized via its graphic user interface (GUI) (Fig. 7.5).

The representation of each substance at an atomic rather than a molecular level allows focus on structural elements of each metabolite. Results of automated pathway reconstruction are based on computed movements in the location of each atom by the tracing engine of the software. The construction of a mapping database, which enables atomic tracers and search functions for certain structural elements, serves to close the gap between pathway databases and computer simulations. Though the coverage of the mapping database is important for reliable reconstruction, once a reasonable set of metabolome data from secondary metabolism is obtained, hypothetical pathways and relationships can be computed using the ARM software. The ARM software and its current data can be downloaded free from http://www.metabolome.jp/.

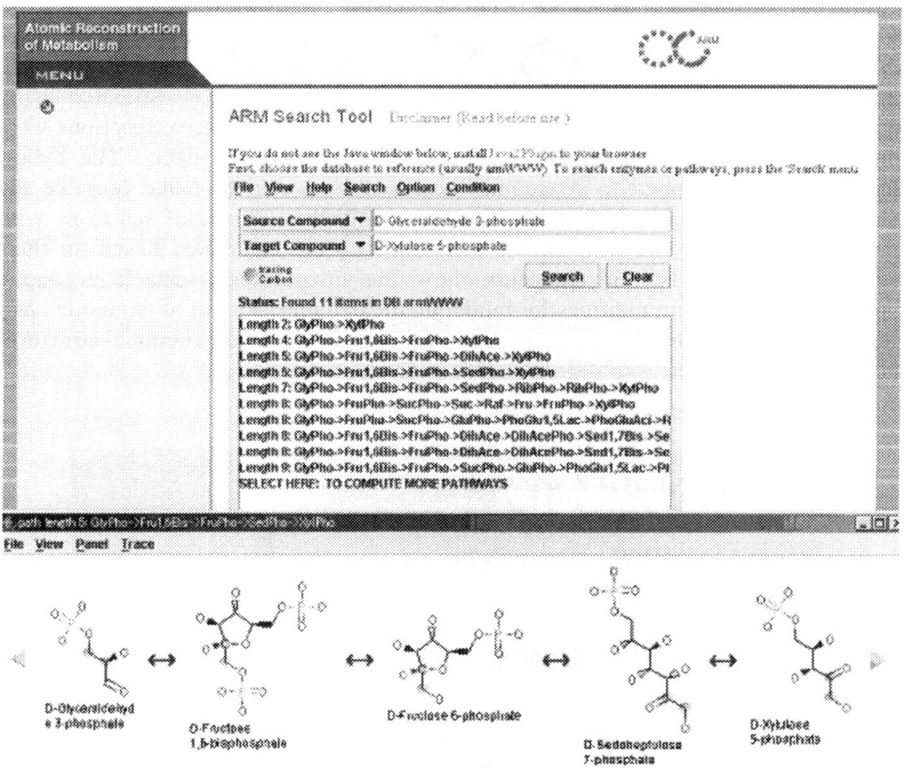

Figure 7.5: Screenshot of the online ARM Database. A list of all logically possible pathways from Glyceraldehyde-3-phosphate to Xylulose-5-phosphate is shown. Selection one of the pathways from the list will display the corresponding metabolite structure and conserved carbon atoms in that pathway.

THE HYBRID STATIC/DYNAMIC SIMULATION ALGORITHM

The Hybrid Static/Dynamic Algorithm, developed by Yugi et al. at the Institute for Advanced Biosciences, Keio University,[19] allows dynamic models to be integrated with stoichiometry-based static models (Fig. 7.6). Previously, in cases where kinetic parameters or rate equations were unavailable, reaction kinetics were either approximated using parameters from other organisms, or estimated using parameter estimation techniques or arbitrary values. In many cases, reactions were simply left out of the model due to the lack of quantitative data. The hybrid algorithm was developed to overcome these problems and to make possible the calculation of approximate reaction flux systematically. Fluxes of reactions with unknown parameters but with known stoichiometry are calculated based on flux-based methods. The hybrid algorithm allows the integration of data from kinetic databases and S-System matrices for modeling the dynamic section of the model, and integration with the stoichiometric section of the model based on chemical structure- and pathway databases.

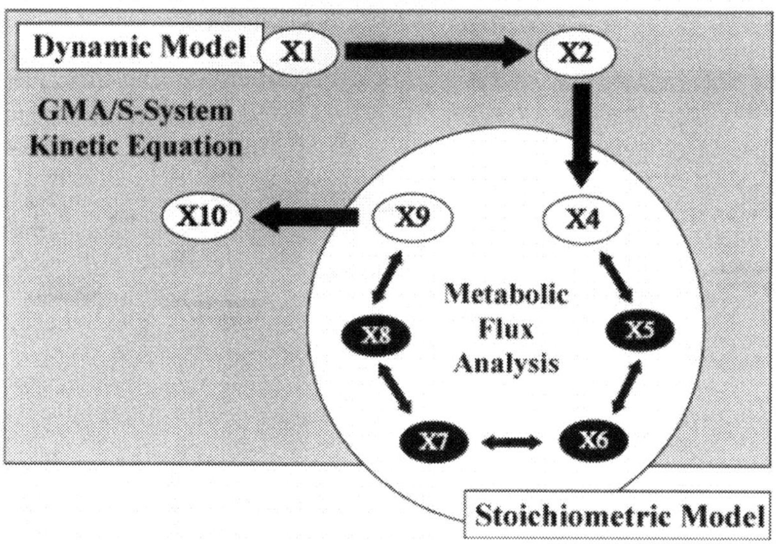

Figure 7.6: Hybrid dynamic/static simulation algorithm. Using flux-based methods, the algorithm allows dynamic models to be integrated with stoichiometry-based models.

SUMMARY AND FUTURE DIRECTIONS

Traditional modeling methods using rate equations and enzyme kinetics alone are insufficient for large-scale plant modeling. As the development of methods to address the challenges posed by plant systems biology is a priority, plant systems biology should take advantage of current modeling strategies and advances in system-wide analysis that have been developed in efforts to achieve both unicellular and multi-cellular modeling.

There are currently two large-scale modeling projects undertaken by E-Cell: e2coli, which aims to simulate the *E. coli* bacterium, and e-Rice, which aims at simulating the rice plant. e-Rice is one of the first attempts to simulate a whole plant organism. Our preliminary goal is to create a generic model for the basic metabolism in a plant cell that can then be adapted to cells residing at different sites in a rice plant. The current version of the basic model consists of reactions that take place in the chloroplast, mitochondrion, peroxisome, and cytosol.

The reconstruction of even a prokaryotic cell poses many challenges.[3, 20, 21] With plants, a huge amount of data and many data types are involved, complicating the process even further.[22 - 24] Even in the same genome, these data vary depending on the type of organelles, cells, tissues, and organs involving multiple development phases. However, by obtaining cell-wide data at two different levels - the genome for global properties and the metabolome for expressed physical properties, a basic dataset for cell simulation can be prepared for computer simulations. Our endeavor to simulate a whole plant is still in its initial stages where we are developing the tools and algorithms necessary for integrating vast amounts of available data and information. The tools and algorithms being developed are indispensable for reliable and comprehensive large-scale simulations, and can be adapted to various organisms including plants and mammals.

In summary, our integrative approach and methods addresses large-scale modeling that is being developed at the Institute for Advanced Biosciences. Our approach for constructing the cell model consists of using 1) the Genome to E-Cell System (GEM System) for modeling pathways based on genomic information, 2) rice metabolome research using capillary electrophoresis mass spectrometry (CE-MS), 3) Atomic Reconstruction of Metabolism (ARM), and 4) the hybrid static/dynamic algorithm for integrating static models based on pathway stoichiometry and dynamic models using kinetics. These applications provide a systematic method for acquiring and integrating various data types for whole cell modeling using the E-Cell simulation environment.

ACKNOWLEDGEMENTS

The e-Rice project is being developed by the bioinformatics and metabolome research units at the Institute for Advanced Biosciences. We wish to acknowledge in particular the work of the following members of the institute: Kazuharu Arakawa, Masanori Arita, Tomoyoshi Soga, Shigeru Sato, Katsuyuki Yugi, Ayako Kinoshita, Nobuyoshi Ishii, and Kotaro Ishii. This work is being supported by the Ministry of Agriculture, Forestry and Fisheries of Japan (Rice Genome Project), a grant from the New Energy and Industrial Technology Development Organization (NEDO) of the Ministry of Economy, Trade and Industry of Japan (Development of a Technological Infrastructure for Industrial Bioprocesses Project), and the following three grants from the Ministry of Education, Culture, Sports, Science and Technology (MEXT): Leading Project for Biosimulation, Grant-in-Aid for the 21st Century Center of Excellence (COE) Program entitled "Understanding and Control of Life's Function via Systems Biology (Keio University)", and Grant-in-Aid for Scientific Research on Priority Areas.

REFERENCES

1. TOMITA, M., HASHIMOTO, K., TAKAHASHI, K., SHIMIZU, T.S., MATSUZAKI, Y., MIYOSHI, F., SAITO, K., TANIDA, S., YUGI, K., VENTER, J.C., HUTCHISON, C.A., E-Cell: software environment for whole cell simulation, Bioinformatics, 1999, **15**, 72-84
2. TAKAHASHI, K., KAIZU, K., HU, B., TOMITA, M., A multi-algorithm, multi-timescale method for cell simulation, *Bioinformatics*, in press.
3. TAKAHASHI, K., YUGI, K., HASHIMOTO, K., YAMADA, Y., PICKETT, C.J.F., TOMITA, M., Computational challenges in cell simulation: A software engineering approach, *IEEE Intelligent Systems*, 2002, **17**, 64-71
4. ARAKAWA, K., MORI, K., IKEDA, K., MATSUZAKI, T., KOBAYASHI, Y., TOMITA, M., G-Language Genome Analysis Environment: A workbench for nucleotide sequence data mining., *Bioinformatics*, 2003, **19**, 305-306
5. OVERBEEK, R., KARSEN, N., PUSCH, G. D., D'SOUZA, M., SELKOV, E. JR., KYRPIDES, N., FONSTEIN, M., MALTSEV, N., SELKOV, E., WIT: Integrated system for high-throughput genome sequence analysis and metabolic reconstruction, *Nuc. Acids Res.*, 2000, **28**, 123-125
6. NAKAI, K., HORTON, P., PSORT: A program for detecting sorting signals in proteins and predicting their subcellular localization, *Trends Biochem. Sci.* 1999, **24**, 34-36
7. GYGI, S.P., ROCHON, Y., FRANZA, B.R., AEBERSOLD, R., Correlation between protein and mRNA abundance in yeast, *Mol. Cell Biol.*, 1999, **19**, 1720-1730
8. BACHMAIR, A., NOVATCHKOVA, M.O., POTUSCHAK, T., EISENHABER, F., Ubiquitylation in plants: A post-genomic look at the post-translation modification, *Trends Plant Sci.*, 2001, **26**, 463-470

9. REDDY, A.S.N., Nuclear pre-mRNA splicing in plants, *Crit. Rev. Plant. Sci.*, 2001, **20**, 523-571
10. SUMNER, L.W., MENDES, P., DIXON, R.A., Plant metabolomics: Large-scale phytochemistry in the functional genomics era., *Phytochemistry*, 2003, **62**, 817-836
11. SATO, S, SOGA, T., TOMITA, M., Time-course analysis of rice metabolome, *in*: Proceedings of the 4th International Conference on Systems Biology, 2003, p. 227
12. SOGA, T., UENO, Y., NARAOKA, H., MATSUDA, K., TOMITA, M., NISHIOKA, T., Pressure-assisted capillary electrophoresis electrospray ionization mass spectrometry for analysis of multivalent anions, *Analy. Chem.*, 2002, **74**, 6224-6229
13. SOGA, T., UENO, Y., NARAOKA, H., OHASHI, Y., TOMITA, M., NISHIOKA, T., Simultaneous determination of anionic intermediates for *Bacillus subtilis* metabolic pathways by capillary electrophoresis electrospray ionization mass spectrometry, *Analytical Chemistry*, 2002, **74**, 2233-2239
14. SOGA, T., OHASHI, Y., UENO, Y., NARAOKA, H., TOMITA, M., NISHIOKA, T., Quantitative metabolome analysis using capillary electrophoresis mass spectrometry, *J. Pro. Res*, 2003, **2**, 488-494
15. WINK, M., Evolution of secondary metabolites from an ecological and molecular phylogenetic perspective, *Phytochemistry*, 2003, **64**, 3-19
16. FIEHN, O., WECKWERTH, W., Deciphering metabolic networks, *Eur. J. Biochem.*, 2003, **270**, 579-588.
17. ARITA, M., In silico atomic tracing by substrate-product relationships in *Escherichia coli* intermediary metabolism, *Genome Res.*, 2003, **13**, 2455-2466
18. MICHAL, G., Biochemical Pathways: An atlas of biochemistry and molecular biology. Wiley & Spektrum, 1999
19. YUGI, K., NAKAYAMA, Y., TOMITA, M., A hybrid static/dynamic simulation algorithm: Towards large-scale pathway simulation, *in*: Proceedings of the 3rd International Conference on Systems Biology (E. Aurell, J. Elf, J. Jeppsson, eds.). 2002, p. 235
20. TOMITA, M., Whole cell simulation: a grand challenge of the 21st century, *Trends Biotechnol*, 2001, **19**, 205-210
21. TOMITA, M., Towards computer aided design (CAD) of useful microorganisms, *Bioinformatics*, 2001, **17**, 1091-1092
22. GIRKE, T., OZKAN. M., CARTER, D., RAIKHEL, N.V., Towards a modeling infrastructure for studying plant cells, *Plant Physiol.*, 2003, **132**, 410-414
23. MINORSKY, P.V., Frontiers of plant cell biology: signals and pathways, system-based approaches 22nd symposium in plant biology (University of California - Riverside), *Plant Physiol.,*, 2003, **132**, 428-435
24. SWEETLOVE, L.J., LAST, R.L., FERNIE, A.R., Predictive metabolic engineering: A goal for systems biology, *Plant Physio.*, 2003, **132**, 420-425

Chapter Eight

METABOLIC ENGINEERING OF SOYBEAN FOR IMPROVED FLAVOR AND HEALTH BENEFITS

Carl A. Maxwell, Maria A. Restrepo-Hartwig, Aideen O. Hession, Brian McGonigle*

E. I. Du Pont de Nemours and Company
DuPont Crop Genetics
PO Box 80402
Wilmington, DE 19880-0402

**Author for correspondence, e-mail* brian.mcgonigle@usa.dupont.com

INTRODUCTION

We choose to eat certain foods based on a complex interaction of factors. Among the most important are safety, environmental impact, cost, religious dictates, perceived health benefits, and flavor. Foods containing soybean (*Glycine max* L.) have positive attributes concerning almost all of these factors, and as such the amount of soyfoods consumed in the US has increased dramatically in recent years. The US market for soyfood products grew to US$ 3.65 billion in 2002 and is expected to continue to grow at a rate of 15-25% over the next several years (www.soyatech.com). Although not consumed as often in Europe as in the US, there as well soyfoods are increasing in popularity with a 2002 market of 1.3 billion Euros and further double-digit growth predicted (www.prosoy.org). Staple soyfood favorites such as tofu are not gaining in popularity as consumers opt for easy to eat foods such as meat alternatives. Categories currently experiencing strong growth are cold cereals, cheese alternatives, non-dairy frozen desserts, soymilk, yogurt, and frozen green soybeans.

Soybeans may represent a safer alternative to some other sources of protein. According to the Center for Disease Control (www.cdc.gov), the most commonly recognized food-borne infections are those caused by the bacteria *Campylobacter*, *Salmonella*, and *E. coli* O157:H7, and by a group of viruses called calicivirus, also known as the Norwalk and Norwalk-like viruses. Raw foods of animal origin are the most likely to be contaminated, including raw meat and poultry, raw eggs, unpasteurized milk, and raw shellfish. Other foods that may contain food borne pathogens are fruits and vegetables that are eaten raw. Proper handling and cooking of foods can protect consumers from food borne pathogens. However because of worries about these and other food borne pathogens, particularly bovine spongiform encephalopathy (BSE), some consumers have chosen to modify their diets and limit the amount of protein obtained from animal sources.

Another reason that some consumers choose to obtain protein from non-animal sources is to lessen their environmental impact. Since at least the 1965 publication of *Diet For a Small Planet*,[1] individuals have become conscious that their individual food choices have global consequences. The production of plant-derived protein has a lesser environmental impact than does the production of similar amounts of animal derived proteins. Although soy protein is somewhat lacking in sulfur-containing amino acids such as methionine and cysteine, it is considered a complete source of protein containing all essential amino acids necessary for the building and maintenance of human body tissues.

Perhaps the strongest driver for the increase in consumption of soyfoods is perceived health benefits. Consumption of soyfoods has been linked to a number of benefits: most prominently prevention of heart disease, prevention of colon cancer and hormone related cancers (*e.g.*, breast and prostate), and alleviation of women's post-menopausal health problems (menopausal symptoms and osteoporosis). For the most part, these effects have not been rigorously proven. The strongest evidence for the health benefits derived from the consumption of soy protein is related to the preventative effects of soy on heart disease. As such, the FDA (www.fda.gov) allows food manufacturers to label products containing a minimum of 6.25 g of soy protein per serving with the statement, "25 grams of soy protein a day, as part of a diet low in saturated fat and cholesterol, may reduce the risk of heart disease". A more scientifically rigorous understanding of the health benefits of consuming soyfoods is necessary for the continued growth of, and indeed the maintenance of, the current soyfoods market.

However, in the end, research by food scientists suggests that although many factors play a role in food choices, a given food will not remain a part of most individuals' diets unless there is an acceptable flavor. In this one category, soyfoods do not rate high. Food companies have spent a significant amount of time and money in formulating soyfoods to be acceptable to Western consumers, and a significant amount of progress has been made. However, according to a recent study by the Center for Food Reformulation at TIAX (www.tiax.biz), many manufacturers are still struggling to formulate a balanced, good tasting product. In many cases, manufacturers are trading off some of the health benefits of soyfoods by including significant amounts (4-16 grams per serving) of sugar.

Metabolic engineering of soybeans can make significant contributions to both the understanding of the health benefits that may be obtained from the consumption of soyfoods, as well as increasing health benefits and improving flavor.

SUPPRESSION OF DAIDZEIN BIOSYNTHESIS

Isoflavonoids, compounds derived from a branch of the phenylpropanoid pathway, are a class of secondary metabolites produced predominantly in legumes. In legumes, these compounds are known to be involved in interactions with other organisms. Isoflavonoid-derived compounds are involved in symbiotic relationships in soybean between roots and rhizobial bacteria that eventually result in nodulation and nitrogen-fixation, [2] and participate in the defense responses of legumes against phytopathogenic microorganisms. [3] Additionally, they have been shown to act as antibiotics, repellents, attractants, and signal compounds. [4]

Isoflavonoids also have been reported to have physiological activity in animal and human studies. Besides acting as an estrogen mimic, it was reported that the isoflavones found in soybean seeds possess antihemolytic, [5] antifungal, [6] tumor-suppressing, [7,8] and serum cholesterol-lowering[9] effects. In addition, both epidemiological and dietary-intervention studies indicate that when isoflavones in soybean seeds and in subsequent protein products prepared from the seeds are part of the human dietary intake, those products provide many significant health benefits.[10-13]

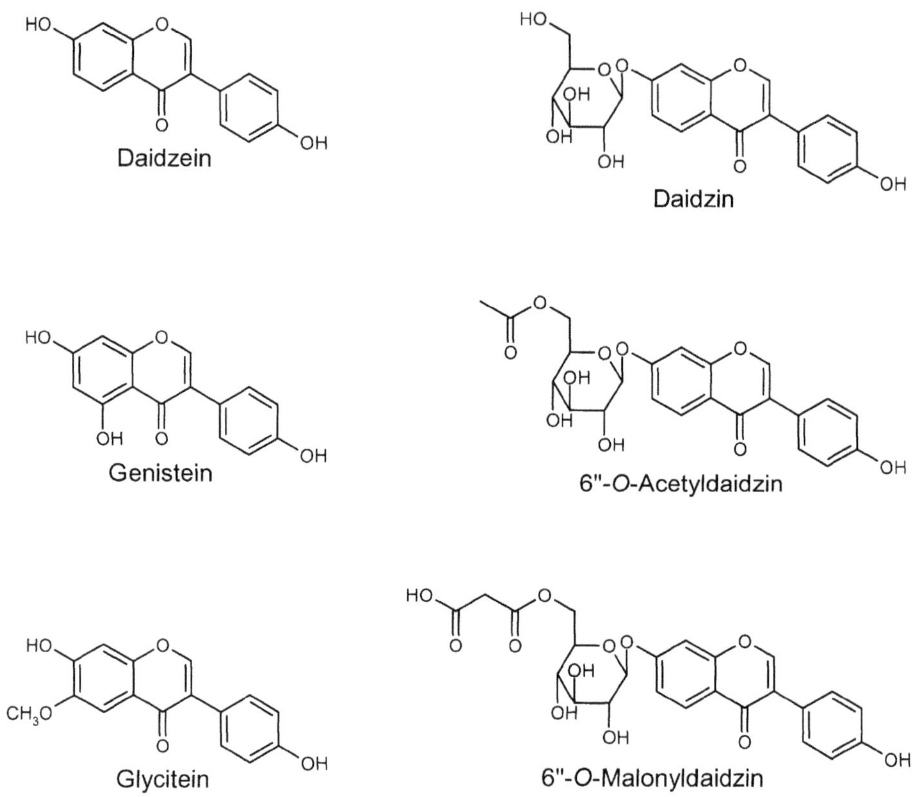

Figure 8.1: A diagram of the chemical structures of the isoflavone aglycones found in soybean as well as conjugates of daidzein. Similar conjugates are formed of genistein and glycitein.

Soybean seeds contain three types of isoflavone aglycones: daidzein, genistein, and glycitein (Fig. 8.1). However, free isoflavones rarely accumulate to high levels in soybeans; instead they are usually conjugated to carbohydrates with or without organic acids. [14] Each aglycone can be found in three different forms: glucoside conjugates known as daidzin, genistin, and glycitin; malonylglucoside conjugates known as 6"-O-malonyldaidzin, 6"-O-malonylgenistin, and 6"-O-malonylglycitin; and acetyl glucoside conjugates. The acetyl conjugates are thought to be formed during processing from the degradation of malonyl glucoside conjugates,[15, 16] and are known as 6"-O-acetyldaidzin, 6"-O-acetylgenistin, and 6"-O-acetylglycitin.

Isoflavonoid content in legumes can be increased by pathogen attack, wounding, high UV light exposure, and pollution. [17] More specifically, the total isoflavone levels, as well as the distribution among different aglycones, is quite variable in soybean seeds and is affected by both genetics and environmental conditions such as growing location and temperature during seed fill. [18, 19] Foods made from soybeans typically reflect the endogenous isoflavone composition, and as such genistein-derived isoflavone forms are the most abundant in most food products, while the daidzein-derived and the glycitein-derived forms are present in lower levels. [20]

The biosynthetic pathway for isoflavonoids in soybean and the relationship of the isoflavonoids to several other classes of phenylpropanoids is presented in Fig. 8.2. Production of *p*-coumaryl-CoA from phenylalanine requires phenylalanine ammonia lyase to convert phenylalanine to cinnamate, cinnamic acid hydroxylase to convert cinnamate to *p*-coumarate, and coumarate:CoA ligase to convert *p*-coumarate to *p*-coumaroyl-CoA. Lignins may be produced from *p*-coumaroyl-CoA or from *p*-coumarate. Chalcone synthase catalyzes the condensation of three molecules of malonyl CoA with *p*-coumaroyl-CoA to form 4, 2', 4', 6'-tetrahydroxychalcone, which is subsequently isomerized in a reaction catalyzed by chalcone isomerase to naringenin, the precursor to genistein, flavones, flavonols, condensed tannins, anthocyanins, and others.

Alternatively, chalcone reductase[21] (CHR; also known as deoxychalcone synthase) together with chalcone synthase and NADPH as a cofactor act in the formation of isoliquiritigenin, which is then isomerized, again by the enzyme chalcone isomerase, to form liquiritigenin, the precursor to daidzein, and the pterocarpan phytoalexins. A type II chalcone isomerase that seems to be found exclusively in the legumes catalyzes this isomerization reaction. [22] Glycitein synthesis is not yet clearly defined, but is likely derived from liquiritigenin via flavonoid 6-hydroxylase, [23] and an unidentified methyltransferase.

In all cases, the unique aryl migration reaction to create the isoflavones is mediated by 2-hydroxyisoflavanone synthase, also known as isoflavone synthase (IFS). Sequences encoding the IFS gene have been identified,[24-26] and the encoded

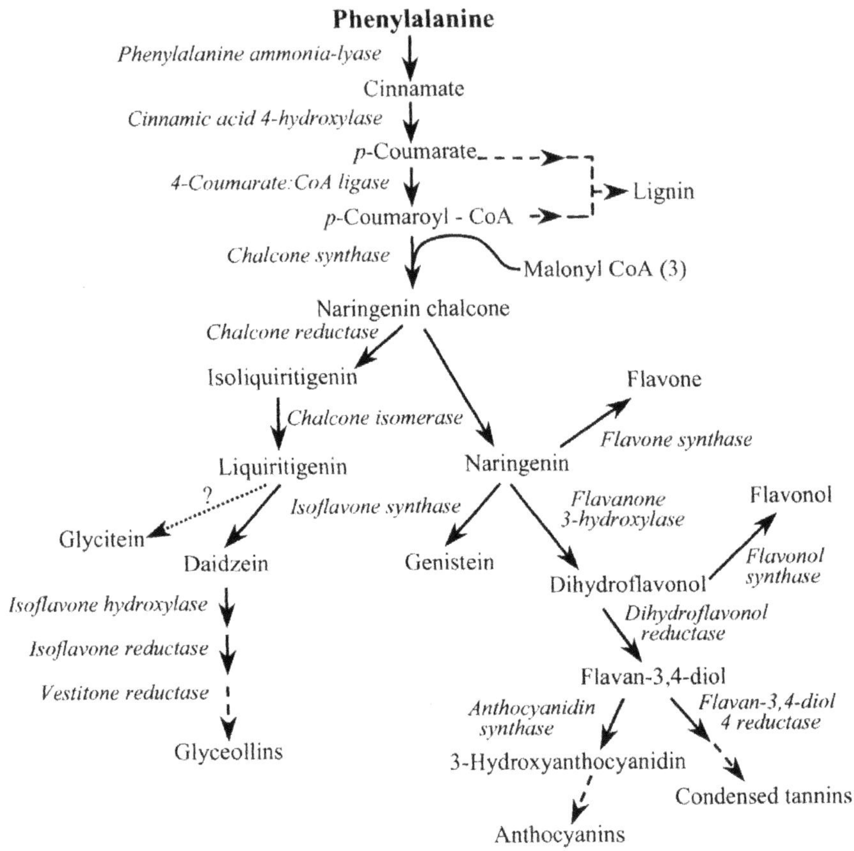

Figure 8.2: A partial diagram of the phenylpropanoid pathway. Intermediates and enzymes involved in isoflavone synthesis, as well as some branch pathways, are shown. Dashed arrows represent multiple steps; the dotted arrow represents a speculative step.

protein is a typical member of the cytochrome P450 superfamily. The reaction requires NADPH and molecular oxygen and forms the compound 2-hydroxyisoflavanone. *In vitro*, the dehydration of the 2-hydroxyisoflavanone to form the isoflavone occurs spontaneously. However, *in planta*, this reaction may by enzymatically catalyzed, and a protein that carries out this reaction has been purified from *Pueraria lobata*.[27]

The physiological benefits associated with isoflavonoids in both plants and humans make the manipulation of their contents in crop plants highly desirable. There have been attempts to produce isoflavones in a number of non-legume species, namely *Arabidopsis*, corn, and tobacco, via the expression of IFS.[26, 28] However, accumulation of isoflavone conjugates was very low. More recently, Liu *et al.*[29] have combined the expression of IFS in *Arabidopsis* with a tt6/tt3 double mutant blocked in flavonol and anthocyanin synthesis, and have shown accumulation of genistein conjugates up to fifty times greater than when IFS is expressed in a wild-type *Arabidopsis* background.

There is currently no recommended level of daily isoflavone consumption. Some of the levels thought to be efficacious would be difficult to obtain from a typical Western diet. This has caused some consumers to turn to supplements to obtain isoflavones. However, this may not be an optimal method of obtaining isoflavones, both because of the difficulty of obtaining good quality supplements,[30] and because it has been suggested that the maximal benefits of isoflavone consumption are obtained through a synergistic effect with either soy protein or other compounds found in soybean.[31] Furthermore, the safety of isoflavone consumption at levels available from supplements is not reported. Additionally, depending upon the processing method of soybeans to protein isolate or protein concentrate (typical ingredients for Western consumption), a significant amount of isoflavones may be lost.

For these reasons, we attempted to increase the levels of isoflavones in soybean.[32] By combining expression of a chimeric transcription factor (CRC), which acts to increase flux through the phenylpropanoid pathway with the suppression of a competing pathway via RNAi mediated silencing of flavanone 3-hydroxylase, levels of isoflavones were shown to accumulate in the seed to levels that were 3-4 times higher than in wild type soybeans.

It is not known through what molecular mechanisms isoflavones exert their beneficial effects although they have been shown to bind to estrogen receptors, inhibit DNA topoisomerase, act as antioxidants, and effect cell signaling pathways through a number of mechanisms including inhibition of tyrosine-specific protein kinases.[33] Recent *in vitro* work has shown that there are different physiological and enzymatic responses to the isoflavones derived from individual aglycones. While both daidzein and genistein act as phytoestrogens, some of the activities attributed to isoflavones are, in fact, specific to genistein, such as inhibition of tyrosine-specific protein kinases.[34] Characterizing the transcriptome of human gut epithelial cell-lines challenged with either daidzein or genistein shows further evidence that the biochemical activities of genistein and daidzein are distinct. Although there is some degree of overlap in the transcriptomes, independent and, in some instances, opposite responses are also found. [35] Because of this, at times, it may be desirable for some individuals to consume genistein and not daidzein. We show a method in which soybeans can be produced in which the level of liquiritingenin-derived isoflavones, namely the aglycones daidzein and glycitein and their conjugates, is significantly reduced.

Vector Construction to Suppress Chalcone Reductase

A cDNA, known as src3c.pk009.e4, identical to one identified as encoding CHR[36] was identified from the DuPont EST collection[37] using BLAST searching[38] with NCBI Accession Number X55730 as a query. A fragment corresponding to a portion of the CHR coding sequence was obtained by PCR amplification using clone src3c.pk009.e4 as template and primers CHR-Not1-sense (5' GCGGCCGCATGGCTGCTGCTATTGAAATC) and CHR-Not1-antisense (5' GCGGCCGCCCTGCTCGCACCTTTCCTCAG). The amplification reaction was performed using advantage 2 polymerase and GC melt reagent (1mM final concentration) following the manufacturer's (Clontech, Palo Alto, CA) protocol. The resulting amplified DNA fragment was first cloned into TopoTA vector (Invitrogen, Carlsbad, CA). The fragment was then liberated from the TopoTA vector by Not I digestion and purified from an agarose gel using the Qiagen Gel Purification Kit (Qiagen, Valencia, CA). The purified DNA fragment was ligated into the Not I site of vector pKS151 to produce the plasmid pAC23.

Vector pKS151 has been described in PCT Publication WO 02/00904, published 03 January 2002, and is derived from the commercially available vector pSP72 (Promega, Madison, WI). Briefly, pKS151 contains the seed-specific expression promoter KTi3[39] followed by nucleotides that promote formation of a stem structure flanking a Not I site, which is followed by a transcription termination signal. Expression of HPT by two different promoters allows the selection for growth in the presence of hygromycin in bacterial and plant systems. The stem structure is formed by two copies of 36 nucleotides at the 5' end of the Not I site and an inverted repeat of the same two 36-nucleotide copies at the 3' end. It has been shown that sequences inserted within the Not I site may be silenced, presumably via RNAi. [40]

Generation of Soybean Transformants and Isoflavone Analysis

Soybean (cv Jack) embryogenic suspension cultures were transformed with plasmid pAC23 by particle gun bombardment[41] following previous protocols. [42] Lines containing the pAC23 construct were identified by PCR using primers: R1 5' CACGGGACGGATGGTAGCAACA and R2 5' CCGATTCTCCCAACATTGCTTATTC. Transformed embryos were germinated and grown to maturity according to above protocols. All plants were allowed to self. Seed of individual plants containing pAC23 were ground in batches of five to eight per plant. A 1-gram sample was extracted with MeOH-H_2O (4:1) and then incubated with 2 N NaOH at room temperature to hydrolyze malonyl and acetyl esters to the corresponding glucosides. Acetic acid was used to adjust the pH to 7, samples were filtered, and then assayed by HPLC (model 2690, Waters, Milford, MA) equipped with UV detector (model 486, Waters) and with a Luna C18 column (3 micron, 4.6 mm X 50 mm, Phenomenex) maintained at 30°C. The column was eluted with 90% A and 10% B (A as 1% acetic acid in water and B as 1% acetic acid in acetonitrile) for 5 min at 1 ml/min, 10%B to 22%B from 5 to 11 min at 1 ml/min, 22% B to 100% B from 11 to 12 min at 1 ml/min to 2 ml/min, and 100% B from 12 to 14.5 min at 2 ml/min. The amounts of daidzin, glycitin, and genistin were calculated by comparison with standard curves prepared from authentic compounds (Indofine Chemical Co., Somerville, NJ; Fujico Co., Japan) at 262 nm. Although this method does not measure the amount of aglycones present, their levels are so low as to not affect the results.

Results of Genetic Modification of Chalcone Reductase

Seed specific suppression of CHR in soybean resulted in soybean seeds with lower content of daidzein-derived and glycitein-derived isoflavones relative to wild-type soybeans. The levels of isoflavones in R1 seed from 88 plants derived from 46 independent transformation events were assayed. Each independent transformation event is labeled with a unique six-digit descriptor; *e.g.*, 3226-6-2, and individual plants from the same event are differentiated by the addition of another digit, *e.g.*, 3226-6-2-1 or 3226-6-2-2. Fig. 8.3 depicts the percentage of total isoflavones that the sum of daidzin and glycitin (the glucoside of daidzein and glycitein, respectively) represents in bulk R1 seeds from transformed plants containing the CHR RNAi construct (pAC23). The sum of daidzin and glycitin was between 10% and 20% of the total isoflavones in two plants derived from one transformation event positive for plasmid pAC23 (Fig. 8.3). The sum of daidzin and glycitin was between 20% and 30% of the total isoflavones in eight plants derived from six independent transformation events, while in a number of other events the amounts of daidzin and glycitin is decreased as compared to wild-type plants although to a lesser extent.

These results (Fig. 8.3) may be compared with those shown in Wang *et al.*,[43] who describe analysis of the agronomic characteristics of 210 soybean cultivars grown in the State of South Dakota in the United States and conclude that the isoflavone content and the distribution of the different aglycones in non-transgenic

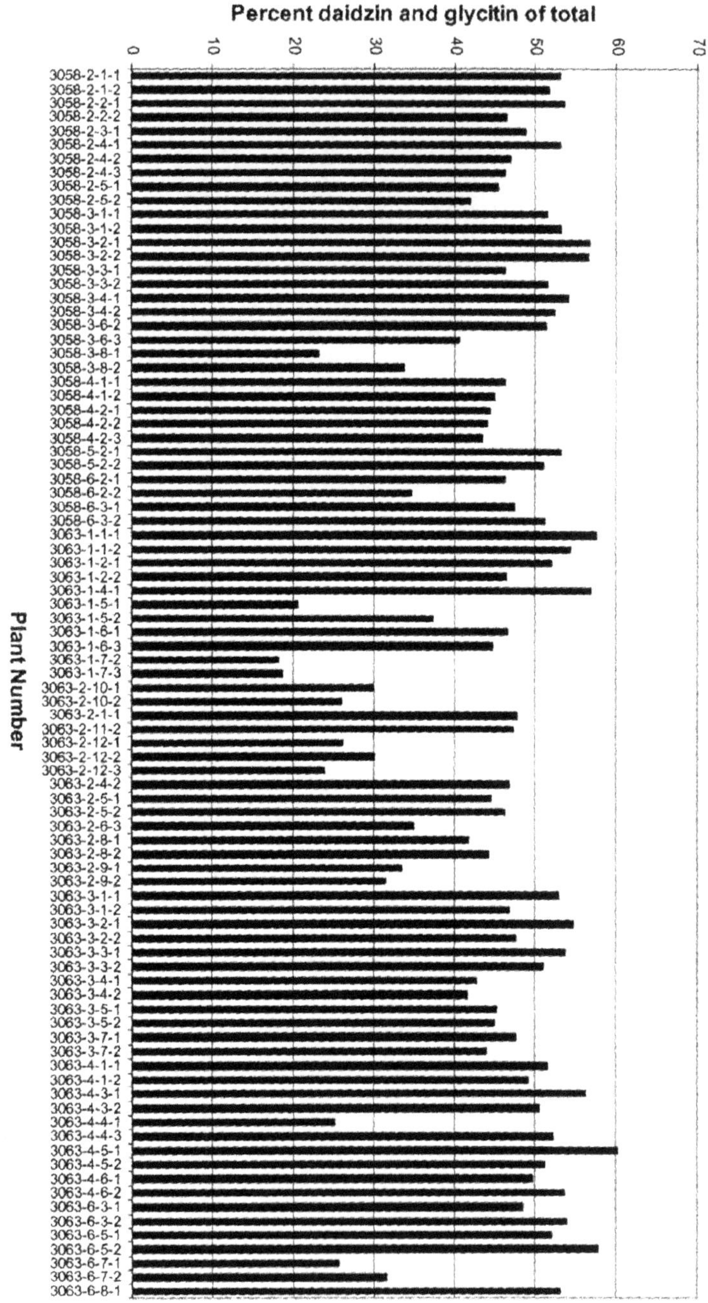
Percent daidzin and glycitin of total
0
10
20
30
40
50
60
70
Plant Number
3058-2-1-1
3058-2-1-2
3058-2-2-1
3058-2-2-2
3058-2-3-1
3058-2-4-1
3058-2-4-2
3058-2-4-3
3058-2-5-1
3058-2-5-2
3058-3-1-1
3058-3-1-2
3058-3-2-1
3058-3-2-2
3058-3-3-1
3058-3-3-2
3058-3-4-1
3058-3-4-2
3058-3-6-2
3058-3-6-3
3058-3-8-1
3058-3-8-2
3058-4-1-1
3058-4-1-2
3058-4-2-1
3058-4-2-2
3058-4-2-3
3058-5-2-1
3058-5-2-2
3058-6-2-1
3058-6-2-2
3058-6-3-1
3058-6-3-2
3063-1-1-1
3063-1-1-2
3063-1-2-1
3063-1-2-2
3063-1-4-1
3063-1-5-1
3063-1-5-2
3063-1-6-1
3063-1-6-3
3063-1-7-2
3063-1-7-3
3063-2-10-1
3063-2-10-2
3063-2-1-1
3063-2-11-2
3063-2-12-1
3063-2-12-2
3063-2-12-3
3063-2-4-2
3063-2-5-1
3063-2-5-2
3063-2-6-3
3063-2-8-1
3063-2-8-2
3063-2-9-1
3063-2-9-2
3063-3-1-1
3063-3-1-2
3063-3-2-1
3063-3-2-2
3063-3-3-1
3063-3-3-2
3063-3-4-1
3063-3-4-2
3063-3-5-1
3063-3-5-2
3063-3-7-1
3063-3-7-2
3063-4-1-1
3063-4-1-2
3063-4-3-1
3063-4-3-2
3063-4-4-1
3063-4-4-3
3063-4-5-1
3063-4-5-2
3063-4-6-1
3063-4-6-2
3063-6-3-1
3063-6-3-2
3063-6-5-1
3063-6-5-2
3063-6-7-1
3063-6-7-2
3063-6-8-1

soybean plants varies greatly. Table 1 of the Wang report shows the total isoflavones (in μg/g), and the total percent of genistein, daidzein, and glycitein for all 210 soybean cultivars. Addition of the total daidzein percent with the total glycitein percent for each cultivar shows that it varies from a low of 35% (for Golden Harvest H-1263 and Newton 1006) to a high of 54% (for Prairie Brand 227EXP). As shown here, suppression of CHR results in transgenic plants having even lower levels of daidzein-derived and glycitein-derived isoflavones than can be found in a survey of a wide range of commonly grown soybean varieties.

The isoflavone analyses (Fig. 8.3) were performed using samples containing from five to eight R1 seeds. It is expected that a bulk sample of R1 seeds will contain a combination of wild type and transgenic seed with, on average, 1/4 of the seeds being wild type. Thus, the levels of daidzin plus glycitin in the transgenic seeds alone will be even lower than what was measured. The data above demonstrate that seed-specific suppression of CHR in soybean results in plants having seeds with reduced levels of daidzein and glycitein derived isoflavones. Further generations of these plants are being characterized and, if the phenotype is stable, will be used for dietary intervention studies to understand the role these individual isoflavones play in the health effects derived from the consumption of soy protein.

SUPPRESSION OF SAPONIN BIOSYNTHESIS

The terpenoids, which are composed of the five-carbon isoprenoids, constitute the largest family of natural products with over 22,000 individual compounds in this class having been described.[44] The terpenoids (hemiterpenes, monoterpenes, sesquiterpenes, diterpenes, triterpenes, tetraterpenes, polyterpenes, and the like) play diverse functional roles in plants as hormones, photosynthetic pigments, electron carriers, mediators of polysaccharide assembly, structural components of membranes, and defense compounds. Many compounds used by man including resins, latex, waxes, and oils contain plant terpenoids.

Two molecules of farnesyl pyrophosphate are joined head-to-head to form squalene, a triterpene, in the first dedicated step towards sterol biosynthesis (Fig. 8.4). Squalene is then converted to 2,3-oxidosqualene, which next can be cyclized to the 30 carbon, 4-ring structure cycloartenol by the enzyme cycloartenol synthase (EC 5.4.99.8). Cycloartenol can be further modified by reactions such as desaturation or demethylation to form the common sterol backbones such as

Figure 8.3: The sum of daidzin and glycitin as a percent of the total amount of isoflavones in plants containing a chalcone reductase RNAi silencing construct.

campesterol, stigmasterol, and sitosterol among others. These compounds, which can be modified further, serve both structural roles in the plant membrane and, when modified to form brassinosteroids, functional roles in plant development. Cycloartenol also serves as the precursor for the steroidal saponins, which are not found in soybean and will not be discussed further.

Alternatively, oxidosqualene cyclases catalyze the cyclization of 2,3-oxidosqualene to form 30 carbon, five ring structures including lupeol, isomultiflorenol, β-amyrin, and α-amyrin. The non-cycloartenol-producing oxidosqualene cyclase enzymes are different, although evolutionarily related, to cycloartenol synthases. [45] β-amyrin synthase, an example of an oxidosqualene cyclase, catalyzes the cyclization of 2,3-oxidosqualene to β-amyrin. [45] The basic β-amyrin ring structure may be modified to give sapogenols, which are further glycosylated to form the triterpenoid saponins. None of the genes that encode enzymes involved in the modification of β-amyrin have been characterized, although an enzyme has been characterized for an UDP-glucoronic acid: soyasapogenol glucuronosyltransferase involved in saponin biosynthesis in germinating soybean seeds. [46]

Saponins have a role in plant defense. Antifungal saponins have been found in a number of plants, and some phytopathogenic fungi have saponin-detoxifying enzymes. [47] For example, saponins in oat roots, avenacins, offer protection against *Gaeumannomyces graminis* var *tritici*, the causative agent of "take-all" in wheat and barley. Mutants of a diploid oat (*Avena strigosa*) with reduced levels of avenacins are susceptible to the fungus. [48] In contrast, *Gaeumannomyces graminis* var *avenae* does infect oat roots and is resistant to avenacin largely due to the saponin-detoxifying enzyme avenacinase. [47] Allelopathic saponins are released from some plants,[49] and other saponins appear to be involved in resistance to insect herbivores. [50,51]

Soybean seeds contain two main classes of saponins, examples of which are shown in Fig. 8.4; group A saponins and the DDMP saponins. Group A saponins are bidesmosidic with two ether linked sugar chains attached to positions 3 and 22. DDMP saponins have a 2,3-dihydro-2, 5-dihydroxy-6-methyl-4H-pyran-4-one (DDMP) moiety attached via an ether linkage to the C-22 hydroxyl residue and an ether linked sugar chain attached to position 22. The DDMP moiety is easily removed during extraction, and the DDMP saponins are converted into either Group B or E saponins. [52] The composition, length and degree of acylation of the attached sugar chains varies greatly depending upon the genotype and can be further modified during the processing of soybeans to food.[53] Removal of the sugar chain leads to the formation of sapogenols.

Total saponin content varies somewhat by soybean cultivar,[54] but is in the range of 0.25% of the seed dry weight. [55] The physiological function of saponins in soybean seeds is not clear, but saponins and sapogenols purified from soybean seeds

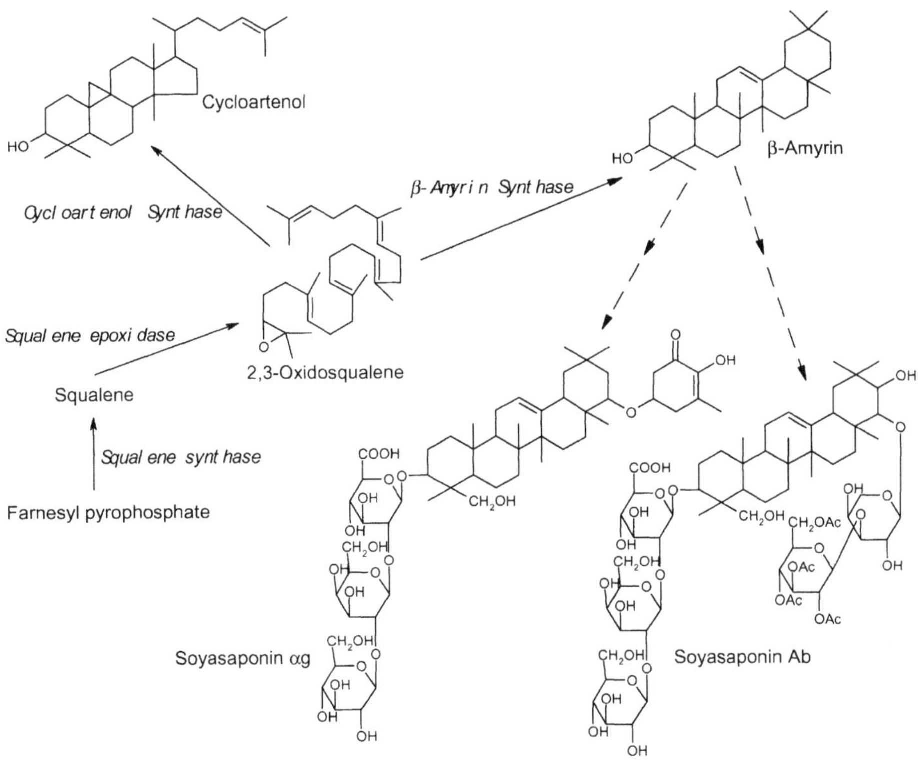

Figure 8.4: A partial diagram of the biosynthetic pathway to saponins; dashed arrows represent multiple steps. Shown here are examples of the two main classes of saponins found in soybeans. Soyasaponin Ab is an example of the bidesmosidic A saponins. Soyasaponin αg is an example of the DDMP-conjugated saponins that lead to B and E soyasaponins upon hydrolysis of the DDMP moiety from the triterpenoid backbone.

have been described as having bitter or astringent taste characteristics when consumed by humans.[56] In an attempt to find the compound(s) possessing undesirable taste characteristics in dried pea, a natural products fractionation approach was taken leading to the purification of soyasponin I (a type of B group saponin).[57] However, the role that saponins play in the undesirable taste characteristics of soyfood products, *e.g.*, not as purified compounds, is still under investigation.

In recent years, there has been interest in quinoa (*Chenopodium quinoa*) as an alternative food crop, in part because of its ability to grow in marginal conditions. Although widely used by the Incas, quinoa requires extensive post-harvest preparation in order to remove undesirable taste characteristics. Some of these characteristics have been removed by the development of sweet quinoa, which has significantly decreased levels of saponins and, thus, a decreased need for extensive post-harvest preparation.[58, 59] It seems likely that saponins will contribute to the undesirable taste characteristics of soyfood products, and reducing the saponin content of soybeans will result in better flavored food products derived from soybean. We show a method in which soybeans can be modified to produce reduced levels of both A saponin and DDMP-saponins when compared to wild-type soybeans.

Vector Construction to Suppress β-amyrin Synthase

Clones sah1c.pk002.n23 and src3c.pk024.m11 were previously identified from the DuPont EST collection, using BLAST homology to known sequences, as encoding oxidosqualene cyclases (PCT publication No. WO01/66773, published 13 September 2001). Furthermore, the cDNA insert in clone src3c.pk024.m11 was demonstrated to be a β-amyrin synthase due to its ability to produce β-amyrin when expressed in yeast as measured by liquid chromatography/mass spectrometry (LC/MS). By using the same methods, we were unable to regularly detect production of a cyclized oxidosqualene (*e.g.*, β-amyrin or α-amyrin) when sah1c.pk002.n23 was expressed in yeast. A portion of the cDNA insert from clone sah1c.pk002.n23 was amplified using primers P2 (5' GCGGCCGCCAACAATTTAGAAGAGGCTCGG) and P3 (5' TTCTTGGAGAAGGACCTAATGGAGGTCATG). A portion of the cDNA insert from clone src3c.pk024.m11 was amplified using primers P4 (5' GCGGCCGCATGTGGAGGCTGAAGATAGCAG) and P5 (5' GTCATGACCTCCATTAGGTCCTTCTCCAAG).

Amplifications were carried out as described above. Primers P3 and P4 were designed in such a way that the amplification products of the two reactions hybridize to form a chimeric recombinant DNA fragment, and a second round of PCR was performed using as template a mixture of 0.01 μL of product from each reaction and primers P2 and P5. The amplification products resulting from using clone src3c.pk024.m11 (β-amyrin synthase) as the template, and from using the mixed amplification products as template were cloned into plasmid pCR2.1 using the TOPO TA Cloning Kit (Invitrogen). The fragments were then liberated from the TopoTA vector by Not I digestion and purified from an agarose gel using the Qiagen Gel Purification Kit. The purified DNA fragments were ligated into the Not I site of vector pKS151 (described above) to create plasmids pAC16 and pAC18, respectively.

Generation of Soybean Transformants and Sapogenol Analysis

Soybean (cv Jack) embryogenic suspension cultures were transformed with plasmid pAC16 or pAC18 by particle gun bombardment[41] following previous protocols.[42] Lines containing the pAC16 or pAC18 construct were identified by PCR using primers: P6 5' ATTTCGTTGGGAGGCAGACATGG and P7 5' CCGATTCTCCCAACATTGCTTATTC, which will give a 643 bp band for pAC16 and a 1099 bp band for pAC18 or P8 5' CCCATCCTCCGTCTTCATTCTGG, and P8 5' ACGGATATAATGAGCCGTAAACAAA, which will give a 778 bp band for both pAC16 and pAC18. Transformed embryos were germinated and grown to maturity according to above protocols. All plants were allowed to self. Five to eight seeds per plant were combined and ground using an Adsit grinder (Adsit Co., Inc., Ft. Meade, FL). About 100-mg ground soybean was weighed into a beater vial, and a ¼ inch steel bead was added along with 1 mL of 60% acetonitrile. The mixture was agitated on a Geno/Grinder™ Model 2000 (SPEX Certiprep, Metuchen, NJ) for 1 minute with the machine set at 1500 strokes per minute, and then placed on an end-over-end tumbler for 1 hour. The vial was placed in the Geno/Grinder™ for 1 minute with the machine set at 1500 strokes per minute, and the sediment removed by centrifugation at 12,000 rpm for 4 minutes. The supernatant was transferred to a 13 x 100-mm glass test tube fitted with a Teflon® cap. The extraction procedure was repeated once, and the supernatants were combined into the same 13 x 100-mm glass test tube. To the tube containing the combined supernatants, 0.4-mL of 12N HCl were added. After mixing, the tube was placed into an 80°C heating block overnight.

After overnight incubation, the tube was removed from the heating block and allowed to cool to room temperature. At that point, 0.5 mL of 30% ammonium hydroxide were added and the solution was mixed. Next, 2 mL of acetonitrile, 100 µL DMSO, and 1.5 mL of methanol were added, and the solution was mixed. The liquid in the tubes was sonicated for 10 minutes, and the volume was measured and recorded. Sediment was removed by centrifuging the tubes for 10 minutes at 3500 rpm at 20°C, and an aliquot of the supernatant was placed into an HPLC vial to analyze the sapogenols using LC/MS. The amount of saponin in a sample is proportional to the amount of measured sapogenols. Thus, a relative saponin content may be calculated by measuring the total sapogenols resulting from removing the sugar moieties from the saponin.

LC/MS was performed with a Waters™ (Waters Corp., Milford, MA) 2690 Alliance HPLC interfaced with a ThermoFinnigan (San Jose, CA) LCQ™ mass spectrometer. Samples were maintained at 25°C prior to injection. A 10 μl sample was injected onto a Phenomenex® (Torrance, CA) Luna™ C18 column (3 μm, 4.6 mm X 50 mm), equipped with a guard cartridge of the same material, and maintained at 40°C. Compounds were eluted from the column at a flow rate of 0.8 mL/minute by using a solvent gradient. For the first two minutes, the eluent was a 50/50 mixture of solvent A (0.1% formic acid in water) and solvent B (0.1% formic acid in acetonitrile). From 2 to 5 minutes, the eluent was a linear gradient from 50% solvent B to 100% solvent B. From 5 to 8 minutes, the eluent was 100% solvent B, and from 8 to 11 minutes, the eluent was a 50/50 mixture of solvent A and solvent B. The mass spectrometer was equipped with an APCI source set to scan m/z of 250 to 500 in positive ion mode. The vaporizer temperature was set to 400°C, the capillary temperature was at 160°C, and the sheath gas flow was at 60 psi. Identification and quantification of sapogenol A and B were based on m/z and co-chromatography of authentic standards (Apin Chemicals, LTD, Oxon, UK).

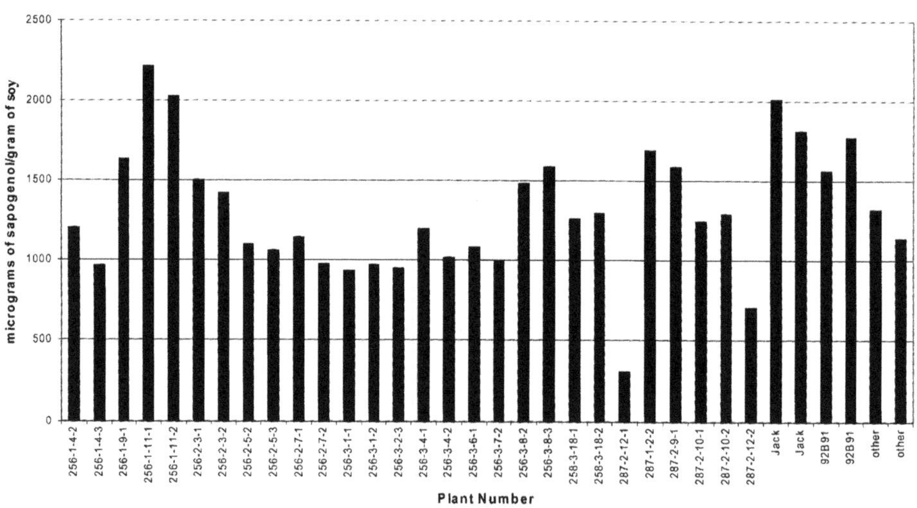

Figure 8.5: The total sapogenol per gram of soy obtained in control plants (Jack, 92B91, or unrelated transgenics) and in soybean plants containing a β-amyrin RNAi silencing construct AC16.

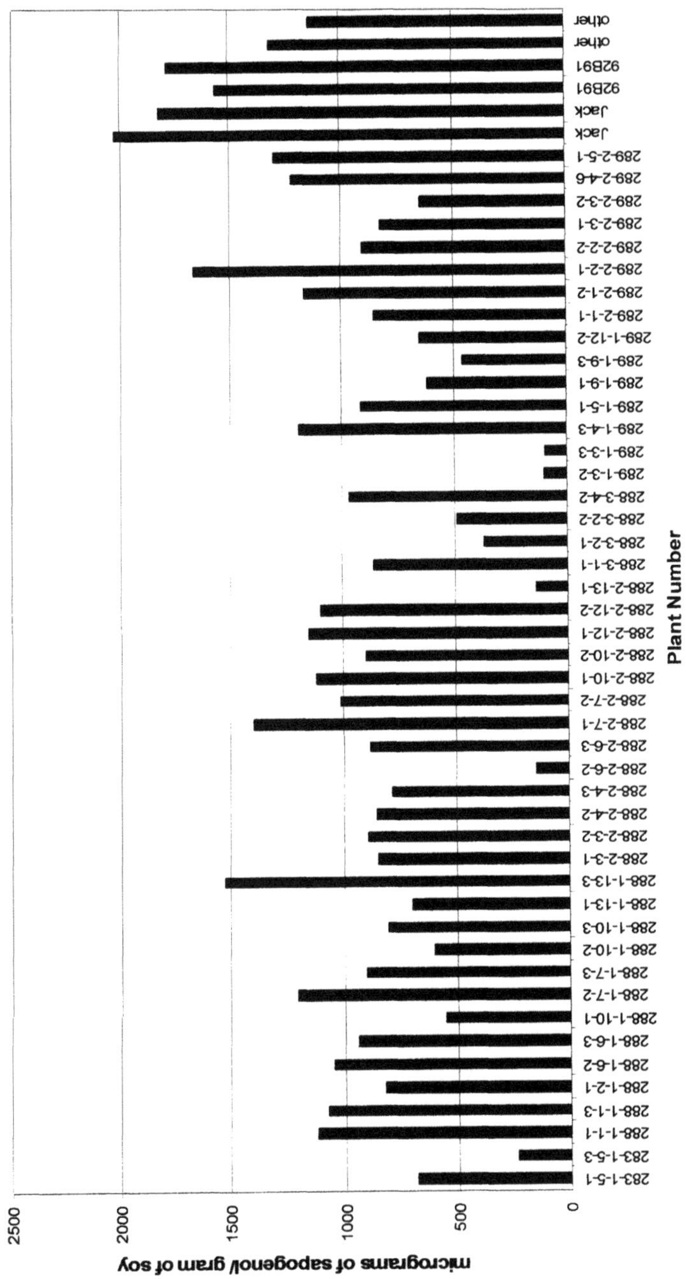

Figure 8.6: The total sapogenol per gram of soy obtained in control plants (Jack, 92B91, or unrelated transgenics) and in soybean plants containing a β-amyrin/oxidosqualene cyclase RNAi silencing construct AC18.

Results of Genetic Modification of β-amyrin Synthase

The sapogenol levels of some of the transgenic plants having pAC16 or pAC18 inserts are much lower than those found in control plants (Figs. 8.5 and 8.6). Wild-type cv Jack, cv 92B91 and Jack plants transformed with recombinant DNA fragments not having DNA sequences derived from oxidosqualene cyclases typically produce seeds with sapogenol levels between 1500 and 2000 ppm (Fig. 8.5). Thirty-two plants representing eighteen independent events transformed with pAC16 were analyzed (Fig. 8.5). One of these plants, 287-2-12-1, showed sapogenol levels below 500 ppm, while seven additional plants derived from six independent events showed soyasapogenol levels between 500 ppm and 1000 ppm. Forty-five plants representing twenty-eight independent events transformed with pAC18 were analyzed (Fig. 8.6). Eight plants derived from six independents events (numbers 283-1-5-3, 288-2-6-2, 288-2-13-1, 288-3-2-1, 288-3-2-2, 289-1-3-2, 289-1-3-3, and 289-1-9-3) showed sapogenol levels below 500 ppm, while an additional 23 plants derived from seventeen additional independent transformation events showed sapogenol levels between 500 ppm and 1000 ppm. It is expected that a bulk sample of R1 seeds will contain a combination of wild type and transgenic seed with, on average, 1/4 of the seeds being wild type. Thus, the levels of saponin in the transgenic seeds alone will be even lower than what was found (Figs. 8.5 and 8.6).

pAC16, an RNAi construct containing a portion of a ß-amyrin synthase gene, suppresses the sapogenol levels in soybean. Further, suppression using pAC18, which contains a recombinant DNA having a chimera composed of a partial ß-amyrin synthase and another partial oxidosqualene cyclase sequence, results in proportionately more plants having very low sapogenol levels (less than 500 ppm) when compared to pAC16. There are several possible explanations for the proportionally greater number of silenced events using the plasmid containing the chimeric ß-amyrin synthase/oxidosqualene cyclase. It may be that this longer construct is more efficient at suppressing the ß-amyrin synthase. Several groups have cloned oxidosqualene cyclases that are capable of producing multiple triterpene structures.[60-62] While these cyclases form predictable mixtures of substances after yeast expression, how the production of the individual triterpene structures is regulated *in planta* is still unclear. It may be that the oxidosqualene cyclase, which is non-functional in yeast under some conditions, functions as a ß-amyrin synthase *in planta*.

Further generations of these plants are being characterized and, if the phenotype is stable, will be used for taste evaluations to gain an understanding of the contribution that soy saponins make towards the overall taste within the soyfood matrix.

SUMMARY AND FUTURE DIRECTIONS

The food choices that consumers make are informed by a variety of criteria including cost, safety, environmental impact, and especially perceived health benefits and taste. Over the past several years these criteria have caused consumers to include more soyfood products in their diets. Two things are necessary for the continuation of this trend. First, the scientific community must more rigorously prove the health benefits associated with eating soybean; specifically which components of soybean cause the health benefits and the physiological mechanisms under which these benefits are obtained. These are difficult studies to carry out. In the past, some of these studies have been conducted using soyfoods that have been treated in such a way as to remove a given compound, although this typically results in the removal of many classes of compounds. Metabolic engineering of soybeans will allow the creation of beans with specific compounds (or lack there of) that can then be used to test the role these compounds play in human health.

Second, soyfoods must be developed that individual consumers consider a tasty part of their diet. Only foods that appeal to an individual are likely to continue to remain a part of that individual's diet, no matter how many other good properties are associated with that food. Progress in formulation of soyfoods has created foods with significantly greater acceptance. However, it may be that specific changes resulting from metabolic engineering will be required to produce new generations of tasty and healthy soyfoods.

ACKNOWLEDGEMENTS

We would like to thank Jan Hazebroek and his team for isoflavone measurements; Christine Hazel, Cheryl Caster, and the soybean transformation team for production of transgenic soybeans; and Joan Odell, Bill Hitz, and Carl Falco for useful discussion.

REFERENCES

1. LAPPE, F.M. Diet for a Small Planet, Ballantine Press, New York. 1975, p. 411.
2. PHILLIPS, D. A., Flavonoids: Plant signals to soil microbes, *in* Phenolic Metabolism in Plants (H. A.Stafford, and R. K Ibrahim, eds.), Plenum Press, New York. 1992, pp. 201-231.
3. DEWICK, P. M., Isoflavonoids, *in*: The Flavonoids, Advances in Research Since 1986 (J. B. Harborne, ed.), Chapman and Hall, London. 1996, pp. 117-238.
4. BARZ, W., WELLE, R., Biosynthesis and metabolism of isoflavones and pterocarpan phytoalexins in chickpea, soybean and phytopathogenic fungi, *in*:

Phenolic Metabolism in Plants (H. A.Stafford, and R. K Ibrahim, eds.), Plenum Press, New York. 1992, pp. 139-164.

5. NAIM, M., GESTETNER, B., BONDI, A., BIRK, Y., Antioxidative and antihemolytic activities of soybean isoflavones, *J. Agric. Food Chem.*, 1976, **24**, 1174-1177.
6. NAIM, M., GESTETNER, B., ZILKAH, S., BIRK, Y., BONDI, A., Soybean isoflavones. Characterization, determination, and antifungal activity, *J. Agric. Food Chem.*, 1974, **22**, 806-810.
7. MESSINA, M., BARNES, S., The role of soy products in reducing risk of cancer, *J. Natl. Cancer Inst.*, 1991, **83**, 541-546.
8. PETERSON, G., BARNES, S., Genistein inhibition of the growth of human breast cancer cells: Independence from estrogen receptors and the multi-drug resistance gene, *Biochem. Biophys. Res. Commun.*, 1991, **179**, 661-667.
9. SHARMA, R.D., Isoflavones and hypercholesterolemia in rats, *Lipids*, 1979, **14**, 535-539.
10. MESSINA, M. J., Legumes and soybeans: Overview of their nutritional profiles and health effects, *Am. J. Clin. Nutr.*, 1999, **70**, 439S-450S.
11. DAVIS, S.R., DALAIS, F.S., SIMPSON, E.R., MURKIES, A.L., Phytoestrogens in health and disease, *Recent Prog. Horm. Res.*, 1999, **54**, 185-210.
12. WATANABE, S., UESUGI, S., KIKUCHI, Y., Isoflavones for prevention of cancer, cardiovascular diseases, gynecological problems and possible immune potentiation, *Biomed. Pharmacother.*, 2002, **56**, 302-312.
13. CLARKSON, T.B., Soy, soy phytoestrogens and cardiovascular disease, *J. Nutr.*, 2002, **132**, 566S-569S.
14. GRAHAM, T.L., Flavonoid and isoflavonoid distribution in developing soybean seedling tissues and in seed and root exudates, *Plant Physiol.*, 1991, **95**, 594–603.
15. GRIFFITH, A.P., COLLISON, M.W., Improved methods for the extraction and analysis of isoflavones from soy-containing foods and nutritional supplements by reversed-phase high-performance liquid chromatography and liquid chromatography-mass spectrometry, *J. Chromatogr. A.*, **913**, 397-413.
16. WANG, H.J., MURPHY, P.A., Isoflavone content in commercial soybean foods, *J. Agric. Food Chem.*, 1994, **42**, 1666-1673.
17. TSUKAMOTO, C., SHIMADA, S., IGITA, K., KUDOU, S., KOKUBUN, M., OKUBO, K., KITAMURA, K., Factors affecting isoflavone content in soybean seeds: Changes in isoflavones, saponins, and composition of fatty acids at different temperatures during seed development, *J. Agric. Food Chem.*, 1995, **43**, 1184-1192.
18. WANG, H.J., MURPHY, P.A., Isoflavone composition of American and Japanese soybeans in Iowa: Effects of variety, crop year, and location, *J. Agric. Food Chem.*, 1994, **42**, 1674-1677.
19. DIXON, R.A., PAIVA, N.L., Stress-induced phenylpropanoid metabolism, *Plant Cell*, 1995, **7**, 1085-1097.
20. MURPHY, P.A., SONG, T., BUSEMAN, G., BARUA, K., BEECHER, G.R., TRAINER, D., HOLDEN, J., Isoflavones in retail and institutional soy foods, *J. Agric. Food Chem.*, 1999, **47**, 2697-2704.

21. WELLE, R., GRISEBACH, H., Induction of phytoalexin synthesis in soybean: Enzymatic cyclization of prenylated pterocarpans to glyceollin isomers, *Arch. Biochem. Biophys.*, 1988, **263**, 191-198.
22. SHIMADA, N., AOKI, T., SATO, S., NAKAMURA, Y., TABATA, S., AYABE, S., A cluster of genes encodes the two types of chalcone isomerase involved in the biosynthesis of general flavonoids and legume-specific 5-deoxy(iso)flavonoids in *Lotus japonicus*, *Plant Physiol.*, 2003, **131**, 941-951.
23. LATUNDE-DADA, A.O., CABELLO-HURTADO, F., CZITTRICH, N., DIDIERJEAN, L., SCHOPFER, C., HERTKORN, N., WERCK-REICHHART, D., EBEL, J., Flavonoid 6-hydroxylase from soybean (*Glycine max* L.), a novel plant P-450 monooxygenase, *J. Biol. Chem.*, 2001, **276**, 1688-1695.
24. AKASHI, T., AOKI, T., AYABE, S., Cloning and functional expression of a cytochrome P450 cDNA encoding 2-hydroxyisoflavanone synthase involved in biosynthesis of the isoflavonoid skeleton in licorice, *Plant Physiol.*, 1999, **121**, 821-828.
25. STEELE, C.L., GIJZEN, M., QUTOB, D., DIXON, R.A., Molecular characterization of the enzyme catalyzing the aryl migration reaction of isoflavonoid biosynthesis in soybean, *Arch. Biochem. Biophys.*, 1999, **367**, 146-150.
26. JUNG, W., YU, O., LAU, S.M., O'KEEFE, D.P., ODELL, J., FADER, G., MCGONIGLE, B., Identification and expression of isoflavone synthase, the key enzyme for biosynthesis of isoflavones in legumes, *Nat. Biotechnol.*, 2000, **18**, 208-212. Erratum in: *Nat. Biotechnol.*, 2000, **18**, 559.
27. HAKAMATSUKA, T., MORI, K., ISHIDA, S., EBIZUKA, Y., SANKAWA, U., Purification of 2-hydroxyisoflavanone dehydratase from the cell cultures of *Pueraria lobata*, *Phytochemistry*, 1998, **49**, 497-505.
28. YU, O., JUNG, W., SHI, J., CROES, R.A., FADER, G.M., MCGONIGLE B., ODELL, J.T., Production of the isoflavones genistein and daidzein in non-legume dicot and monocot tissues, *Plant Physiol.*, 2000, **124**, 781-794.
29. LIU, C.J., BLOUNT, J.W., STEELE, C.L., DIXON, R.A.., Bottlenecks for metabolic engineering of isoflavone glycoconjugates in *Arabidopsis*, *Proc. Natl. Acad. Sci. USA.*, 2002, **99**, 14578-14583.
30. SETCHELL, K.D., BROWN, N.M., DESAI, P., ZIMMER-NECHEMIAS, L., WOLFE, B.E., BRASHEAR, W.T., KIRSCHNER, A.S., CASSIDY, A., HEUBI, J.E., Bioavailability of pure isoflavones in healthy humans and analysis of commercial soy isoflavone supplements, *J. Nutr.*, 2001, 131(4 Suppl), 1362S-1375S.
31. WAGNER, J.D., SCHWENKE, D.C., GREAVES, K.A., ZHANG, L., ANTHONY, M.S., BLAIR, R.M., SHADOAN, M.K, WILLIAMS, J.K., Soy protein with isoflavones, but not an isoflavone-rich supplement, improves arterial low-density lipoprotein metabolism and atherogenesis, *Arterioscler. Thromb. Vasc. Biol.*, 2003, Oct 23 [Epub ahead of print].
32. YU, O., SHI, J., HESSION, A.O., MAXWELL, C.A., MCGONIGLE, B., ODELL, J.T., Metabolic engineering to increase isoflavone biosynthesis in soybean seed, *Phytochemistry*, 2003, **63**, 753-763.
33. FORD, D., Mechanistic explanations for the chemopreventive action of soyabean isoflavones: reducing the possibilities, *Br. J. Nutr.*, 2002, **88**, 439-441.

34. AKIYAMA, T., ISHIDA, J., NAKAGAWA, S., OGAWARA, H., WATANABE, S., ITOH, N., SHIBUYA, M., FUKAMI, Y., Genistein, a specific inhibitor of tyrosine-specific protein kinases, *J. Biol. Chem.*, 1987, **262**, 5592-5595.
35. GILLIES, P., SMITH, H., VAN OMMEN B., STIERUM R., Nutrigenomic profiling of soy isoflavones: A comparison of the effects of genistein and daidzein on a human gut epithelial cell-line transcriptome, 5th International Symposium on the Role of Soy in Preventing and Treating Chronic Disease, September 21-24, 2003, www.aocs.org.
36. WELLE, R., SCHRODER, G., SCHILTZ, E., GRISEBACH, H., SCHRODER, J., Induced plant responses to pathogen attack. Analysis and heterologous expression of the key enzyme in the biosynthesis of phytoalexins in soybean (*Glycine max* L. Merr. cv. Harosoy 63), *Eur. J. Biochem.*, 1991, **196**, 423-430.
37. MCGONIGLE, B., KEELER, S.J., LAU, S.M., KOEPPE, M.K., O'KEEFE D.P., A genomics approach to the comprehensive analysis of the glutathione S-transferase gene family in soybean and maize, Plant Physiol., 2000, **124**, 1105-1120.
38. ALTSCHUL, S.F., MADDEN, T.L., SCHAFFER, A.A., ZHANG, J., ZHANG, Z., MILLER, W., LIPMAN, D.J., Gapped BLAST and PSI-BLAST: A new generation of protein database search programs, *Nuc. Acids Res.*, 1997, **25**, 3389-3402.
39. JOFUKU, K.D., GOLDBERG, R.B., Kunitz trypsin inhibitor genes are differentially expressed during the soybean life cycle and in transformed tobacco plants, *Plant Cell*, 1989, **11**, 1079-1093.
40. YU, H., KUMAR, P.P., Post-transcriptional gene silencing in plants by RNA, *Plant Cell Rep.*, 2003, **22**, 167-174.
41. KLEIN, T.M., WOLF, E.D., WU, R., SANFORD, J.C., High-velocity microprojectiles for delivering nucleic acids into living cells, *Nature*, 1987, **327**, 70-73.
42. KINNEY, A.J., FADER, G.M., Suppression of specific classes of soybean seed protein genes, 2001 U.S. Patent 6362399.
43. WANG, C.Y., SHERRARD, M., PAGADALA, S., WIXON, R., SCOTT, R.A., Isoflavone content among maturity group 0 to II soybeans, *J. Am. Oil Chem Soc.*, 2000, **77**, 483-487.
44. BRAMLEY, P.M., Isoprenoid metabolism, *in*: Plant Biochemistry (P.M. Dey and J.B. Harborne, eds.) Academic Press, San Diego. 1998, pp. 417-437.
45. KUSHIRO, T., SHIBUYA, M., EBIZUKA, Y., Beta-amyrin synthase--cloning of oxidosqualene cyclase that catalyzes the formation of the most popular triterpene among higher plants, *Eur. J. Biochem.*, 1998, **256**, 238-244.
46. KUROSAWA, Y., TAKAHARA, H., SHIRAIWA, M., UDP-glucuronic acid:soyasapogenol glucuronosyltransferase involved in saponin biosynthesis in germinating soybean seeds, *Planta*, 2002, **5**, 620-629.
47. OSBOURN, A.E., Preformed antimicrobial compounds and plant defense against fungal attack, *Plant Cell*, 1996, **8**, 1821-1831.
48. PAPADOPOULOU, K., MELTON, R.E., LEGGET, M., DANIELS, M.J., OSBOURN, A.E., Compromised disease resistance in saponin-deficient plants, *Proc. Natl. Acad. Sci. USA,* 1999, **96**, 12923-12928.

49. WALLER, G.R., JURZYSTA, M., THORNE, R.L.Z., Allelopathic activity of root saponins from alfalfa (*Medicago sativa* L.) on weeds and wheat, *Bot. Bull. Acad. Sin.*, 1993, **34**, 1-11.
50. AGRELL, J., OLESZEK, W., STOCHMAL, A., OLSEN, M., ANDERSON, P., Herbivore-induced responses in alfalfa (*Medicago sativa*), *J. Chem. Ecol.*, 2003, **29**, 303-320.
51. OLESZEK, W., HOAGLAND, R.E., ZABLOTOVICZ, R.M., Ecological significance of plant saponins, *in*: Principles and Practices in Plant Ecology: Allelochemical Interactions (K.M.M. Dakshini and C.L. Foy, eds.), CRC Press, New York. 1999, pp. 451-465.
52. YOSHIKI, Y., KUDOU, S., OKUBO, K., Relationship between chemical structures and biological activities of triterpenoid saponins from soybean, *Biosci. Biotechnol. Biochem.*, 1998, **62**, 2291-2299.
53. GU, L., TAO, G., GU, W., PRIOR, R. L., Determination of soyasaponins in soy with LC-MS following structural unification by partial alkaline degradation, *J. Agric. Food Chem.* 2002, **50**, 6951-6959.
54. SHIRAIWA, M., HARADA, K., OKUBO, K., Composition and content of saponins in soybean seed according to variety, cultivation year and maturity. *Agric. Biol. Chem.*, 1991, **55**, 323-331.
55. RUPASINGHE, H.P., JACKSON, C.J., POYSA, V., DI BERARDO, C., BEWLEY, J.D., JENKINSON, J., Soyasapogenol A and B distribution in soybean (*Glycine max* L. Merr.) in relation to seed physiology, genetic variability, and growing location, J. Agric. Food Chem., 2003, **51**, 5888-5894.
56. OKUBO, K., IIJIMA, M., KOBAYASHI, Y., YOSHIKOSHI, M., UCHIDA, T., KUDOU, S., Components responsible for the undesirable taste of soybean seeds, *Biosci. Biotechnol. Biochem.*, 1992, **56**, 99-103.
57. PRICE, K.R., FENWICK, G.R., Soyasaponin I, a compound possessing undesirable taste characteristics isolated from the dried pea (*Pisum sativum* L.), *J. Sci. Food Agric.*, 1984, **35**, 887-892.
58. ZHU, N., SHENG, S., SANG, S., JHOO, J.W., BAI, N., KARWE, M.V., ROSEN, R.T., HO, C.T., Triterpene saponins from debittered quinoa (*Chenopodium quinoa*) seeds, *J. Agric. Food Chem*, 2002, **50**, 865-7.
59. GEE, J.M., PRICE, K.R., RIDOUT, C.L., WORTLEY, G.M., HURRELL, R.F., JOHNSON, I.T., Saponins of quinoa (*Chenopodium quinoa*): Effects of processing on their abundance in quinoa products and their biological effects on intestinal mucosal tissue, *J. Sci. Food Agric.*, 1993, **63**, 201-209.
60. MORITA, M., SHIBUYA, M., KUSHIRO, T., MASUDA, K., EBIZUKA, Y., Molecular cloning and functional expression of triterpene synthases from pea (*Pisum sativum*). New alpha-amyrin-producing enzyme is a multifunctional triterpene synthase, *Eur. J. Biochem.*, 2000, **12**, 3453-3460.
61. KUSHIRO, T., SHIBUYA, M., MASUDA, K., EBIZUKA, Y., A novel multifunctional triterpene synthase from *Arabidopsis thaliana*, *Tetrahedron Lett.*, 2000, **41**, 7705-7710.

62. HUSSELSTEIN-MULLER, T., SCHALLER, H., BENVENISTE, P., Molecular cloning and expression in yeast of 2,3-oxidosqualene-triterpenoid cyclases from *Arabidopsis thaliana*, *Plant Mol. Biol.*, 2001, **45**, 75-92.

Chapter Nine

MINING SOYBEAN EXPRESSED SEQUENCE TAG AND MICROARRAY DATA

Martina V. Strömvik,[a,b] Françoise Thibaud-Nissen,[a] and Lila O. Vodkin[a*]

[a]*Department of Crop Sciences*
1201 West Gregory Drive
University of Illinois
Urbana, Illinois 61801, USA

[b]*Department of Plant Science*
McGill University
Macdonald campus
21111 Lakeshore Road
Ste-Anne-de-Bellevue
Québec, Canada H9X 3V9

**Author for correspondence, email:* l-vodkin@uiuc.edu

INTRODUCTION

Prior to the availability of genomics resources, plant scientists could only employ the technologies of 'single gene' molecular biology including DNA blots (*i.e.*, 'southern blots') and RNA blots (*i.e.*, 'northern blots'). Because of gene duplications and divergence, most crop plants will have well over 26,000 genes, the number found in the model plant *Arabidopsis*. To examine expression of 30,000 to 50,000 genes would be nearly impossible if such experiments are conducted with pre-genomics technologies and resources. Fortunately, the data from large scale EST (expressed sequence tag) projects and from microarray experiments now provide a way for individual researchers to examine the expression of thousands of genes simultaneously in multiple tissue and organ systems and under differing physiological or environmental conditions. Data mining is the process of extracting as much information as possible about the genes and pathways likely to be expressed during growth and development or challenged by stress or pathogens.

As a result of two recent soybean genomics projects, the availability of public genomics tools for expression analysis in soybean has increased dramatically. The "Public EST Project", supported by the soybean grower associations, produced a collection of over 300,000 5' ESTs (expressed sequence tags) that represent 80 diverse cDNA libraries.[1] As a result of the National Science Foundation Project "A Functional Genomics Program for Soybean", a "unigene" set of 27,000 cDNAs has been processed, verified by 3' sequencing, and used in cDNA microarrays.[2-3]

The cDNA libraries represented in the soybean EST databases and on the microarrays have been constructed from many stages and tissues of soybean development and tissue and organ systems (see http://129.186.26.94/soybeanest.html for a detailed list). From the 5' EST data of these libraries, cDNAs have been selected by sequence clustering and reracked to form low redundancy sets of over 27,500 tentatively unique cDNAs, *i.e.*, unigenes, that were used in the construction of three microarrays containing 9,216 PCR amplified inserts on each array. Set 1 (designated Gm-r1070) is highly representative of cDNAs expressed in the developing seed coats, immature cotyledons, developing flowers and buds, and young pods. Set 2 (Gm-r1021 + Gm-r1083) is highly representative of cDNAs expressed in the roots of seedlings and adult plants, and roots infected with *Bradyrhizobium japonicum*. Subset 3 (Gm-r1088) is highly representative of cDNAs selected from libraries made from the germinating cotyledons, germinating seedlings and young plants under various stresses, and leaves of two-week old plants under various pathogen challenges.

In this chapter, we present examples of the use of these data resources in order to gain novel information on gene expression that will increase our understanding of genes and pathways that operate during plant growth and development. We illustrate mining of the soybean EST collection for expression data exhibited by a small family of lectin-related proteins. This approach known as 'electronic northern' can be applied to any of the thousands of gene families in soybean using the data available in the public EST collection. In the second section, we review how microarrays can be used to obtain quantitative data on the simultaneous expression of thousands of genes and many pathways operational during the process of induction of regenerable soybean somatic embryos during culture of the embryos on media with exogenously applied hormones.

Although we present only two examples, the use and applications of these genomics resources are essentially unlimited. For example, researchers will use soybean microarrays to compare how the plant responds to various nutritional changes during growth, as well has how it responds to temperature stress, and to challenges by various pathogens. In addition, microarrays can be used to probe genetic stocks that contain single gene mutations as well as QTLs (quantitative trait loci) in order to gain clues to the nature of the genetic variation and identify pathways responsible for the traits.

EXPLOITING THE SOYBEAN EST COLLECTION

Lectins as an Example

Mining the extensive soybean EST data using the "electronic Northern" approach illustrates how information can be inferred about the possible sites of expression of gene family members. Here, we illustrate this approach for a small family of soybean lectins and present new information on the expression patterns of two non-seed lectins.

Lectins are collectively a large, widespread family of interesting proteins with various functions involving protein-protein, protein-carbohydrate, and/or protein-RNA (adenine) binding activities. In mammals they are known to function in non-immune pathogen defense.[4] In plants, lectins have been widely used as model systems for protein storage, pathogen and herbivore defense, and for root nodulation. One group of lectins, the classical legume lectins, is prevalent in particular among the members of the legume (*Fabaceae*) family, although one member was recently reported from ivy, a non-legume species.[5] Like many gene families, the legume lectins have differentially expressed homologues. They are most expressed in the seeds but also are found in the vegetative parts of the plant. They may have different functions in each tissue, but they are known to serve as storage proteins.[6-9] Other groups of lectins that are less related to the classical legume lectins are also studied in legumes, such as the phloem lectins and the root apyrases.[10-12] In general, the seed lectins are thought to play a role in protein storage whereas in addition to protein storage, the vegetative and phloem lectins are thought to have a sugar transport and/or pathogen defense role. Lectins (apyrases) may also play a vital role in host-rhizobia recognition during the early stages of root nodulation by binding to the bacterial exopolysaccharides.[11]

In the legume plant *Dolichos biflorus*, two different classical legume lectins have been described, one of which is seed specific and the other is expressed in leaves and stem.[8,13] In soybean, the well known seed lectin *Le1* is highly expressed in cotyledons (ca 2%).[6-7,14] This seed specific expression was well studied *in planta* and by transgenic studies with the *Le1* promoter.[15-18] A soybean vegetative lectin (SVL), expressed in leaves, stems, petioles, and seedling cotyledons has also been described.[9,19] The SVL protein was observed in tissues where *Le1* (SBA) was not present. In addition, a related lectin homolog, *Le2,* that hybridizes to soybean DNA blots was isolated from genomics libraries, but clearly it is not expressed in the developing seed as determined by northern blots.[14]

Contig Analysis and Electronic Northerns

The sequences for the *Le1* gene were used to retrieve other lectin-like soybean EST sequences by nucleotide similarity from a collection of 303,149 public soybean ESTs. A BLASTN comparison yielded 304 EST sequences with significant similarity. These 304 sequences were contigged together using the standard approaches including Phrap software.[20-23] Examples of two of the contigs that represent the non-seed vegetative lectins and contain 111 and 4 ESTs, respectively, are shown in Figure 9.1. The largest contig containing 172 EST (not shown on Figure 9.1) represents the well characterized seed lectin, *Le1*. A detailed discussion of the contig analysis follows in order to illustrate how the process of mining EST data is conducted as well as to present the overall conclusions on the expression of the lectin homologues.

Three multiple-read contigs resulted: Contig 5 with 172 ESTs, Contig 4 with 111 ESTs, and Contig 3 with just four EST members. Contigs 1 and 2 were classified as "singleton contigs," and 10 singletons also resulted from the analysis. One first examines the few singletons and singleton contigs to determine their origin, which can include chimeras or low quality sequences. For example, the singletons (GenBank accession nos. AI941218, AI748013, AI940866, AI940981, AI941174, AI941268, AW397471, AW568999, AW317941, AW472598) were short and of lower quality. Upon visual inspection, they could be assigned to either Contig 4 or Contig 5, and they were thus left out of the rest of the analysis. Likewise, Contig 1, which consists of one sequence (sg46a07.y1, GenBank accession no. AW317250), would have been clustered with Contig 4, and Contig 2 (sg72a06.y1 GenBank accession no. AW395503) would have clustered with Contig 5, but both seem to be chimeric and have been left out of the remainder of the analysis.

Contigs 3, 4, and 5 consensus sequences were analyzed with BLAST against nr GenBank to find potential genes for the ESTs. The contigging of these ESTs revealed the consensus sequences of the *Le1* gene, and two additional and distinct *Le1*-like genes, neither of which are *Le2*, nor are they represented in nr with a (genomic) gene sequence.

(A)

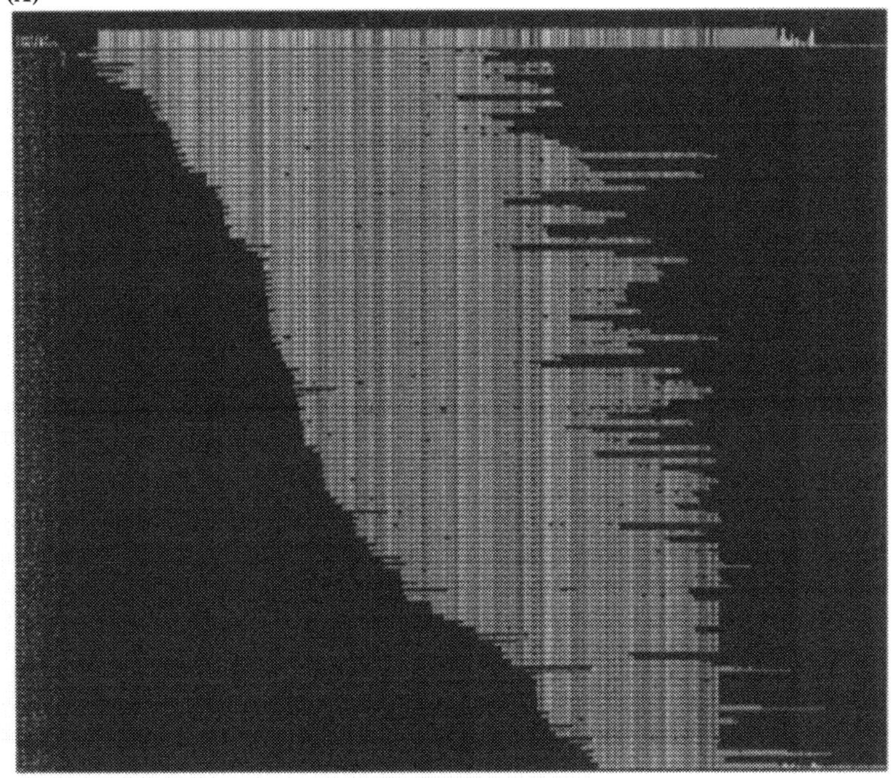

(B)

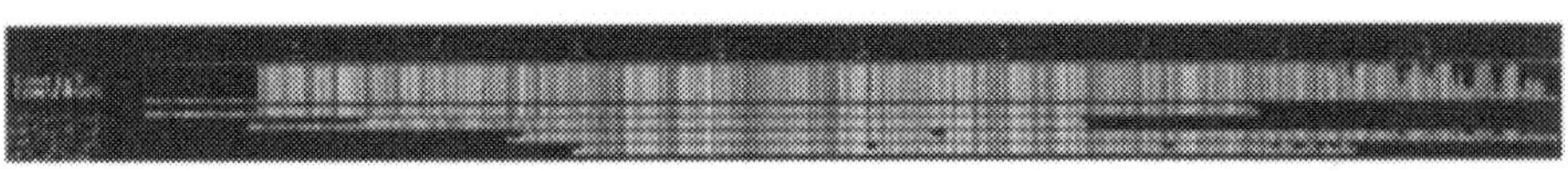

Figure 9.1: Examples of contigs resulting from clustering all sequences in the soybean public EST collection that are related to the soybean seed lectin, *Le1*. (A) A contig consisting of 111 ESTs representing the *Le3* gene (see cover) and (B) a contig with four members representing the *Le4* gene. The largest-most contig consisting of 172 members is that of *Le1* and is not showns. Contig images were created using the contigimage software from the bioData suite (http://ccgb.umn.edu).

Contig 5 contains 172 ESTs (data not shown). This is the largest contig and it represents transcripts from the *Le1* (SBA/SBL) gene. These ESTs come from cotyledon containing tissues only (cotyledons, seed, germinating seedlings, 1-2 cm pods containing very young seeds). Only three members of Contig 5 are chimeras of *Le1* and other sequences. Nucleotides 1-261 of the Gm-c1055-4407clone (sae07d08.y1, GenBank accession no. BI944757) are similar to chloroplast RNA, while the rest of the sequence is similar to *Le1* (annotated as lectin). Nucleotides 1-138 of Gm-c1007-912 (sg61d12.y1, GenBank accession no. AW318064) are similar to *Le1*, while the rest of the sequence appears to be of the ribosomal kind (annotated as ribosomal). The most interesting chimera is Gm-c1007-1663 (sg69c04.y1, GenBank accession no. AW397865) which is a Kunitz trypsin inhibitor (1-216) *Le1* chimera (212-430) and annotated as trypsin inhibitor. All four sequences, *Le1*, chloroplast RNA, ribosomal and Kunitz trypsin inhibitor are all highly abundant transcripts, and chimeras with these are more likely to be formed during library construction.

The second largest contig, Contig 4 with 111 ESTs, is shown in Figure 9.1a. Contig4 represents a gene which we call *Le3*. At the peptide level, it is most similar to the *Vigna linearis* leaf lectin (GenBank accession no. CAD43280) (68% identical), and the leaf and stem lectin DB58 from *Dolichos biflorus* (GenBank accession no. P19588) (64% identical). It is also 63% identical to the *Phaseolus vulgaris* leucoagglutinatig phytohemagglutinin precursor (GenBank accession no. P05087), a lectin toxic to the cowpea weevil. Contig4/Le3 is ca 56% identical to the soybean *Le1* at the protein level. Though only a sequence of 21 amino acids have been published for the soybean vegetative lectin (SVL),[9,19] the alignment of this sequence with the predicted peptide sequences of *Le3*, yields a 100% identity (alignment not shown). The tissues from which the ESTs are derived are much more widespread than those for *Le1*, ranging from flowers and pods, to leaves, stems, and vegetative buds, but not cotyledons, which is in accordance with previous SVL protein localization results.[9]

The third contig with multiple reads, Contig 3, shown in Figure 9.1b, has only four members from two different cultivars: Gm-c1062-5244 (GenBank accession no. CA799172) and Gm-c1062-5319 (GenBank accession no. CA799234) from a one-month old stem library made from cultivar Raiden and Gm-c1052-3368 (GenBank accession no. BQ094785) and Gm-c1053-2094 (GenBank accession no. BU091370) from two seedling tissue libraries from the cultivar Harosoy. This sequence, *Le4*, when compared by its predicted peptide sequence, is most similar to *Le3*, and also highly similar to the *Vigna linearis* lectins GenBank accession nos. CAD43280 and CAD43279.1 (nucleotide GenBank accession no. AJ504725) from leaf (67% and 65% identical, respectively), and also 60% identical to the *Phaseolus vulgaris* leucoagglutinating phytohemagglutinin precursor (GenBank accession no. P05087). In contrast, *Le4* is only ca 61% identical to the soybean seed *Le1*. We believe *Le4* to be a vegetative lectin because of the expression of *Le4* in stem and seedlings, together with the higher homology to the vegetative lectins.

None of the 304 EST sequences matched the *Le2* gene better than they match *Le1*, *Le3*, or *Le4*. Based on the contigging results, we draw the conclusion that in soybean, there are at least four genes with high nucleotide homology to the *Le1* seed lectin, although their expression pattern and also likely their function varies. *Le1* is the seed specific lectin, *Le3* and *Le4* are vegetative lectins, and *Le2* seems to be expressed at very low levels under special conditions. No ESTs for *Le2* were found in the 303,149 public soybean ESTs that we searched. The only occurrence of an *Le2* cDNA was found during screening of high density filters containing clones from Gm-c1005, an etiolated hypocotyl cDNA libaray (data not shown). The full sequence of the cDNA clone, designated *b22*, and representing *Le2*, has been entered as accession AY342213.

It has previously been shown that *Le1* gene expression is in general high and confined to cotyledons (seeds).[6-7,14-15] In contrast, *Le2* gene expression appears to be of low abundance. Four additional soybean cDNA libraries (Gm-c1012 [2286 ESTs sequenced] Williams, Gm-c1024 [548 ESTs sequenced] Williams, Gm-c1044 [1920 ESTs sequenced] Williams, Gm-c1045 [5432 ESTs sequenced] from Williams 82) are constructed from tissues similar to the Gm-c1005 (Williams 82 etiolated hypocotyls). In total, there are 10,382 ESTs sequenced from apical shoots from 9-10 day old etiolated seedlings in the cultivars Williams and Williams 82, but the *b22* cDNA clone from the Gm-c1005 [196 ESTs sequenced] is the only one representing *Le2*.

More soybean EST libraries are constructed from etiolated tissues of cultivars other than Williams, resulting in a total of 17,433 ESTs that have been sequenced from these tissues: Gm-c1058 [1580 ESTs sequenced] from "hypocotyl, 2 week old seedlings, etiolated (*G. soja*)", Gm-c1059 [4413 ESTs sequenced] from "whole seedling, 2 week old, etiolated (*G. soja*)", Gm-c1069 [6703 ESTs sequenced] from "degenerating cotyledons, 9-10 day old etiolated seedling (Williams 82)" and Gm-c1084 [4737 ESTs sequenced] from "etiolated hypocotyls, inoculated with *Phytophthora sojae* race 1 (Williams 82)". The only Le like sequences in these libraries are five ESTs representing *Le3*. Thus, etiolation alone may not be the key reason for *Le2* expression in the seedling shoot.

The libraries in which either *Le1* or *Le3* were present were grouped by tissues to aid in the "electronic northern". The occurrences of *Le1* and *Le3* sequences were counted for each tissue and a normalized value (ESTs per million) was calculated with the following formula: (Le ESTs in Library A + Le ESTs in Library B + ...+ Le ESTs in Library n / ESTs in Library A + ESTs in Library B + ... + ESTs in Library n) x (1.0×10^6) = Expression level in ESTs per million (EPM).

Figure 9.2 shows an 'electronic northern' of the *Le1* and *Le3* expression patterns. Upon observing the different tissues where *Le1* (Contig5) and the *Le3* (Contig 4) sequences are expressed, it is clear that *Le1* is specific to cotyledon tissues, whereas the *Le3* is more widely expressed in the plant. The highest expression of *Le3* is in floral meristem. Only in two libraries were *Le1* and *Le3* expressed simultaneously, Gm-c1056 (seedling) and Gm-c1071 (pods and seeds). Both of these libraries are considered 'mixed' tissues since the seedling still has the cotyledons attached, and the pod libraries were constructed using young 1-2 cm pods containing very young seeds. *Le1* was previously shown to be expressed early in seed development,[13] as well as being very highly expressed in seed development. We show that *Le3* appears to have a relatively high expression in floral meristem tissue and is also present in flowers, vegetative buds, leaf, and seedling. *Le4*, being expressed at very low levels, has only been observed in stem and in seedling. It is interesting to note that none of the four *Le1*-like sequences were present in the root libraries as determined by the computational analysis of EST data.

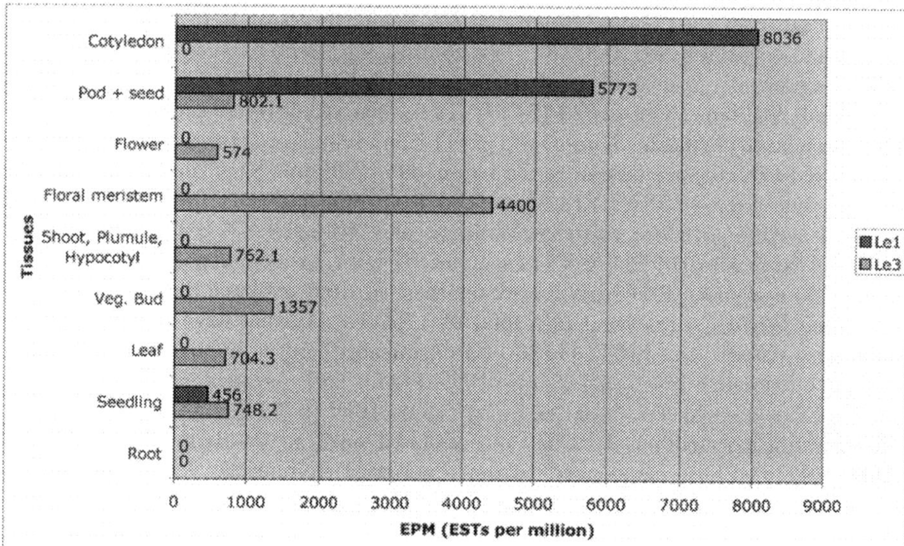

Figure 9.2: "Electronic northern" histogram of *Le1* and *Le3* expression patterns. *Le1* (first darker bar), is expressed in cotyledons of the seeds, and *Le3* (second, lighter bar) in non-seed tissues. Root is included to illustrate that neither *Le1* nor *Le3* are expressed in roots.

In summary, retrieval and subsequent contigging of 304 *Le1* like EST-sequences revealed that there are likely four *Le1*-like soybean genes, the well known *Le1* seed lectin (SBA), *Le3* that is likely the soybean vegetative lectin (SVL), and a stem and shoot *Le4* whose occurrence has not been reported before. Only one occurrence of *Le2* has been found, and that was in etiolated shoots. *Le1* and *Le2* are more homologous to each other than to either of *Le3* or *Le4*, and *Le3* and *Le4* are more homologous to each other than to either *Le1* or *Le2*.

EXPLOITING MICROARRAYS FOR GLOBAL ANALYSIS OF PATHWAYS

The electronic northerns are a semi-quantitative, first approximation of expression levels in various tissues. They are more accurate when the input data are from non-normalized cDNA libraries, the database is large, and most libraries have been sequenced relatively deeply. Microarrays provide a method to quantify relative gene expression in RNA samples by dual labeling with fluorescent dyes. They also enable the simultaneous comparisons of expression levels of thousands of genes at the same time. Genes that respond in a similar developmental manner and/or to the same stress or environmental conditions can be determined. Below, we illustrate the power of microarrays to reveal information on networks of pathways that operate during the induction of somatic embryos on tissue culture media.

Induction of Somatic Embryos as a System

Somatic embryos follow the same general pattern of development as zygotic embryos, but the progression from one stage to the next is induced externally by changes in the culture medium. In soybean, somatic embryos are initiated from immature cotyledons on high levels of the synthetic auxin 2,4-dichlorophenoxyacetic acid (2,4-D, 40 mg).[24] Within 30 days, embryos appear from the epidermal or subepidermal layers of the upper side (away from the medium), while the rest of the cotyledon degenerates into a brown callus mass.[24] Auxin inhibits the differentiation of embryo cells beyond the globular stage in the soybean system, while in many species, such as the carrot, auxin inhibits the organization of the callus into embryos. In soybean, embryos can be maintained indefinitely at the globular stage on 20 mg/L 2,4-D.[25] The heart through the cotyledon stages occur on MS medium free of auxin and are followed by several days of desiccation.[26] The mature embryo can then be placed on a germination medium and grown into plants.

A critical difference with zygotic embryos is that somatic embryos are never surrounded by an endosperm and that they do not develop a suspensor.[27] In addition, somatic embryo are often larger than zygotic embryos. The morphological diversity and the lower rates of germination seen in somatic embryos compared to zygotic embryos are generally attributed to *in vitro* culture. However, the extent to which the zygotic and somatic developmental routes are molecularly similar is unclear.

Indoleacetic acid (IAA) is the main form of natural auxin in plants and is involved in many aspects of embryo development through the regulation of cell expansion and cell division. Genes transcribed within minutes following exposure to auxin have been identified. These genes fall into three large families: the *Aux/IAAs*, the *GH3s*, and the *SAURs*. Aux/IAA proteins are encoded by 29 genes in *Arabidopsis*[28] and at least two genes in soybean.[29] The *GH3s* have been identified by differential screening in soybean and constitute a 7-gene family in *Arabidopsis*.[30] There is some evidence that some Aux/IAAs are phosphorylated by phytochrome A in vitro and that GH3s are part of the phytochrome A transduction pathway.[30] A group of 5 clustered *SAUR* genes were identified in soybean,[31] and a total of 70 are present in *Arabidopsis*.[30] However, the function of SAURs is still unknown.

Soybean cotyledons can produce somatic embryos when subjected to high levels (180 μM) of the synthetic auxin 2,4-D. Little is known of the cellular and molecular events underlying the transition from differentiated epidermal cells to globular embryos in the presence of large amounts of auxin. Several lines of evidence suggest that the auxin response is mediated by reactive oxygen species (ROS). ROS mediate the response to numerous stresses, including pathogen challenge in the hypersensitive response, mechanical wounding, and osmotic shock.[32-35] Accumulation of stress-induced transcripts is commonly observed upon auxin treatment. In addition, increased free radicals during auxin-induced cell expansion were reported in maize coleoptiles,[36] and the measuring of ROS using the probe 2-7-dihydrofluorescein have provided evidence for an auxin-induced oxidative burst in cells of *Chenopodium rubrum*.[35]

Relatively few genes have been identified, that are expressed in somatic embryos but not in callus cultures. Based on the number of auxin mutants in *Arabidopsis* that are arrested early in their development, it is likely that this hormone plays an important role during embryo development. However, the molecular events linking auxin to the formation and development of embryos are still mostly unknown. Somatic embryos, shown to be morphologically similar to zygotic embryos, have been used to advance our knowledge of the early events in development, but multiple studies using traditional differential screening techniques have led to the identification of a small number genes and have provided few clues on the molecular or physiological aspects of the developmental process.

Transcript Profiles During Early Somatic Embryogenesis

In this section, we review our approach using microarrays to gain the global view of gene expression changes during the development of somatic embryos through the globular stage.[3] We used a 9,728-element microarray consisting of 9,216 single-spotted soybean cDNA clones (Gm-r1070 library), and of 64 choice clones each printed eight times. Each cDNA clone was chosen as a representative of a unigene as illustrated in Figure 9.3. The estimated redundancy is between 15 and 20%, so the 9,216 cDNA clones represent about 8,000 unique genes from flowers, pods, developing cotyledons, and seed coats.

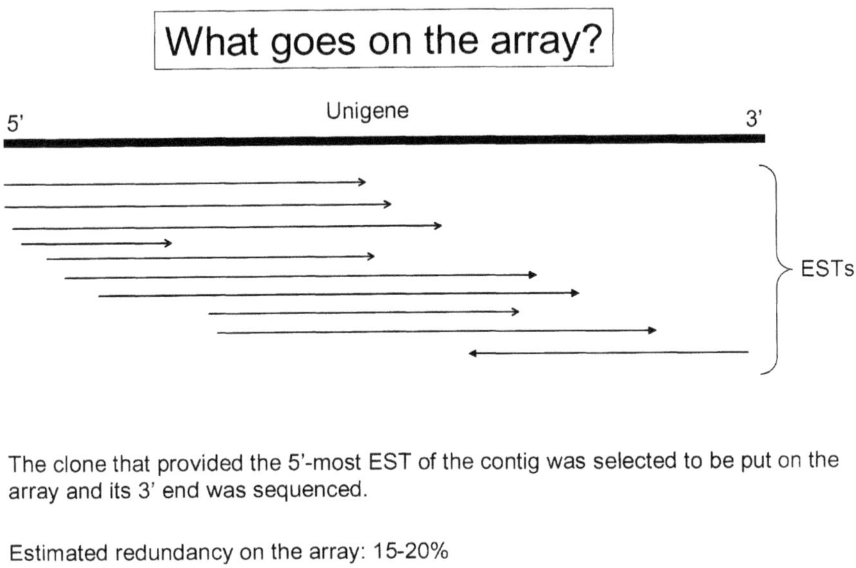

Figure 9.3: Diagram illustrating selection of the tentatively unique cDNA clones to include the microarrays. After contigging ESTs from a selection of flower, pod, seed, and seed coat libraries, the singletons or 5'-most members of each contig were re-racked into a new library, Gm-r1070. The 3' ends of the cDNA clones in Gm-r1070 were then sequenced.

In order to determine the global expression patterns during somatic embryogenesis, we sampled adaxial and abaxial sides of the cotyledons separately, at 7-day intervals during the 4-week induction, and obtained RNA from the small amounts of tissue. Expression in the adaxial side (on which embryos form) was compared to expression in the abaxial side (that becomes a callus tissue) collected at the same time point by hybridization of the corresponding labeled cDNAs to a soybean microarray representing the 9,216 cDNA clones in the Gm-r1070 unigene set. In addition, transcript profiles of the genes expressed in the adaxial side on which the embryos form were obtained by comparing each time point to the previous one. In that manner, we determined a time course of the global expression profiles during the first four weeks of somatic embryogenesis.

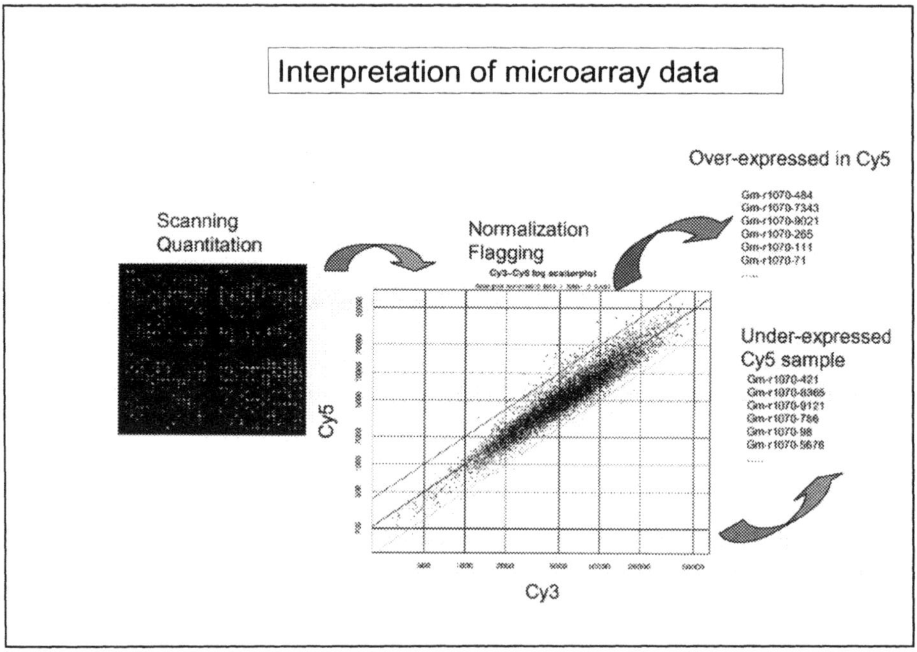

Figure 9.4: Diagram illustrating the microarray process and initial steps in data analysis.

Of the approximately 8,000 unique genes represented on the array, 495 cDNAs (5.3% of the cDNAs on the array) showed a difference in their levels of expression above or below the two-fold level. Figure 9.4 diagrams the general process of interpreting microarray data to obtain the differentially expressed cDNAs. The relative levels of RNA expression in two samples are compared by incorporating different fluorescent dyes into the RNA by reverse transcription. RNA extracted from one tissue or time point that is labeled with cy5 and another is labeled with cy3. The two labeled samples are pooled and hybridized simultaneously to the microarray. After scanning and quantification of the array, it is important to flag spots that may be bad and to normalize the two images. The center image shows a scatter plot of the comparison of the log values of each RNA sample. Equal expression is represented by the center line, and the two outer lines represent two-fold higher or two-fold lower expression. Each dot represents the value for one of the 9,728 elements on the array. The cDNAs that have differential expression of over or under two-fold can be identified by their location on the array and its corresponding clone ID number (Gm-r1070-1, etc).

One of the major conclusions from the data is that polarity between the abaxial and adaxial side in gene expression is removed within the first 7 days on auxin media as the cotyledons dedifferentiate. Further analysis of the 495 cDNAs that were differentially expressed during the time course included statistical clustering using a non-hierarchical method (K-means) to reveal cDNAs with similar profiles during the time course of somatic embryo development on the adaxial side. A sample of the interpretation of the gene lists from two of the eleven clusters is shown in Figures 9.5 and 9.6. Auxin induces dedifferentiation of the cotyledon and provokes a surge of cDNAs involved in the oxidative burst. For example, glutathione-S-transferases (GST) are prominent in set 6, as shown which is characterized by a peak in the newly forming embryos at 14 days and then a decline. GSTs are induced by reactive oxygen species as hydrogen peroxide[32] and they detoxify byproducts of membrane lipid hydroperoxides.[37] Many members of the flavonoid pathway including chalcone synthases (CHS), chalcone isomerase (CI), flavonoid hydroxylase (F3'5'H), and isoflavone synthase (IFS) are also found in this set. Among other compounds, the flavonoid pathway is needed to produce phytoalexinins which are induced upon stress.[33]

Our data indicate that the formation of somatic globular embryos is accompanied by the accumulation of storage protein transcripts. Figure 9.6 illustrates that transcripts for the synthesis of gibberellic acid also accompany the formation of somatic embryos. These microarray hybridizations experiments have been entered into the GEO (Gene Expression Omnibus) database of NCBI at http://www.ncbi.nlm.nih.gov/geo. For complete details on interpretation of the data, see Thibaud-Nissen *et al.*[3]

2,4-D induces an oxidative burst

Function	Annotation	Function	Annotation
cgm	glucose-6-P 1-dehydrogenase (cytoplasm.)	cgm	glucose-6-P 1-dehydrogenase (cytoplasm.)
cgm	probable coatomer complex subunit	cgm	beta-glucosidase
cgm	glycosyl hydrolase family 17		
oth	F3'5'H (flavonoid-3',5'-hydroxylase)	oth	CI (chalchone isomerase)
oth	IFR1 (isoflavone reductase 1)	oth	F3'5'H (flavonoid-3',5'-hydroxylase)
oth	IFS2 (isoflavone synthase 2)	oth	CHS2 (chalcone synthase gene 2)
oth	IFS1 (isoflavone synthase 1)	oth	CHS7 (chalcone synthase gene 7)
oth	CYP93A1 (dihydroxypterocarpan-6a-hydroxy)	oth	CHS* (chalcone synthase homologue)
oth	CHS6 (chalcone synthase gene 6)		
ox	putative NtPRp27-like protein	ox	glutathione S-transferase GST 8
ox	glutathione S-transferase GST 16	ox	NtPRp27
ox	probable glutathione transferase	ox	glutathione S-transferase GST 11
ox	endo-beta-1 4-glucanase	ox	glutathione S-transferase GST 10
ox	glutathione S-transferase GST 7	ox	In2-1 protein
ox	probable disease resistance response protein	ox	expansin
ox	glutathione S-transferase GST 19	ox	probable glutathione transferase
ox	probable glutathione transferase		
sig	calcium-binding protein-like	sig	auxin-responsive GH3 product
to	homeodomain-leucine zipper protein 56	to	NtWRKY2
to	DNA-binding protein WRKY1		
u	cytochrome P450 82A4	u	cytochrome P450
u	cytochrome P450	u	none
u	none	u	unknown protein
u	none	u	putative ripening-related protein
u	none	u	unknown protein
u	none	u	hypothetical protein
u	unknown protein	u	unknown
u	cytochrome P450 82A4	u	none
u	none	u	conserved hypothetical protein
u	cytochrome P450	u	putative ripening-related protein

Figure 9.5: Illustration of genes expressed similarly during a time course of somatic embryogenesis in one of the K-means cluster sets, set 6, of the microarray data. Adapted from Thibaud-Nissen, *et al.*[3] The annotations matching clone IDs on the microarray sets represent the top Blastx hit[21] in the public databases at a threshold of E - 6. The putative function in cell metabolism is shown: oth, other; ox, oxidative stress/defense; sig, signaling; to, transcription; u, unknown.

Genes involved in gibberellin synthesis and photosynthesis increase steadily in the developing embryos but not in the subtending callus

Function	Annotation	Function	Annotation
cgm	argonaute (zwille, pinhead) like protein	cgm	tyrosyl-tRNA synthase
cgm	putative xyloglucan endotransglycolase	cgm	putative epimerase/dehydratase
cgm	ATP-dpdent Clp protease proteolytic subunit	cgm	SMT3 ubiquitin-like protein
cgm	ribosomal protein S2	cgm	histone H2A.F/Z
cgm	phosphatidylserine decarboxylase	cgm	acetyltransferase
cgm	histone H2A.F/Z		
en	chlorophyll a/b bdg prot. LHCII type I prec. 3	en	photosystem II type I chlorophyll a/b-binding protein
en	chlorophyll a/b bdg prot. LHCII type I prec. 3	en	chlorophyll a/b-binding protein type I precursor
en	RuBisCO small subunit RBCS1		
oth	4-coumarate-CoA ligase-like protein		
ox	glutathione reductase chloroplast precursor		
sig	ent-kaurenoic acid hydroxylase	sig	zing finger protein like, Ser/Thr protein kinase like
sig	ent-kaurenoic acid hydroxylase	sig	nodulin-like protein
sig	gibberellin 20-oxidase	sig	ent-kaurene oxidase
sig	protein kinase C inhibitor-like	sig	gibberellin 20-oxidase
sig	receptor-like kinase 2		
sp	Kunitz-type trypsin inhibitor KTI2 precursor	sp	lectin precursor (agglutinin), (SBA)
sp	beta-amylase	sp	lectin precursor
to	putative ribonucleoprotein chloroplast prec.	to	filamentous flower protein FIL or YABBY1
u	gibberellin-regulated protein GAST1-like	u	putative protein
u	gibberellin-regulated protein	u	none
u	none	u	hypothetical protein
u	none	u	putative senescence-associated protein
u	none	u	none
u	none	u	putative protein
u	hypothetical protein	u	none
u	cytochrome P-450	u	cytochrome P-450
u	none	u	unknown protein
u	none	u	none
u	putative protein	u	LTCOR11

Figure 9.6: Illustration of genes expressed with similar patterns in cluster set 11 of the microarray data of somatic embryogenesis. Adapted from Thibaud-Nissen *et al.*[3] The annotations matching clone IDs on the microarray sets represent the top Blastx hit[21] in the public databases at a threshold of E -6. The putative function in cell metabolism is shown: cgm, cell growthm and maintenance; en, energy; sig, signaling; sp, seed protein; to, transcription; u, unknown.

SUMMARY

In summary, we have illustrated the approach of mining the large soybean EST collection to deduce knowledge about expression of individual gene family members using the soybean lectins as an example. Plants have many different sets of genes that are homologous at the sequence level, but which may have very different biological functions in different cells, tissues, or organs in which the gene product is active. The sequence alone provides little information about function. By adding the information about spatial and temporal gene expression (as from 'electronic northerns'), we get a first view of gene expression profiles that puts us one step closer to understanding their functions and the chemical pathways in which they participate.

In the case of lectins, we can be relatively certain the seed lectin product of the *Le1* has a role as a seed storage protein because of the abundance of ESTs representing *Le1* in only the seed tissues. In addition, we see that there are three homologs closely related to *Le1*, but none of these is expressed in seeds. These related lectins may have function as vegetative storage proteins or defend against pathogens in vegetative tissues. In the future, the different lectins can be investigated for insecticidal properties and possibly used in pathogen defense strategies using genetic engineering. The more distantly related apyrases, whose ESTs are found in root libraries, likely have a function in recognition of bacteria.

As opposed to ESTs, microarrays provide a quantitative approach to global gene expression, and microarray data can be subjected to advanced statistical clustering analysis. Clustering the cDNAs by similarity of expression profile derived from microarray data over the course of early somatic embryogenesis allowed a determination of the timing of the molecular events taking place during that phase of development. For example, several genes involved in polarity (adaxial versus abaxial) of the tissue culture soybean embryos were found,[3] and these could be targets for improving the process of regeneration of embryos from tissue culture.

Of course, mRNA abundance data alone do not ensure that the metabolic products of a pathway are present since control can be exerted at multiple levels, including, transcriptional, post-transcriptional, translational, and post-translational. However, mining the soybean EST databases and determining transcript profiles of many thousands of genes simultaneously using microarrays was not feasible for soybean until recently. These approaches will stimulate many avenues of research into the complex physiology and metabolic systems operational in this important crop. For example, we are currently determining the expression profiles of 27,000 soybean cDNAs during normal seed development in order to elucidate the developmental profiles of genes that produce compositional traits as protein, oil, and secondary compounds. In addition, germplasm lines that vary in protein, oil, or isoflavone content can be examined by microarrays to determine key control points in production of these compounds under normal or stress conditions. These studies will yield information that can be applied toward breeding or genetic engineering approaches to improve seed composition. Finally, the cDNA microarrays will enable similar studies in related legumes that do not have genomics resources yet available.

REFERENCES

1. SHOEMAKER, R., KEIM, P., VODKIN, L., RETZEL, E., CLIFTON, S., WATERSTON, R., SMOLLER, D., CORYELL, V., KHANNA, A., ERPELDING, J., GAI, X., BRENDEL, V., RAPH-SCHMIDT, C., SHOOP, E.G., VIELWEBER, C.J., SCHMATZ, M., PAPE, D., BOWERS, Y., THEISING, B., MARTIN, J., DANTE, M., WYLIE, T., GRANGER, C., A Compilation of soybean ESTs: generation and analysis, *Genome*, 2002, **45**, 329-338..
2. VODKIN, L.O., KHANNA, A., SHEALY, R.T., CLOUGH, S.J., GONZALEZ, D.O., PHILIP, R., ZABALA, G., THIBAUD-NISSEN, F., SIDAROUS, M., STROMVIK, M.V., SHOOP, E., SCHMIDT, C., RETZEL, E., ERPELDING, J., SHOEMAKER, R.C., RODRIQUEZ-HEUTE, A., POLACCO, J.C., CORYELL, V., KEIM, P., GONG, G., LIU, L., PARDINAS, J., SCHWEITZER, P., Microarrays for Global Expression Using 27,500 Sequenced cDNAs Representing an Array of Developmental Stages and Physiological Conditions of the Soybean Plant, http:www.soybeangenomics.cropsci.uiuc.edu.
3. THIBAUD-NISSEN, F., SHEALY, R.T., KHANNA, ANUPAMA, VODKIN, L.O., Clustering of microarray data reveals transcript patterns associated with somatic embryogenesis in soybean, *Plant Physiol.* 2003, **132**, 118-136.
4. HAKANSSON, K., REID, K.B.M., Collectin structure – a review, *Protein Science*, 2000, **9**, 1607-1617.
5. WANG, W., HAUSE, B., PEUMANS, W.J., SMAGGHE, G., MACKIE, A., FRASER, R., VAN DAMME, E.J.M., The Tn antigen-specific lectin from ground ivy is and insecticidal protein with and unusual physiology, *Plant Physiol.*, 2003, **132**, 1322-1334.
6. VODKIN, L.O., Isolation and characterization of messenger RNAs for seed lectin and Kunitz trypsin inhibitor in soybeans, *Plant Physiol.*, 1981, **68**,766-771.
7. VODKIN, L.O., RHODES, P.R., GOLDBERG, R.B., A lectin gene insertions has the structural features of a transposable element, *Cell*, 1983, **34**, 1023-1031.
8. HARADA, J.J., SPADORO-TANK, J., MAXWELL, J.C., SCHNELL, D.J., ETZLER, M.E., Two lectin genes differentially expressed in Dolichos biflorus differ primarily by a 116-base pair sequence in their 5' flanking regions, *J. Biol. Chem.*, 1990, **265**, 4997-5001.
9. SPILATRO, S.R., COCHRAN, G.R., WALKER, R.E., CABLISH, K.L., BITTNER, C.C., Characterization of a new lectin of soybean vegetative tissues, *Plant Physiol.*, 1996, **110**, 825-834.
10. DINANT, S., CLARK, A.M., ZHU, Y., VILAINE, F., PALAUQUI, J.-C., KUSIAK, C., THOMPSON, G.A., Diversity of the superfamily of phloem lectins (Phloem Protein 2) in angiosperms, *Plant Physiol.*, 2003, **131**, 114-128.

11. ETZLER, M.E., KALSI, G., EWING, N.N., ROBERTS, N.J., DAY, R.B., MURPHY, J.B., A nod factor binding lectin with apyrase activity from legume roots, *Proc. Natl. Acad. of Sci. USA*, 1999, **96**, 5856-5861.
12. DAY, R.B., MCALVIN, C.B., LOH, J.T., DENNY, R.L., WOOD, T.C., YOUNG, N.D., STACEY, G., Differential expression of two soybean apyrases, one of which is an early nodulin, *Molec. Plant Microbe Interactions*, 2000, **13**, 1053-1070.
13. BUNKER, T.W., ETZLER, M.E., The stem and leaf lectin of Dolichos biflorus L., previously thought to be cell wall associated, is sequestered in vacuoles, *Planta*, 1994, **192**, 144-147.
14. GOLDBERG, R.B., HOSCHEK, G., VODKIN, L.O., An insertion sequence blocks the expression of a soybean lectin gene, *Cell*, 1983, **33**, 465-475.
15. OKAMURO, J.K., JOFUKO, D.K., GOLDBERG, R.B., Soybean seed lectin gene and flanking nonseed protein genes are developmentally regulated in transformed tobacco plants, *Proc. Natl. Acad. Sci. USA*, 1986, **83**, 8240-8244.
16. LINDSTROM, J.T., VODKIN, L.O., HARDING, R.W., GOEKEN, R.M., Expression of soybean lectin gene deletions in tobacco, *Devel. Genet.*, 1990, **11**, 160-167.
17. CHO, M.-J., WIDHOLM, J.M., VODKIN, L.O., Cassettes for seed-specific expression tested in transformed embryogenic cultures of soybean, *Plant Molec. Biol. Rept.*, 1995, **13**, 255-269.
18. PHILIP, R., DARNOWSKI, D.W., MAUGHAN, P.J., VODKIN, L.O., Processing and localization of bovine beta-casein expressed in transgenic soybean seeds under control of a soybean lectin expression cassette, *Plant Sci.*, 2001, **161**, 323-335.
19. SPILATRO, S.R., ANDERSON, J.M., Characterization of a soybean leaf protein that is related to the seed lectin and is increased with pod removal, *Plant Physiol.*, 1989, **90**, 1387-1393.
20. ALTSCHUL, S.F., MADDEN, T.L., SCHAFFER, A.A., ZHANG, J., ZHANG, Z., MILLER, W., LIPMAN, D.J., Gapped BLAST and PSI-BLAST: a new generation of protein database search programs, *Nuc. Acids Res.*, 1997, **25**, 3389-3402.
21. BENSON, D.A., KARSCH-MIZRACHI, I., LIPMAN, D.J., OSTELL, J., RAPP, B.A., WHEELER, D.L., GenBank, *Nuc. Acids Res.*, 2002, **30**, 17-20.
22. THOMPSON, J.D., HIGGINS, D.G., GIBSON, T.J., ClustalW: improving the sensitivity of progressive multiple sequence alignment through sequence weighing, position-specific gap penalties and weight matrix choice, *Nuc. Acids Res.*, 1994, **22**, 4673-4680.
23. GREEN, P., Documentation for Phrap (http://www.phrap.org), University of Washington, Seattle. 1996.
24. FINER, J.J., Apical proliferation of embryonic tissue of soybean [*Glycine max* (L.) Merrill], *Plant Cell Rep.*, 1988, **7**, 238-241.
25. WRIGHT, M.S., LAUNIS, K.L., NOVITZKY, R., DUESING, J.H., HARMS, C.T., A simple method for the recovery of multiple fertile plants from individual somatic embryos of soybean [*Glycine max* (L.) Merrill], *In Vitro Cell Dev. Biol.*, 1991, **27**, 153-157.
26. FINER, J.J., MCMULLEN, M.D., Transformation of soybean *via* particle bombardment of embryogenic suspension culture tissue, *In Vitro Cell Dev. Biol.*, 1991, **27**, 175-182.

27. ZIMMERMAN, J.L., Somatic embryogenesis: a model for early development in higher plants, *Plant Cell*, 1993, **5**, 1411-1423
28. LISCUM, E., REED, J.W., Genetics of Aux/IAA and ARF action in plant growth and development, *Plant Mol. Biol.*, 2002, **49**, 387-400.
29. AINLEY, W.M., WALKER, J.C., NAGAO, R.T., KEY, J.L., Sequence and characterization of two auxin-regulated genes from soybean, *J. of Biol. Chem.*, 1988, **263**, 10658-10666.
30. HAGEN, G., GUILFOYLE, T., Auxin-responsive gene expression: genes, promoters and regulatory factors, *Plant Mol. Biol.*, 2002, **49**, 373-385.
31. MCCLURE, B.A., HAGEN, G., BROWN, C.S., GEE, M.A., GUILFOYLE, T.J., Transcription, organization, and sequence of an auxin-regulated gene cluster in soybean, *Plant Cell*, 1989, **1**, 229-239.
32. LEVINE, A., TENHAKEN, R., DIXON, R., LAMB, C., H_2O_2 from the oxidative burst orchestrates the plant hypersensitive disease resistance response, *Cell*, 1994, **79**, 583-593.
33. WOJTASZEK, P., Oxidative burst - an early plant response to pathogen infection, *Biochem. J.*, 1997, **322**, 681-692.
34. GUS-MAYER, S., NATON, B., HAHLBROCK, K., SCHMELZER, E., Local mechanical stimulation induces components of the pathogen defense response in parsley, *Proc. Natl. Acad. Sci. USA*, 1998, **95**, 8398-8403.
35. PFEIFFER, W., HÖFTBERGER, M., Oxidative burst in *Chenopodium rubrum* suspension cells: Induction by auxin and osmotic changes, *Physiol. Plantarum*, 2001, **111**, 144-150.
36. SCHOPFER, P., LISZKAY, A., BECHTOLD, M., FRAHRY, G., WAGNER, A., Evidence that hydroxyl radicals mediate auxin-induced extension growth, *Planta*, 2002, **214**, 821-828.
37. BERHANE, K., WIDERSTEN, M., ENGSTROM, A., KOZARICH, J.W., MANNERVIK, B., Detoxication of base propenals and other alpha, beta-unsaturated aldehyde products of radical reactions and lipid peroxidation by human glutathione transferases, *Proc. Natl. Acad. Sci. USA*, 1994, **91**, 1480-1484.

Chapter Ten

ASPERGILLUS NIDULANS AS A MODEL SYSTEM TO STUDY SECONDARY METABOLISM.

Lori A. Maggio-Hall, Thomas M. Hammond, Nancy P. Keller*

Department of Plant Pathology
University of Wisconsin
Madison, WI 53706

**Author for correspondence:* npk@plantpath.wisc.edu

INTRODUCTION

Aspergillus as a genus is of intense biological, industrial, agricultural, and medicinal importance. This genus represents a large family of fungi with over 185 recognized species.[1] Members are distributed world-wide and occupy diverse ecological niches.[1,2] Most species are saprophytes that grow on a large number of substrates from plant and animal waste to pesticides and platicizers, and thus are important in nutrient cycling and detoxification.[2] The inherent properties associated with degradation of diverse substrates have lent themselves well to industrial applications and food fermentation. Industrially important *Aspergillus* spp., including *A. oryzae, A. terreus,* and *A. niger,* generate such products as citric acid, lovastatin, and penicillin,[3-5] and *A. oryzae* and *A. sojae* are intensively used in the production of a series of world enjoyed oriental condiments.[6] Because members of this species can grow over a wide range of temperatures and colonize substrates with relatively low water activity (some can grow at water activity as low as – 40 Mpa), they are well suited to colonize a number of grain and nut crops in the field and in storage. They represent the major class of fungi involved in the deterioration of grain. Further, *Aspergillus* species are known to contaminate grain and prepared food with an array of mycotoxins including, aflatoxins, sterigmatocystins, ochratoxins, versicolorins, gliotoxin, citrinin, cyclopiazonic acid, cietrovirdin, and tremogens.[7,8] With increasing frequency, *Aspergillus* spp. are also implicated in a number of human and animal diseases. No other fungal genus has such a diverse membership and plays such an important role in industry, agriculture, medicine, and soil ecology.

Whereas a diverse number of *Aspergillus* spp. are credited with the above biological properties, efforts to elucidate the genetics of *Aspergillus* development and metabolism have centered on the model *Aspergillus nidulans.* This fungus is one of the best described eukaryotic genetic systems and has been used to decipher the biology of the cell cycle, pathogenicity, drug resistance, human disease, primary and secondary metabolism among other topics. The available genomic sequence (http://www-genome.wi.mit.edu/annotation/fungi/aspergillus/index.html), useful vectors and DNA libraries (www.fgsc.net), and existence of a sexual cycle (rare in *Aspergillus*) have contributed to the ease of genetic manipulation of this spp. This amenability of *A. nidulans* to genetic analysis provides a powerful tool for examining important questions on the development and metabolism of *Aspergillus* species. In this chapter we review the contributions *A. nidulans* has made to the understanding of fungal secondary metabolism.

STERIGMATOCYSTIN

Sterigmatocystin (ST) has received considerable attention due to its biosynthetic relationship to the better-known mycotoxin, aflatoxin (AF). Both compounds are teratogenic, mutagenic and carcinogenic,[7] but AF is the more prevalent contaminant in agricultural settings. The AF producing Aspergilli (primarily *A. flavus* and *A. parasiticus*) are common seed infecting fungi and produce copious amounts of AF in infested seed. ST, produced by *A. nidulans*, is the penultimate precursor in the AF biosynthetic pathway. Our current understanding of AF biosynthesis has been greatly enhanced through genetic studies of ST biosynthesis in *A. nidulans*.

Biosynthetic Pathway

The discovery of an approximately 60 kb ST gene cluster was a crowning achievement in the field of mycotoxin genetics.[9] Since then, the genetics and biochemistry of ST and AF synthesis have been largely worked out and are the subjects of several extensive reviews.[10-14] This section will summarize the pathway with particular attention to the enzymes involved in ST production by *A. nidulans*. All of the enzymatic genes required for ST synthesis are found within the 60 kb cluster (Fig. 10.1A). Additionally, there are genes in the cluster for which there is currently no described function.

In the late 1960s, Biollaz et al.[15,16] used ^{14}C-labeled acetate to show that the carbon skeleton of AF (and thus ST) was formed by linking acetate precursor molecules into a polyketide chain. Over a decade later, Townsend et al.[17,18] found that intact ^{13}C-labeled hexanoic acid was also incorporated into the AF/ST carbon skeleton. With these results, Townsend et al. proposed that the early stages of AF/ST biosynthesis involved the use of a hexanoic acid primer synthesized by a distinct fatty acid synthase (FAS). This proposal was supported with genetic evidence from two labs a dozen years later. Mahanti et al.[19] identified an *A. parasiticus* FAS subunit, and Brown et al.[20] identified two *A. nidulans* FAS subunits (StcJ and StcK) that were required for production of AF/ST at early points in the biosynthetic pathway (Fig. 10.2). In fact, hexanoic acid rescues ST production in *A. nidulans stcJ* and *stcK* mutants.[20] In *A. parasiticus*, evidence exists to suggest that the AF polyketide synthase responsible for elongation of the hexanoic acid primer forms a complex with the AF-specific FAS subunits, leading to an efficient shuttling of the starter unit to the polyketide synthase (PKS).[21] By analogy, *A. nidulans* StcJ and StcK may form a complex with StcA,[22] the *A. nidulans* ST PKS.

Completion of the ST carbon skeleton (Fig. 10.2) requires that StcA add 7 acetate subunits (as malonyl-CoA) to the FAS-produced hexanoic acid, forming an

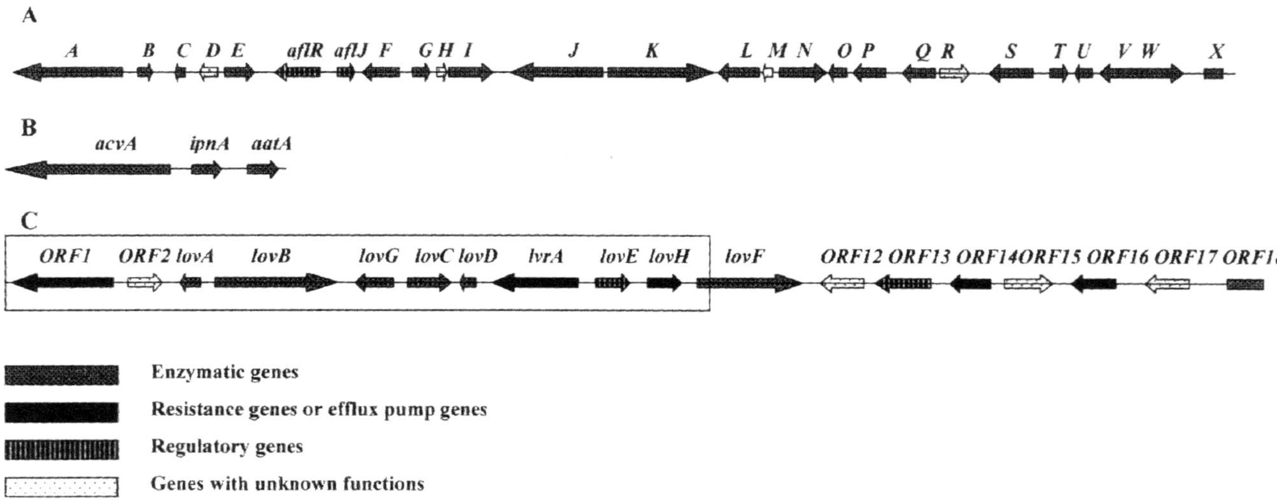

Figure 10.1: Architecture of *Aspergillus* secondary metabolite gene clusters. Shown are the *A. nidulans* Sterigmatocystin (ST) (**A**) and Penicillin (PN) (**B**) gene clusters as well as the *A. terreus* Lov gene cluser (**C**). The rectangle indicates the portion of the Lov cluster that has been heterologously expressed in *A. nidulans*, as discussed in the text.

unstable intermediate called noranthrone. Once created, noranthrone is then oxygenated to the pathway's first stable biosynthetic intermediate, norsolorinic acid, by a mechanism that is not well understood. Three possible mechanisms have been postulated for this step of the pathway, two involving specific enzymes and one involving a spontaneous conversion event.[10,23,24] In *A. nidulans*, the next step of the pathway is characterized by conversion of norsolorinic acid to averantin by a dehydrogenase (StcE).[9,25] Genetic studies suggest that averantin is then oxidized by StcF to 5'-hydroxyaverantin,[26] an intermediate between averantin and averufin.[27] 5'-Hydoxyaverantin is then dehydrogenated by StcG (Sim and Keller, unpublished results) to an open chain form of averufin. The open chain form of averufin is then thought to undergo spontaneous ring closure.[27,28] Averufannin was previously thought to be an AF/ST intermediate occurring between averantin and averufin,[10] however, it is currently believed to be a non-enzymatic by-product of 5'-hydroxyaverantin[28] with an ability to re-enter the AF/ST biosynthetic pathway.[29]

The oxygenation of averufin to versiconal hemiacetal acetate is believed to be a two-step process. Yabe et al.[30] recently used a cell free assay derived from *A. parasiticus* to provide evidence that averufin is first converted to hydroxyversicolorone by an enzyme that is located in microsomes, before conversion to versiconal hemiacetal acetate by a cytosolic enzyme. One of these enzymes may be encoded by *stcO*, as the *A. parasiticus* homolog (*avfA*) was found to complement averufin-accumulating mutants of that species.[31] Additionally, disruption of *stcB* and *stcW*, genes encoding two putative monooxygenases, was shown to cause the accumulation of averufin in *A. nidulans*.[26] Further studies with *A. nidulans stcO*, *stcB*, and *stcW* mutants should determine which of these genes is required for each of the two conversion steps. Next in the pathway is versiconal hemiacetal acetate esterfication to versiconal, followed by ring closure to form versicolorin B (VB). The *A. nidulans* enzymes responsible for these steps are thought to be *stcI* and *stcN*, respectively.[14]

The final steps of the AF/ST biosynthesis pathway (Fig. 10.2) occur soon after the formation of VB. *A. nidulans* StcL dehydrates the bisfuran moiety of VB producing Versicolorin A (VA).[32] Genetic evidence indicates that a ketoreductase (StcU)[33] and a p450 monooxygenase (StcS)[34] convert VA and VB to demethylsterigmatocystin (DMST) and dihydro-demethylsterigmatocystin (DH-DMST), respectively. Disruption of either gene leads to the accumulation of VA and VB.[33,34] The final conversion to ST (or DMST) in *A. nidulans* requires an *O*-methyltransferase, StcP. StcP mutants accumulate DMST and DH-DMST.[35] In *A. flavus* and *A. parasiticus*, at least two more enzymatic steps are required to form AF from ST.

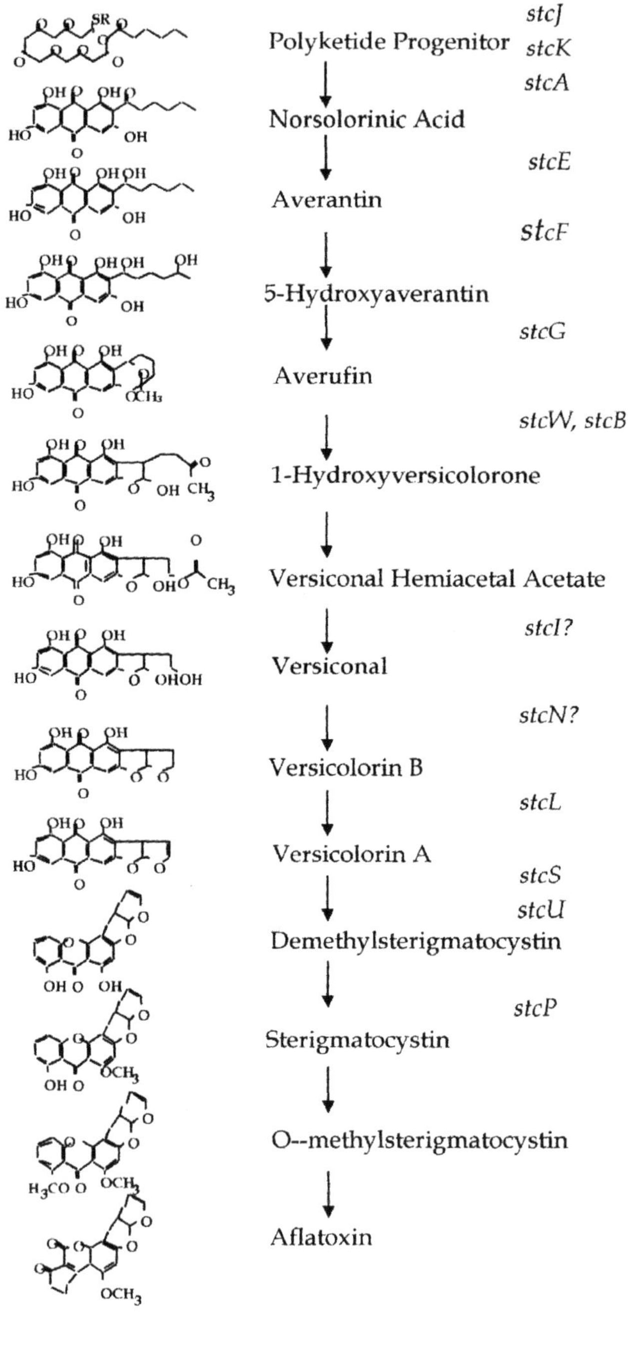
stcJ
stcK
Polyketide Progenitor
stcA
Norsolorinic Acid
stcE
Averantin
stcF
5-Hydroxyaverantin
stcG
Averufin
stcW, stcB
1-Hydroxyversicolorone
Versiconal Hemiacetal Acetate
stcI?
Versiconal
stcN?
Versicolorin B
stcL
Versicolorin A
stcS
stcU
Demethylsterigmatocystin
stcP
Sterigmatocystin
O--methylsterigmatocystin
Aflatoxin

In contrast to the function of the *stc* genes described above, several ST cluster genes have been deleted with no or little effect on ST production. These genes include *stcC,* encoding a putative chloroperoxidase (Hammond and Keller unpublished results), *stcM,* encoding a protein of unknown function (Maggio-Hall and Keller unpublished results), *stcT,* encoding a putative glutathione S-transferase (Zhang and Keller unpublished results), and *stcQ* and *stcV,* encoding putative dehydrogenases (Kelkar, Adams and Keller unpublished results). These mutants have not been studied extensively, and it is possible they may play a role in ST synthesis under conditions not yet examined.

Regulation

Transcription Factors

Within the ST gene cluster lie two genes, *aflR* and *aflJ*, important for the expression of ST and AF enzymatic genes. *aflR* encodes a zinc binuclear transcription factor that positively regulates *stc* genes in the ST cluster and AF genes in the AF cluster.[36,37] Biochemical studies in *A. nidulans* have shown that AflR binds to the palindromic sequence $TCGN_5CGA$,[38] and additional studies in *A. parasiticus* have identified a second binding site TTAGGCCTAA.[39] However this site has not been shown to function in *A. nidulans*. Deletion of *aflR* in all species results in lack of cluster gene expression.[37,40,41]

Adjacent to *aflR* but transcribed in the opposite direction is *aflJ* (Fig. 10.1A). There have been no studies of AflJ function in *A. nidulans,* but experiments using *A. flavus* and *A. parasiticus* have demonstrated a regulatory role for this protein. Currently it appears that AflJ forms a complex with AflR that aids in transcriptional regulation of the AF (and presumably ST) cluster genes.[42]

Further insight into transcriptional regulation of the ST cluster has come from an *A. nidulans* mutant hunt.[25] Complementation of a mutant unable to express *aflR* or *stc* genes – but otherwise near wild-type in appearance – revealed the *laeA* (loss of *aflR* expression) gene. This gene encodes a putative protein methyltransferase

Figure 10.2: The sterigmatocystin and aflatoxin biosynthetic pathway. The structures of the intermediates are on the left, the names of the intermediates in the middle and the *A. nidulans* biosynthetic genes which encode the enzymes required to convert one intermediate to the next precursor are indicated on the right. Gene names followed by a question mark (*i.e., stcN*) indicates that the gene is predicted, not proven, to function at this particular step in the ST pathway.

required for the expression of the ST gene cluster and other secondary metabolite clusters.[43] Analysis of available fungal databases suggests this protein is conserved in all filamentous fungi.

Signal Transduction

Initial work in the analysis of ST regulation was sparked by studies of *A. nidulans* conidiation mutants.[44-47] These studies found two genes, *fadA* and *flbA*, that play important roles in the regulation of asexual reproduction and ST biosynthesis. The protein products of these two genes function in a G-protein signaling cascade, a well conserved mechanism throughout eukaryotes.[48] *fadA* encodes a heterotrimeric G protein α-subunit.[47] Together with β and γ subunits, FadA forms part of a heterotrimeric protein that is believed to be coupled to an unknown membrane bound receptor. According to G-protein dogma, ligand binding to the receptor should cause attachment of a GTP molecule to FadA. In the GTP bound state, $FadA_{GTP}$ dissociates from the receptor and the $\beta\gamma$-subunits. $FadA_{GTP}$ then activates or inactivates specific downstream effectors, remaining active as long as it is in the GTP-bound state. There are two known mechanisms that can account for $FadA_{GTP}$ hydrolysis to $FadA_{GDP}$: intrinsic FadA GTPase activity or extrinsic GTPase activity provided by enzymes known as regulators of G-protein signaling (RGS). *flbA* encodes such an RGS protein.[44,47] While FadA is in the active state, the uncoupled $\beta\gamma$-heterodimer may also activate or deactivate downstream effectors.[49]

Hicks et al.[50] performed an extensive analysis of the roles of FadA and FlbA in ST regulation. They found that *A. nidulans* strains with a *flbA* deletion or constitutive FadA activation (*fadA*^{G42R}) did not produce ST, while FlbA overexpression or FadA constitutive deactivation (*fadA*^{G203R}) strains produced ST earlier than normal. Thus, their data support a strong role for $FadA_{GTP}$ in ST inhibition. Analysis of other *A. nidulans* strains also suggested a role for the uncoupled $\beta\gamma$-heterodimer in suppression of ST production. For example, deletion of *fadA* does not give the precocious ST phenotype observed in the constitutive deactivation mutant, $FadA^{G203R}$. Theoretically, the only difference in these two strains is the state of the $\beta\gamma$-heterodimer. The heterodimer would be permanently bound to $FadA^{G203R}$ in a FadA deactivation strain, but would remain free in a FadA deletion strain. This suggests that uncoupled G-$\beta\gamma$ heterodimer may have an inhibitory role in the regulation of ST production. Support for this hypothesis is also found in studies of the *A. nidulans* Gβ-subunit, *sfaD*. Mutations in *sfaD* restore ST production in a *flbA* loss of function mutant.[51,52]

Protein kinase A (PKA) is a well-characterized signaling protein in eukaryotes[53] whose activity is often influenced by G-proteins. It is composed of a regulatory subunit with cyclic AMP (cAMP) binding sites, and a catalytic subunit with kinase activity. cAMP binds the PKA regulatory subunit releasing the catalytic subunit, which then phosphorylates conserved serine or threonine residues in target proteins. Shimizu et al.[54,55] studied the role of *A. nidulans* PkaA (catalytic subunit) in ST regulation. They found that PkaA has important effects on the ST-specific transcription factor AflR. Overexpression of *pkaA* decreased the amount of *aflR* transcript, resulting in less AflR for the activation of *stc* genes. Recent studies suggest that this transcriptional control is mediated through LaeA.[43] PkaA also has an important role in post-transcriptional regulation of AflR. When *aflR* was heterologously expressed in the *pkaA* overexpression background, PkaA still suppressed production of ST. Further analysis revealed that this post-transcriptional control is due to the presence of three PkaA phosphorylation sites in AflR. These sites appear to affect AflR localization in the cell. When PkaA was over-expressed, a heterologously expressed AflR-GFP fusion was found largely in the cytoplasm. However, mutation of the three putative PkaA phosphorylation sites in AflR resulted in increased nuclear localization of AflR-GFP even in the presence of over-expressed levels of PkaA. Thus, it appears that PkaA modulates AflR activity by controlling its access to the nucleus.[55] A model depicting G protein/PkaA control of ST biosynthesis is presented in Fig. 10.3.

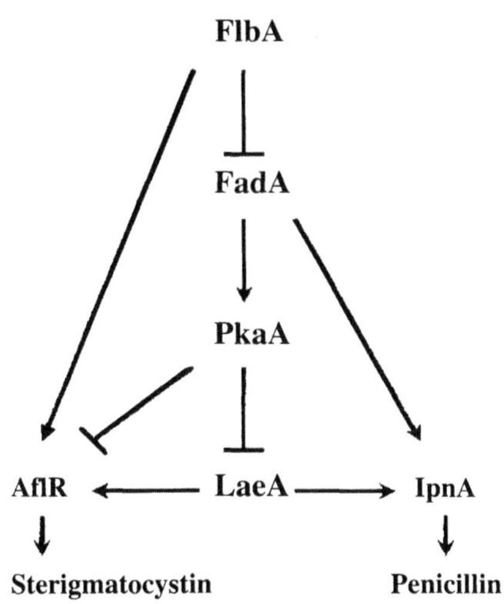

Figure 10.3: Summary of secondary metabolite gene regulation via G protein-mediated signal transduction. The indicated positive and negative regulation (arrows and bars, respectively) has been found to be mediated transcriptionally, post-transcriptionally, or by both mechanisms, as described in the text.

AflR localization studies have also revealed an additional role for the RGS protein FlbA in ST regulation that is independent of its role in deactivating $FadA_{GTP}$. A dominant activating mutation in FadA (*fadA*^{G42R}) inhibits ST production, presumably through activation of PkaA. Overexpression of *aflR* in this background reestablishes ST production. If FlbA were to influence ST production through FadA deactivation only, deletion of *flbA* from the overexpression *aflR*, *fadA*^{G42R} genetic background would be expected to have no effect on ST production (FlbA cannot inactivate FadA^{G42R}). However, deleting FlbA eliminates ST production in this background. Shimuzu et al.[55] suggest that since nuclear localization of AflR does not seem to be influenced by the *flbA* deletion, there must exist an uncharacterized mechanism by which FlbA influences AflR activity while AflR is in the nucleus.

pH

Production of ST and AF is typically higher in acidic pH. This regulation is likely mediated by the well-characterized PacC regulator.[39,56] PacC is active in alkaline pH, and its role in ST and AF production is predicted to be as a negative regulator. Although several putative PacC binding sites have been found in the promoters of several AF/ST genes, the molecular details of pH regulation of this pathway remain unknown. Further discussion of this protein will be covered in the next section.

PENICILLIN

The discovery of penicillin (PN) is arguably one of the greatest achievements of the twentieth century. While *Penicillium chrysogenum* has served as the industrial production strain, significant discoveries of the intricate regulation of PN biosynthesis have been made using *A. nidulans*. This includes regulation by pH, carbon source and amino acids via global and possibly pathway-specific regulatory elements. The biosynthesis of penicillins is the subject of a number of excellent reviews.[57-59] The goal of this section is to highlight the contributions of *A. nidulans* to this endeavor.

Biosynthetic Pathway

PN is synthesized by three enzymes encoded by a cluster of three genes (Fig. 10.1B). The arrangement of PN genes in a cluster was first demonstrated in *A. nidulans*.[60] The first enzyme is a nonribosomal peptide synthetase, ACV synthetase (ACVS), catalyzing the condensation of L-α-aminoadipic acid (L-α-AAA), L-cysteine, and L-valine into a tripeptide, δ-(L-α-aminoadipyl)-L-cysteinyl-D-valine (ACV) (Fig. 10.4).[61] In *A. nidulans*, ACV synthetase is encoded by the intronless,

11.5 kb *acvA* gene.[60] ACV is cyclized to form isopenicillin N by isopenicillin N synthase (IPNS, encoded by *ipnA*).[60,62] The last enzyme, acyl Co-A:isopenicillin N acetyltransferase (IAT, encoded by *aatA*), catalyzes the exchange of the hydrophilic L-α-AAA group with a variety of hydrophobic acyl groups.[57,63] While the identity of the acyl group can be controlled by the addition of exogenous compounds (for example the addition of phenoxyacetic and phenylacetic acid lead to the production of penicillin V and G, respectively), a variety of short chain fatty acids (hexenoic, Δ3-hexenoic and octanoic acid) are typically found in nature.[59] It should be pointed out that the first two steps of the PN pathway are common to other β-lactam biosynthetic pathways of fungi and bacteria (for example cephalosporin and clavulanic acid pathways), and IAT is only found in PN producing organisms.

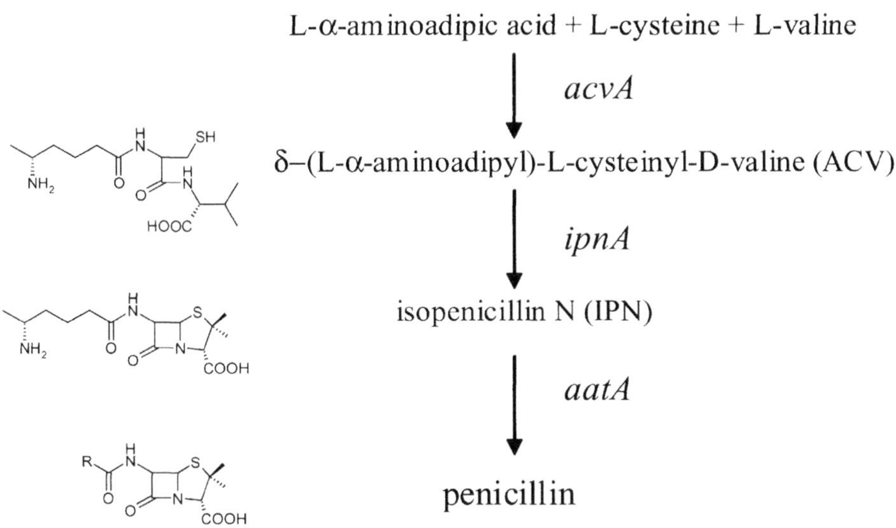

Figure 10.4: The penicillin biosynthetic pathway. AcvA, IpnA, and AatA indicate ACV synthase, IPN synthase, and acyl-CoA : IPN acyltransferase, respectively. R-COOH represents a large variety of aliphatic and aromatic acid side chains, such as phenylacetic (Penicillin G), phenoxyacetic (V), octanoic (K), hexenoic (DF), and Δ3-hexenoic (F) acids.

While it has served primarily as the genetic model in elucidating the structure and regulation of the PN gene cluster, *A. nidulans* has also contributed to our knowledge of the enzymology of PN biosynthesis. Indeed, ACVS was first purified from this organism after unsuccessful attempts with other β-lactam-producing bacteria and fungi.[64] Until this isolation, it was unclear whether the ACV tripeptide was synthesized by a single enzyme or a two-enzyme complex. The crystal structure of the *A. nidulans* IPNS with bound substrate was determined to 1.3Å resolution, providing information on the unique enzymatic mechanism that produces the characteristic 4-membered β-lactam ring.[65] The activation of potential acyl side chains to their CoA thioesters and their subsequent incorporation into penicillins was demonstrated *in vitro* using acetyl-CoA synthetase from *A. nidulans* coupled with IAT.[66]

Regulation

Transcription Factors

While the PN gene cluster appears to lack a pathway-specific transcription factor, a host of global factors contribute to PN regulation. Indeed, the study of PN gene regulation has simultaneously benefited from and contributed to our knowledge of global regulatory mechanisms in *A. nidulans*. In many of these studies, β-galactosidase (encoded by *lacZ*) or β-glucuronidase (encoded by *uidA*) fusions to the three PN biosynthetic genes have been incorporated into strains carrying mutations in known global regulatory pathways.[57] Many environmental factors that had previously been shown to alter PN titers could now be more easily analyzed to see if their effects were mediated by changes in expression of the PN structural genes. These same fusions have been used in mutant hunts to discover new modes of regulation as well.

Despite the common location of the PN biosynthetic genes, each is expressed at different levels.[67,68] The low expression of ACVS makes this the rate-limiting step in the pathway. Overexpression of ACVS using the inducible *alcA* promoter led to a 30-fold increase in PN titer, whereas overexpression of the other two enzymes resulted in only modest gains.[69,70] Deletion analysis of the *ipnA* promoter using an *ipnA*::*lacZ* fusion suggested the existence of multiple negative acting elements and showed that at least one promoter element was located within the coding region of the divergently transcribed *acvA* gene.[71] Significant overlap between the promoter regions of *acvA* and *ipnA* was also found in an analysis of the 872 bp intergenic region between these genes using *lacZ* and *uidA* fusions to *ipnA* and *acvA*, respectively.[71] This design enabled the simultaneous monitoring of the expression of both genes, and revealed the existence of a *cis*-acting element that is now known to be regulated by the CCAAT-binding AnCF complex.[68,72] AnCF was found to positively regulate *ipnA* and *aatA*. This eukaryotic multimeric transcription factor complex may regulate as many as 200 genes in *A. nidulans*.[73] Due to the wide range of expression conditions of the various genes that have been identified, additional transcription factors may modulate AnCF regulation. Indeed, a novel transcription factor has been found to bind a site that overlaps the CCAAT box of *aatA* and, therefore, functions as a repressor.[74] While deletion of the CCAAT box from the *acvA/ipnA* intergenic region led to reduced expression of *ipnA::lacZ*, it increased expression of *acvA::uidA* by 4-fold.[72] However, *acvA* expression was not increased in an AnCF mutant strain (Δ*hapC*), suggesting the existence of an additional, negative regulator that binds to an overlapping site here as well.[75]

The G-protein signaling pathway known to regulate AF/ST production (Fig. 10.3) also appears to be involved in the regulation of PN biosynthesis. Increased levels of PN and *ipnA* mRNA were detected in a strain carrying the dominant activating *fadA* allele,[76] although this same allele suppressed ST formation and asexual development. In contrast to ST regulation, PKA does not appear to have a major effect on PN synthesis[77] and, thus, FadA-downstream factors responsible for transduction of this signal specifically to the PN pathway have not been identified. However, LaeA, an apparent global regulator of secondary metabolism (described above), has been found to be necessary for PN gene expression.[43]

The *lacZ* reporter fusion has also been used in mutant hunts to discover additional *trans*-acting regulatory factors. The recovery of *cis*-acting mutants was reduced by using strains carrying a duplication of the fusion. *Cis*-acting mutations would require two independent mutation events to be detected, a statistically unlikely event. In one such hunt, mutation of the *npeE* gene of Chromosome IV reduced PN production by 10-fold and nearly eliminated expression of twin *ipnA::lacZ* fusions located adjacent to the *argB* gene.[78] Another gene isolated by this method was *suAprgA1* (a suppressor of *prgA1*).[79-80] Deletion of the gene led to a 50% reduction in *ipnA::lacZ* expression and a reduction in PN titer to about 60% of wild type.[79] The encoded protein is homologous to others found in humans (p32), *Saccharomyces cerevisiae* (Mam33p) and *Trypanosoma brucei* (p22) and is, therefore, not likely to be a PN pathway-specific regulator.[79] Although Mam33p and p22 are mitochondrial matrix proteins,[81,82] the human homolog (p32) appears to be a substrate for MAP kinase in the cytoplasm and can be translocated to the nucleus.[83]

Carbon Source

As in the commercial PN strains, glucose and sucrose were found to repress PN production in *A. nidulans*. Compared to lactose-grown cultures, *ipnA::lacZ* expression was reduced with these two carbon sources.[67,84] The glucose-dependent repression of many genes in primary metabolism is mediated by the well-characterized carbon catabolite repression (*cre*) system.[85,86] While CreA mediates repression directly through binding of promoter DNA, CreB and CreC seem to have an indirect role contributing to repression. Surprisingly, PN titers were still subject to glucose repression in *creA*, *creB*, and *creC* mutant strains.[67,84,87] Only the most extreme, morphologically debilitating *creA* alleles led to somewhat derepressed levels of *ipnA*::*lacZ* expression, and deletion of the putative CreA binding site from the carbon responsive region of the *ipnA* promoter did not alleviate glucose-based repression of an *ipnA*::*lacZ* fusion.[84,88] Acetate mediates the repression of some *cre* regulated genes, but was found to increase transcription of *ipnA* and PN production.[88] Therefore, the data do not support a role for *cre* in carbon source regulation of *ipnA*. Interestingly, expression of *acvA* and *aatA* fusions showed little or no change with repressing carbon sources, although the specific activity of IAT was found to be reduced.[67,68] The mechanism of this apparent post-transcriptional regulation is unknown.

pH

The highest titers of PN are achieved in alkaline pH,[88,89] and expression of both *acvA* and *ipnA* is controlled by ambient pH via the PacC regulator.[88,90-92] PacC is active under alkaline conditions, and, therefore, acts as a positive regulator in this system. Hence, a constitutively active version of PacC (PacC5) increased expression of both *acvA* and *ipnA* promoter fusions.[88,90] Three functional PacC-binding sites were identified within the *ipnA* promoter, and all three were required for maximal *ipnA* expression under alkaline conditions.[92] In *A. nidulans*, alkaline pH can even override the repressing affects of glucose and sucrose.[88] It has been noted that the pH of cultures grown with repressing carbon sources is lower than that found with derepressing ones.[88] Is the carbon source regulation observed then just a pH effect? PacC binding sites in the *ipnA* promoter do not overlap with the carbon regulatory element.[91] Also, acidic pH does not prevent derepression by lactose.[88] Therefore, carbon regulation appears to be a real, pH-independent entity that remains to be characterized.

Amino Acids

Since the first step of PN biosynthesis is the formation of a tripeptide from two amino acids (L-cys and L-val) and an amino acid precursor (L-a-AAA), the effects of amino acid supplementation on PN titer and on *acvA* and *ipnA* expression were studied.[90] Of the twenty naturally-occurring amino acids, only serine and arginine had no effect on either of the *acvA* or *ipnA* reporter gene fusions. A subset of amino acids that repressed the expression of both genes (Met, Val, His, and Lys) was tested further by promoter deletion analysis. The repression by Val and His was dependent on promoter regions overlapping with the PacC regulatory elements. Accordingly, repression by these amino acids was abolished in a strain carrying a constitutively active PacC. Lys and Met, however, exhibited pH-independent regulation. Met regulation remains uncharacterized. Lys had previously been shown to repress PN gene expression and titer.[93] The PN precursor L-α-AAA is an intermediate of Lys biosynthesis in fungi.[94] Logically, one would expect Lys to feedback regulate its own biosynthetic pathway, thereby restricting the pool of L-α-AAA available to PN biosynthesis. However, the transcriptional effect on the PN structural genes shows that Lys does more than just control the availability of substrate to the pathway. Lysine biosynthesis is subject to a cross-pathway control (CPC) mechanism that responds to general amino acid starvation.[95] Overexpression of this system led to repression of PN gene expression (*ipnA* and *acvA*) and reduced PN production. The mechanism for PN regulation by CPC is likely indirect. Overexpression of cross-pathway control would increase the level of lysine in the cell while reducing L-α-AAA (Lys biosynthetic steps above the branch point were not as responsive to CPC as those below). While the molecular mechanism for Lys regulation of PN genes remains unknown, it is clear from this study that general amino acid starvation would direct L-α-AAA away from PN biosynthesis and back to Lys biosynthesis.

LOVASTATIN

Biosynthesis

Lovastatin and its chemical derivatives, known more commonly by such trade names as Mevacor (Merck), Pravachol (Bristol-Myers Squibb), Zocor (Merck), Lipitor (Parke-Davis), Lescol (Novartis), and Baycol (Bayer), are powerful cholesterol reducing drugs, generating some $11 billion in U.S. sales annually.[96] Lovastatin is a fungal polyketide produced by *A. terreus*.[97] Moving individual genes and/or portions of the lovastatin gene cluster into the non-producer *A. nidulans* has helped elucidate the biosynthetic pathway, particularly in establishing a role for accessory polypetides in polyketide synthesis (Fig. 10.5). Polyketide synthesis proceeds much like fatty acid synthesis, but with differing degrees of reduction possible with each condensation step. Iterative polyketide synthases (PKSs), such as the lovastatin nonaketide synthase, use the same active site repeatedly but are somehow able to achieve different levels of reduction/hydration at each cycle.[98] By moving portions of the lovastatin gene cluster into *A. nidulans*, Kennedy et al.[3] were able to show LovC was necessary to modulate the LovB PKS to synthesize the precursor dihydromonocolin L. Expression of *lovB* alone lead to the accumulation of abortive polyketide structures that suggested a necessary enoyl reductase (ER) activity was missing. Addition of *lovC*, a lovastatin cluster gene with homology to ER domains of PKSs, to the *lovB* strain resulted in synthesis of the completed polyketide, dihydromonocolin L. LovB and LovC are likely to be in physical contact, as they have been found to copurify with each other from *A. nidulans*.[99]

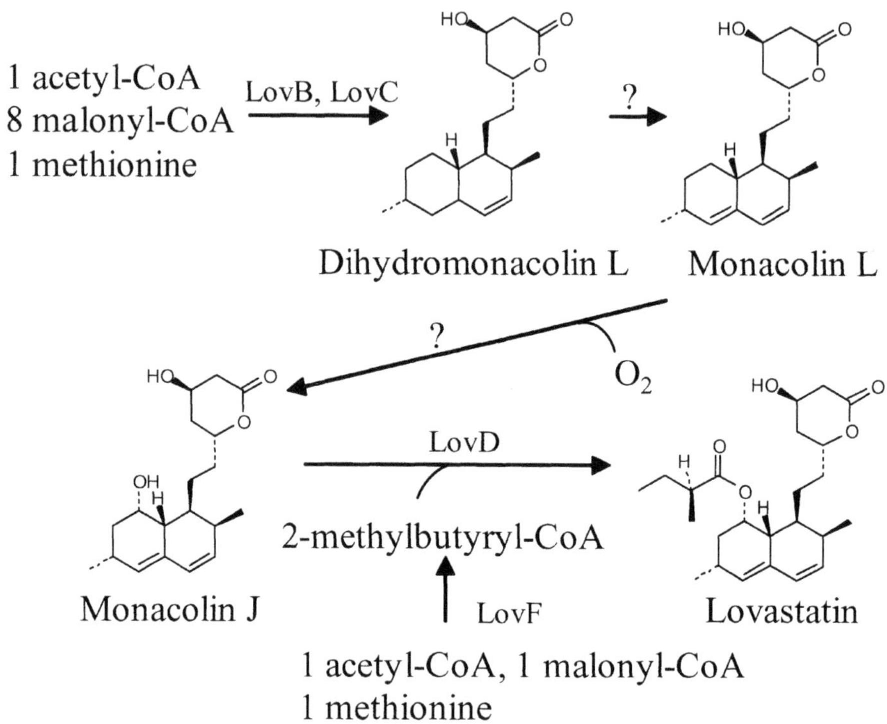

Figure 10.5: Lovastatin biosynthetic pathway. LovB and LovC, a nonaketide synthase and enoyl reductase, respectively, generate the first stable intermediate, dihydromonacolin L. Monacolin L and Monacolin J are recognized pathway intermediates, however enzymes that catalyze these conversions have not been identified. LovF is a diketide synthase generating 2-methylbutyryl-CoA. This molecule is esterified to Monacolin J by LovD, generating Lovastatin.

Indeed, unequal (heterologous) expression of the two proteins may account for the accumulation of small amounts of abortive polyketide structures observed in the engineered *A. nidulans* strain.[100]

Regulation

Lovastatin production in *A. terreus* is under the control of the Zn_2Cys_6 transcription factor, LovE.[3] It is likely that LovE also regulates the lov gene cluster when the cluster is introduced into *A. nidulans*. Bok and Keller have also shown that the Lov gene cluster is regulated by LaeA in *A. nidulans* as well as *A. terreus*, thus suggesting a conserved role of LaeA function in secondary metabolite gene regulation.[43]

SUMMARY AND FUTURE STUDIES

In this review, we have summarized studies illustrating the strides that have been made in understanding secondary metabolism using *A. nidulans* as a model system. This organism produces many natural products including ST and PN and has been used as a heterologous host to study the biosynthesis of other natural products including lovastatin. Critical advances in our understanding of fungal secondary metabolism include the discovery of ST and PN biosynthetic gene clusters and the discovery of a G-protein/cAMP/protein kinase A mediated growth pathway in *A. nidulans* regulating secondary metabolism production. This later pathway coordinates both secondary metabolism and asexual development, similar in spirit, but certainly not in mechanism, to the γ-butyrolactone signaling systems that have been found to simultaneously regulate secondary metabolism and morphological differentiation in bacteria.[101] The interwoven coregulation of these two processes may be unraveled through our discovery of LaeA, which plays no major role in development (Bok and Keller, unpublished results). The molecular details of LaeA regulation, found only in secondary metabolite-producing fungi, is the subject of ongoing work in our lab. Where else will the future take this unique fungal model system?

Another aspect of eukaryotic (fungal or plant) secondary metabolism that differs distinctively from that of bacterial secondary metabolism is the compartmentalization of biosynthetic precursors into various organelles. For example, the final step of PN biosynthesis (catalyzed by IAT) occurs in the peroxisome.[102] Thus, naturally occurring PN side chains must be generated in or, like exogenously provided side chains, be transported into this organelle. The amino acid substrates of PN biosynthesis are sequestered in vacuoles, although ACVS is believed to be cytoplasmic. The synthesis of polyketides, including ST, draws carbon from the heart of primary metabolism (acetyl-CoA). The acetyl-CoA pool is deliberately divided between the cytoplasm, mitochondria, and peroxisomes to strike the proper balance between energy generation and requisite biosynthetic capabilities (*i.e.*, gluconeogenesis, fatty acid synthesis). How polyketide secondary pathways fit into this network is not yet appreciated. Having only certain subpools of precursor molecules available for secondary metabolic processes could represent an import level of regulation. Knowing which pool of a given metabolite is supplying a secondary pathway could give us insights into how pools are coordinated and could create new opportunities for metabolic engineering. We expect future work on ST and PN biosynthesis in *A. nidulans* to elaborate more on this important interface between primary and secondary metabolism.

The genome sequence of *A. nidulans* has recently been completed (http://www-genome.wi.mit.edu/annotation/fungi/aspergillus/index.html) and will be a valuable tool for discovery in all aspects of the physiology of this fungus, including secondary metabolism. Genes required for a given secondary metabolic pathway are invariably clustered in the genome. This is in contrast to other types of genes and likely reflects the importance of horizontal transfer in acquiring these pathways. Thus, genes of secondary metabolic pathways can be predicted just as they are in bacterial genomes. Genes neighboring a PKS- or NRPS-encoding gene are likely required for the same pathway and can be analyzed for coregulation or, if the product of the pathway is known, by deletion analysis. Preliminary BLAST searches of the *A. nidulans* genome sequence suggest the existence of at least two-dozen polyketide pathways and about a dozen non-ribosomal peptide pathways. Despite this diversity only five of these compounds have been identified: ST, PN, the iron chelator ferricrocin,[103] and the polyketides responsible for sexual and asexual spore pigmentation.[104,105] A systematic approach could now be taken to delete putative secondary pathway genes and look for alterations in basic physiology and in the production of extractable compounds. The impact of the deletion or overexpression of identified global regulators or individual pathways on the expression of all of the putative secondary pathways could now be assessed with genome-wide transcriptional profiling. More than fifty years after Guido Pontecorvo and coworkers first championed the use of *A. nidulans* as a genetic model,[106] the completed genome sequence has us primed for the next fifty years.

ACKNOWLEDGMENTS

This work was supported by NIH F32 AI052654 to L.A. M.-H. and NSF MCB-0196233 to N.P.K.

REFERENCES

1. SAMPSON, R.A., Current Taxonomic Schemes of the Genus *Aspergillus* and its Teleomophs. Butterworth-Heinemann, Boston, 1992, pp. 355-390.

2. KLICH, M.A., TIFFANY, L. H., KNAPHUS, G., Ecology of the Aspergilli of Soils and Liter. Butterworth-Heinemann, Boston, 1992, pp. 329-353.
3. KENNEDY, J., AUCLAIR, K., KENDREW, S.G., PARK, C., VEDERAS, J.C., HUTCHINSON, C.R., Modulation of polyketide synthase activity by accessory proteins during lovastatin biosynthesis, *Science,* 1999, **284**, 1368-1372.
4. KIRIMURA, K., YODA, M., SHIMIZU, H., SUGANO, S., MIZUNO, M., KINO, K., USAMI, S., Contribution of cyanide-insensitive respiratory pathway, catalyzed by the alternative oxidase, to citric acid production in *Aspergillus niger, Biosci. Biotechnol. Biochem.,* 2000, **64**, 2034-2039.
5. SWIFT, R.J., KARANDIKAR, A., GRIFFEN, A.M., PUNT, P.J., VAN DEN HONDEL, C.A., ROBSON, G.D., TRINCI, A.P., WIEBE, M.G., The Effect of organic nitrogen sources on recombinant glucoamylase production by *Aspergillus niger* in chemostat culture, *Fungal Genet Biol.*, 2000, **31**, 125-133.
6. O'TOOLE, D.K., Characteristics and use of okara, the soybean residue from soy milk production—a review, *J. Agric. Food Chem.,* 1999, **47**, 363-371.
7. CAST, Mycotoxins: Risks in Plant, Animal, and Human Systems, Council for Agricultural Science and Technology (CAST), Ames, IA., 2003.
8. JELINEK, C.F., POHLAND, A.E., WOOD, G.E., Review of mycotoxin contamination: worldwide occurrence of mycotoxins in foods and feeds-an update, *J. Assoc. Off. Anal. Chem.,* 1989, **72**, 223-230.
9. BROWN, D.W., YU, J.-H., KELKER, H.S., FERNANDES, M., NESBITT, T.C., KELLER, N.P., ADAMS, T.H., LEONARD, T.J., Twenty-five coregulated transcripts define a sterigmatocystin gene cluster in *Aspergillus nidulans*, *Proc. Natl. Acad. Sci. USA,* 1996, **93**, 1418-1422.
10. BHATNAGAR, D., ERHLICH, K.C., CLEVELAND, T.E., Oxidation reduction reactions in biosynthesis of secondary metabolites, *in*: Handbook of Applied Mycology: Mycotoxins in Ecological Systems (D. Bhatnagar, E.B. Lillehoj, and D.K. Arora, eds,), Marcel Dekker, New York. 1992, pp. 255-286
11. MINTO, R.E., TOWNSEND, C.A., Enzymology and molecular biology of aflatoxin biosynthesis, *Chem. Rev.*, 1997, **97**, 2537-2556.
12. PAYNE, G.A., BROWN, M.P., Genetics and physiology of aflatoxin biosynthesis, *Annu. Rev. Phytopathol.,* 1998, **36**, 329-362.
13. YU, J., BHATNAGAR, D., EHRLICH, K.C., Aflatoxin biosynthesis, *Rev. Iberoam Micol.*, 2002, **19**, 191-200.
14. HICKS, J.K., SHIMIZU, K., KELLER, N.P., Genetics and biosynthesis of aflatoxins and sterigmatocystin, *in*: The Mycota XI, (Kempken, ed.), Springer-Verlag, Berlin. 2002, pp. 55-69.
15. BIOLLAZ, M., BUCHI, G., MILNE, G., Biosynthesis of aflatoxins, *J. Am. Chem Soc.*, 1968, **90**, 5017-5019.
16. BIOLLAZ, M., BUCHI, G., MILNE, G., The biosynthesis of the aflatoxins, *J. Am. Chem Soc.*, 1970, **92**, 1035-1042.
17. TOWNSEND, C.A., CHRISTENSEN, S.B., Stable isotope studies of anthraquinone intermediates in the aflatoxin pathway, *Tetrahedron,* 1983, **39**, 3575-3582.

18. TOWNSEND, C.A., CHRISTENSEN, S.B., TRAUTWEIN, K., Hexanoate as a starter unit in polyketide biosynthesis, *J. Am. Chem. Soc.*, 1984, **106**, 3868-3869.
19. MAHANTI, N., Structure and function of fas-1A, a gene encoding a putative fatty acid synthetase directly involved in aflatoxin biosynthesis in *Aspergillus parasiticus*, *Appl. Environ. Microbiol.*, 1996, **62**, 191-195.
20. BROWN, D.W., ADAMS, T.H., KELLER, N.P., *Aspergillus* has distinct fatty acid synthases for primary and secondary metabolism, *Proc. Natl. Acad. Sci. USA,* 1996, **93**, 14873-14877.
21. WATANABE, C.M.H., WILSON, D., LINZ, J.E., TOWNSEND, C.A., Demonstration of the catalytic roles and evidence for the physical association of typeI fatty acid synthase and a polyketide synthase in the biosynthesis of aflatoxin B1, *Chem. Biol.*, 1996, **3**, 463-469.
22. YU, J.-H., LEONARD, T.J., Sterigmatocystin biosynthesis in *Aspergillus nidulans* requires a novel type I polyketide synthase, *J. Bacteriol.*, 1995, **177**, 4792-4800.
23. VEDERAS, J.C., NAKASHIMA, T.T., Biosynthesis of averufin by *Aspergillus parasiticus*: detection of ^{18}O-label by ^{13}C NMR isotope shifts, *J. Chem. Soc. Chem. Commun.*, 1980, **4**, 183-185.
24. DUTTON, M.F., Enzymes and aflatoxin biosynthesis, *Microbiol Rev.*, 1988, **52**, 274-295.
25. BUTCHKO, R.A.E., ADAMS, T.H., KELLER, N.P., *Aspergillus nidulans* mutants defective in *stc* gene cluster regulation, *Genetics,* 1999, **153**, 715-720.
26. KELLER, N.P., WATANABE, C.M.H., KELKAR, H.S., ADAMS, T.H., TOWNSEND, C.A., Requirement of monooxygenase-mediated steps for sterigmatocystin biosynthesis by *Aspergillus nidulans*, *Appl. Environ. Microbiol.*, 2000, **66**, 359-362.
27. YABE, K., NAKAMURA, Y., NAKAJIMA, H., ANDO, Y., HAMASAKI, T., Enzymatic conversion of norsolorinic acid to averufin in aflatoxin biosynthesis, *Appl. Environ. Microbiol.*, 1991, **57**, 1340-1345.
28. CHANG, P.-K., YU, J., EHRLICH, K.C., BOUE, S.M., MONTALBANO, B.G., BHATNAGAR, D., CLEVELAND, T.E., *adhA* in *Aspergillus parasiticus* is involved in conversion of 5'-hydroxyaverantin to averufin, *Appl. Environ. Microbiol.*, 2000, **66**, 4715-4719.
29. MCCORMICK, S.P., BHATNAGAR, D., LEE, L.S., Averufanin is an aflatoxin B_1 precursor between averantin and averufin in the biosynthetic pathway, *Appl. Environ. Microbiol.*, 1987, **53**, 14-16.
30. YABE, K., CHIHAYA, N., HAMAMATSU, S., SAKUNO, E., HAMASAKI, T., NAKAJIMA, H., BENNETT, J.W., Enzymatic conversion of averufin to hydroxyversicolorone and elucidation of a novel metabolic grid involved in aflatoxin biosynthesis, *Appl. Environ. Microbiol.*, 2003, **69**, 66-73.
31. YU, J., WOLOSHUK, C.P., BHATNAGAR, D., CLEVELAND, T.E., Cloning and characterization of avfA and omtB genes involved in aflatoxin biosynthesis in three *Aspergillus* species, *Gene,* 2000, **248**, 157-167.
32. KELKAR, H.S., SKLOSS, T.W., HAW, J.F., KELLER, N.P., ADAMS, T.H., *Aspergillus nidulans stcL* encodes a putative cytochrome P-450 monooxygenase

required for bisfuran desaturation during aflatoxin/sterigmatocytin biosynthesis, *J. Biol. Chem.*, 1997, **272**, 1589-1594.
33. KELLER, N.P., KANTZ, N.J., ADAMS, T.A., *Aspergillus nidulans ver*A is required for production of the mycotoxin sterigmatocystin, *Appl. Environ. Microbiol.*, 1994, **60**, 1444-1450.
34. KELLER, N.P., SEGNER, S., BHATNAGAR, D., ADAMS, T.H., *stcS*, a putative P-450 monooxygenase, is required for the conversion of versicolorin A to sterigmatocystin in *Aspergillus nidulans*, *Appl. Environ. Microbiol.*, 1995, **61**, 3628-3632.
35. KELKAR, H.S., KELLER, N.P., ADAMS, T.H., *Aspergillus nidulans stcP* encodes an O-methyltransferase that is required for sterigmatocystin biosynthesis, *Appl. Environ. Microbiol.,* 1996, **62**, 4296-4298.
36. WOLOSHUK, C.P., FOUNT, K.R., BREWER, J.F., BHATNAGAR, D., CLEVELAND, T.E., PAYNE, G.A., Molecular characterization of *aflR*, a regulatory locus for aflatoxin biosynthesis, *Appl. Environ. Microbiol.*, 1994, **60**, 2408-2414.
37. YU, J.H., BUTCHKO, A.E., FERNANDES, M., KELLER, N.P., LEONARD, T.J., ADAMS, T.H., Conservation of structure and function of the aflatoxin regulatory gene *aflR* from *Aspergillus nidulans* and *A. flavus*, *Curr. Genet.*, 1996, **29**, 549-555.
38. FERNANDES, M., KELLER, N.P., ADAMS, T.H., Sequence-specific binding by *Aspergillus nidulans* AflR, a C_6 zinc cluster protein regulating mycotoxin biosynthesis, *Mol. Microbiol.*, 1998, **28**, 1355-1365.
39. EHRLICH, K.C., CARY, J.W., MONTALBANO, B.G., Characterization of the promoter for the gene encoding the aflatoxin biosynthetic pathway regulatory protein AflR, *Biochim. Biophys. Acta,* 1999, **1444**, 412-417.
40. CHANG, P.-K., CARY, J.W., BHATNAGAR, D., CLEVELAND, T.E., BENNETT, J.W., LINZ, J.E., WOLOSHUK, C.P., PAYNE, G.A., Cloning of the *Aspergillus parasiticus apa-2* gene associated with the regulation of aflatoxin biosynthesis, *Appl. Environ. Microbiol.,* 1993, **59**, 3273-3279.
41. PAYNE, G.A., NYSTROM, G.J., BHATNAGAR, D., CLEVELAND, T.E., WOLOSHUK, C.P., Cloning of the *afl-2* gene involved in aflatoxin biosynthesis from *Aspergillus flavus*, *Appl. Environ. Microbiol.*, 1993, **59**, 156-162.
42. CHANG, P.K., The *Aspergillus parasiticus* protein AFLJ interacts with the aflatoxin pathway-specific regulator AFLR, *Mol. Genet. Genomics*, 2003, **268**, 711-719.
43. BOK, J.W., KELLER, N.P., LaeA, a regulator of secondary metabolism in *Aspergillus*, *Eukary. Cell*, 2004, in press.
44. LEE, B.N., ADAMS, T.H., Overexpression of *flbA,* an early regulator of *Aspergillus* asexual sporulation leads to activation of *brlA* and premature initiation of development, *Mol. Microbiol.*, 1994, **14**, 323-334.
45. LEE, B.N., ADAMS, T.H., The *Aspergillus nidulans fluG* gene is required for production of an extracellular developmental signal, *Genes Dev.*, 1994, **8**, 641-651.
46. WIESER, J., ADAMS, T.H., *flbD* encodes a myb-like DNA binding protein that regulates initiation of *Aspergillus nidulans* conidiophore development, *Genes Dev.*, 1995, **9**, 491-502.

47. YU, J., WIESER, J., ADAMS, T.H., The *Aspergillus* FlbA RGS domain protein antagonizes G-protein signaling to block proliferation and allow development, *EMBO J.,* 1996, **15**, 5184-5190.
48. DOHLMAN, H.G., THORNER, J.W., Regulation of G protein-initiated signal transduction in yeast: paradigms and principles, *Annu. Rev. Biochem.*, 2001, **70**, 703-754.
49. VANDERBELD, B., KELLY, G.M., New thoughts on the role of the beta-gamma subunit in G-protein signal transduction, *Biochem. Cell Biol.*, 2000, **78**, 537-550.
50. HICKS, J.K., YU, J.-H., KELLER, N.P., ADAMS, T.H., *Aspergillus* sporulation and mycotoxin production both require inactivation of the FadA Ga protein-dependent signaling pathway, *EMBO J.*, 1997, **16**, 4916-4923.
51. YU, J.-H., ROSEN, S., ADAMS, T.H., Extragenic suppressors of loss-of-function mutations in the *Aspergillus* FlbA regulator of G-protein signaling domain protein, *Genetics,* 1999, **151**, 97-105.
52. ROSÉN, S., YU, J.-H., ADAMS, T.H., The *Aspergillus nidulans sfaD* gene encodes a G protein β subunit that is required for normal growth and repression of sporulation, *EMBO J.*, 1999, **18**, 5592-5600.
53. THEVELEIN, J.M., DE WINDE, J.H., Novel sensing mechanisms and targets for the cAMP-protein kinase A pathway in the yeast *Saccharomyces cerevisiae*, *Mol. Microbiol.,* 1999, **33**, 904-918.
54. SHIMIZU, K., KELLER, N.P., Genetic involvement of a cAMP-dependent protein kinase in a G protein signaling pathway regulating morphological and chemical transitions in *Aspergillus nidulans*, *Genetics,* 2001, **157**, 591-600.
55. SHIMIZU, K., HICKS, J., HUANG, T.-P., KELLER, N.P., Pka, Ras and RGS protein interactions regulate activity of AflR, a Zn(II)2Cys6 transcription factor in *Aspergillus nidulans*, *Genetics,* 2003, in press.
56. KELLER, N.P., NESBITT, C., SARR, B., PHILLIPS, T.D., BUROW, G.B., pH regulation of sterigmatocystin and aflatoxin biosynthesis in *Aspergillus* spp., *Phytopathology,* 1997, **87**, 643-648.
57. BRAKHAGE, A.A., Molecular regulation of β-lactam biosynthesis in filamentous fungi, *Microbiol. Mol. Biol. Rev,* 1998, **62**, 547-585.
58. MARTIN, J.F., Molecular control of expression of penicillin biosynthesis genes in fungi: Regulatory proteins interact with a bidirectional promoter region, *J. Bacteriol.*, 2000, **182**, 2355-2362.
59. LUENGO, J.M., PEÑALVA, M.A., Penicillin biosynthesis, *Prog. Ind. Microbiol.*, 1994, **29**, 603-638.
60. MACCABE, A.P., RIACH, M.B.R., UNKLES, S.E., KINGHORN, J.R., The *Aspergillus nidulans npeA* locus consists of three contiguous genes required for penicillin biosynthesis, *EMBO J.*, 1990, **9**, 279-287.
61. KLEINKAUF, H., VON DÖHREN, H., A nonribosomal system of peptide biosynthesis, *Eur. J. Biochem.*, 1996, **236**, 335-351.
62. RAMOS, F.R., LÓPEZ-NIETO, M.J., MARTIN, J.F., Isopenicillin N synthetase of *Penicillium chrysogenum*, an enzyme that converts δ-(L-α-aminoadipyl)-L-cysteinyl-D-valine to isopenicillin N, *Antimicrob. Agents Chemother.*, 1985, **27**, 380-387.

63. WHITEMAN, P.A., ABRAHAM, E.P., BALDWIN, J.E., FLEMING, M.D., SCHOFIELD, C.J., SUTHERLAND, J.D., WILLIS, A.C., Acyl coenzyme A:6-aminopenicillanic acid acyltransferase from *Penicillium chrysogenum* and *Aspergillus nidulans*, *FEBS Lett.*, 1990, **262**, 342-344.
64. VAN LIEMPT, H., VON DOHREN, H., KLEINKAUF, H., δ-(L-α-Aminoadipyl)-L-cysteinyl-D-valine synthetase from *Aspergillus nidulans*, *J. Biol. Chem.*, 1989, **264**, 3680-3684.
65. ROACH, P.L., CLIFTON, I.J., HENSGENS, C.M.H., SHIBATA, N., SCHOFIELD, C.J., HAJDU, J., BALDWIN, J.E., Structure of isopenicillin N synthase complexed with substrate and the mechanism of penicillin formation, *Nature*, 1997, **387**, 827-830.
66. MARTINEZ-BLANCO, H., REGLERO, A., FERNÁNDEZ-VALVERDE, M., FERRERO, M.A., MORENO, M.A., PEÑALVA, M.A., LUENGO, J.M., Isolation and characterization of the acetyl-CoA synthetase from *Penicillium chrysogenum*: Involvement of this enzyme in the biosynthesis of penicillins, *J. Biol. Chem.*, 1992, **267**, 5474-5481.
67. BRAKHAGE, A.A., BROWNE, P., TURNER, G., Regulation of *Aspergillus nidulans* penicillin biosynthesis and penicillin biosynthesis genes *acvA* and *ipnA* by glucose, *J. Bacteriol.*, 1992, **174**, 3789-3799.
68. LITZKA, O., THEN BERGH, K., BRAKHAGE, A.A., Analysis of the regulation of the *Aspergillus nidulans* penicillin biosynthesis gene *aat* (*penDE*), which encodes acyl coenzyme A:6-aminopenicillanic acid acyltransferase, *Mol. Gen. Genet.*, 1995, **249**, 557-569.
69. KENNEDY, J., TURNER, G., δ-(L-α-Aminoadipyl)-L-cysteinyl-D-valine synthetase is a rate limiting enzyme for penicillin production in *Aspergillus nidulans*, *Mol. Gen. Genet.*, 1996, **253**, 189-197.
70. FERNÁNDEZ-CAÑÓN, J.M., PEÑALVA, M.A., Overexpression of two penicillin structural genes in *Aspergillus nidulans*, *Mol. Gen. Genet.*, 1995, **246**, 110-118.
71. PÉREZ-ESTEBAN, B., OREJAS, M., GÓMEZ-PARDO, E., PEÑALVA, M.A., Molecular characterization of a fungal secondary metabolism promoter: Transcription of the *Aspergillus nidulans* isopenicillin N synthetase gene is modulated by upstream negative elements, *Mol. Microbiol.*, 1993, **9**, 881-895.
72. THEN BERGH, K., LITZKA, O., BRAKHAGE, A.A., Identification of a major *cis*-acting DNA element controlling the biodirectionally transcribed penicillin biosynthesis genes *acvA* (*pcbAB*) and *ipnA* (*pcbC*) of *Aspergillus nidulans*, *J. Bacteriol.*, 1996, **178**, 3908-3916.
73. BRAKHAGE, A.A., ANDRIANOPOULOS, A., KATO, M., STEIDL, S., DAVIS, M.A., TSUKAGOSHI, N., HYNES, M.J., HAP-like CCAAT-binding complexes in filamentous fungi: Implications for biotechnology, *Fungal Genet. Biol.*, 1999, **27**, 243-252.
74. CARUSO, M.L., LITZKA, O., MARTIC, G., LOTTSPEICH, F., BRAKHAGE, A.A., Novel basic-region helix-loop-helix transcription factor (AnBH1) of *Aspergillus nidulans* counteracts the CCAAT-binding complex AnCF in the promoter of a penicillin biosynthesis gene, *J. Mol. Biol.*, 2002, **323**, 425-439.

75. LITZKA, O., PAPAGIANNOPOLOUS, P., DAVIS, M.A., HYNES, M.J., BRAKHAGE, A.A., The penicillin regulator PENR1 of *Aspergillus nidulans* is a HAP-like transcriptional complex, *Eur. J. Biochem.*, 1998, **251**, 758-767.
76. TAG, A., HICKS, J., GARIFULLINA, G., AKE JR., C., PHILLIPS, T.D., BEREMAND, M., KELLER, N., G-protein signaling mediates differential production of toxic secondary metabolites, *Mol. Microbiol.*, 2000, **38**, 658-665.
77. MCDONALD, T., DEVI, T., SHIMIZU, K., SIM, S.-C., KELLER, N. P., Signaling events connecting mycotoxin biosynthesis and sporulation in *Aspergillus* and *Fusarium spp.*, 2004, in press.
78. PÉREZ-ESTEBAN, B., GÓMEZ-PARDO, E., PEÑALVA, M.A., A *lacZ* reporter fusion method for the genetic analysis of regulatory mutations in pathways of fungal secondary metabolism and its application to the *Aspergillus nidulans* penicillin pathway, *J. Bacteriol.*, 1995, **177**, 6069-6076.
79. BRAKHAGE, A.A., VAN DEN BRULLE, J., Use of reporter genes to identify recessive *trans*-acting mutations specifically involved in the regulation of *Aspergillus nidulans* penicillin biosynthesis genes, *J. Bacteriol.*, 1995, **177**, 2781-2788.
80. VAN DEN BRULLE, J., STEIDL, S., BRAKHAGE, A.A., Cloning and characterization of an *Aspergillus nidulans* gene involved in the regulation of penicillin biosynthesis, *Appl. Environ. Microbiol.*, 1999, **65**, 5222-5228.
81. SEYTTER, T., LOTTSPEICH, F., NEUPERT, W., SCHWARZ, E., Mam33p, an oligomeric, acidic protein in the mitochondrial matrix of *Saccharomyces cerevisiae* is related to the human complement receptor gC1q-R, *Yeast,* 1998, **14**, 303-310.
82. HAYMAN, M.L., MILLER, M.M., CHANDLER, D.M., GOULAH, C.C., READ, L.K., The trypanosome homolog of human p32 interacts with RBP16 and stimulates its gRNA binding activity, *Nucleic Acids Res.*, 2001, **29**, 5216-5225.
83. MAJUMDAR, M., MEENAKSHI, J., GOSWAMI, S.K., DATTA, K., Hyaluronan binding protein 1 (HABP1)/C1QBP/p32 is an endogenous substrate for MAP kinase and is translocated to the nucleus upon mitogenic stimulation, *Biochem. Biophys. Res. Comm.*, 2002, **291**, 829-837.
84. ESPESO, E.A., PEÑALVA, M.A., Carbon catabolite repression can account for the temporal pattern of expression of a penicillin biosynthetic gene in *Aspergillus nidulans*, *Mol. Micriobiol.*, 1992, **6**, 1457-1465.
85. BAILEY, C., ARST, JR., H.N., Carbon catabolite repression in *Aspergillus nidulans*, *Eur. J. Biochem.*, 1975, **51**, 573-577.
86. HYNES, M.J., KELLY, J.M., Pleiotrophic mutants of *Aspergillus nidulans* altered in carbon metabolism, *Mol. Gen. Genet.*, 1977, **150**, 193-204.
87. ESPESO, E.A., FERNÁNDEZ-CAÑÓN, J.M., PEÑALVA, M.A., Carbon regulation of penicillin biosynthesis in *Aspergillus nidulans*: a minor effect of mutations in *creB* and *creC*, *FEMS Microbiol. Lett.*, 1995, **126**, 63-68.
88. ESPESO, E.A., TILBURN, J., ARST, H.N., PEÑALVA, M.A., pH regulation is a major determinant in expression of a fungal penicillin biosynthetic gene, *EMBO J.*, 1993, **12**, 3947-3956.

89. SHAH, A.J., TILBURN, J., ADLARD, M.W., ARST, J., H.N., pH regulation of penicillin production in *Aspergillus nidulans*, *FEMS Microbiol. Lett.*, 1991, **77**, 209-212.
90. THEN BERGH, K., BRAKHAGE, A.A., Regulation of the *Aspergillus nidulans* penicillin biosynthesis gene *acvA* (*pcbAB*) by amino acids: Implication for involvement of transcription factor PACC, *Appl. Environ. Microbiol.* 1998, **64**, 843-849.
91. TILBURN, J., SARKAR, S., WIDDICK, D.A., ESPESO, E.A., OREJAS, M., MUNGROO, J., PEÑALVA, M.A., ARST JR., H.N., The *Aspergillus* PacC zinc finger transcription factor mediates regulation of both acid- and alkaline-expressed genes by ambient pH, *EMBO J.*, 1995, **14**, 779-790.
92. ESPESO, E.A., PEÑALVA, M.A., Three binding sites for the *Aspergillus nidulans* PacC zinc-finger transcription factor are necessary and sufficient for regulation by ambient pH of the isopenicillin N synthase gene promoter, *J. Biol. Chem.*, 1996, **271**, 28825-28830.
93. BRAKHAGE, A.A., TURNER, G., L-Lysine repression of penicillin biosynthesis and the expression of penicillin biosynthesis genes *acvA* and *ipnA* in *Aspergillus nidulans*, *FEMS Microbiol. Lett.*, 1992, **98**, 123-128.
94. BHATTACHARJEE, J.K., Evolution of α-aminoadipate pathway for the synthesis of lysine in fungi, *in*: The Evolution of Metabolic Function (R.P. Mortlock, ed,), CRC Press, Inc., Boca Raton, FL. 1992, pp. 47-80
95. BUSCH, S., BODE, H.B., BRAKHAGE, A.A., BRAUS, G.H., Impact of the cross-pathway control on the regulation of lysine and penicillin biosynthesis in *Aspergillus nidulans*, *Curr. Genet.*, 2003, **42**, 209-219.
96. CONAWAY, C., Too much of a good thing can be bad., *Regional Review*, 2003, (Qtr. 1).
97. MOORE, R.N., BIGAM, G., CHAN, J.K., HOGG, A.M., NAKASIMA, T.T., VEDERAS, J.C., Biosynthesis of the hypocholesterolemic agent mevinolin by *Aspergillus terreus*. Determination of the origin of carbon, hydrogen and oxygen atoms by 13C NMR and mass spectrometery, *J. Am. Chem. Soc.*, 1985, **107**, 3694-3701.
98. HUTCHINSON, C.R., KENNEDY, J., PARK, C., KENDREW, S., AUCLAIR, K., VEDERAS, J., Aspects of the biosynthesis of non-aromatic fungal polyketides by iterative polyketide synthases, *Antonie van Leeuwenhoek*, 2000, **78**, 287-295.
99. HUTCHINSON, C.R., KENNEDY, J., PARK, C., AUCLAIR, K., KENDREW, S.G., VEDERAS, J., The molecular genetics of lovastatin and compactin biosynthesis, *in*: Handbook of Industrial Microbiology (Zhiqiang An, ed.), Marcel Dekker, Inc., New York. 2004, In press.
100. SORENSEN, J.L., VEDERAS, J.C., Monacolin N, a compound resulting from derailment of type I iterative polyketide synthase function *en route* to lovastatin, *Chem. Comm.*, 2003, **13**, 1492-1493.
101. HORINOUCHI, S., A microbial hormone, A-factor, as a master switch for morphological differentiation and secondary metabolism in *Streptomyces griseus*, *Front Biosci.*, 2002, **7**, 2045-2057.

102. VAN DE KAMP, M., DRIESSEN, A.J.M., KONINGS, W.N., Compartmentalization and transport in β-lactam antibiotic biosynthesis by filamentous fungi, *Antonie van Leeuwenhoek,* 1999, **75**, 41-78.
103. EISENDLE, M., OBEREGGER, H., ZADRA, I., HAAS, H., The siderophore system is essential for viability of *Aspergillus nidulans*: functional analysis of two genes encoding l-ornithine N 5-monooxygenase (sidA) and a non-ribosomal peptide synthetase (sidC), *Mol. Microbiol.*, 2003, **49**, 359-375.
104. BROWN, D., SALVO, J., Isolation and characterization of sexual spore pigments from *Aspergillus nidulans*, *Appl. Environ. Microbiol.*, 1994, **60**, 979-983.
105. FUJII, A.W., MORI, Y., EBIZUKA, Y., Structures and functional analyses of fungal polyketide synthase genes, *Actinomycetology,* 1998, **12**, 1-14.
106. PONTECORVO, G., ROPER, J.A., HEMMONS, L.M., MACDONALD, K.P., BUFTON, A.W.J., The genetics of *Aspergillus nidulans*, *Adv. Gen.*, 1953, **5**, 141-238.

Chapter Eleven

GENETICS AND BIOCHEMISTRY OF AFLATOXIN FORMATION AND GENOMICS APPROACH FOR PREVENTING AFLATOXIN CONTAMINATION

Jiujiang Yu,* Deepak Bhatnagar, and Thomas E. Cleveland

U. S. Department of Agriculture
Agricultural Research Service
Southern Regional Research Center
1100 Robert E. Lee Boulevard
New Orleans, Louisiana, 70124 U.S.A.

**Author for correspondence:* jiuyu@srrc.ars.usda.gov

INTRODUCTION

Mycotoxins and Aflatoxins

Mycotoxins are toxic, small molecular weight secondary metabolites produced by fungi. Research on mycotoxins gained worldwide attention after the notorious "Turkey X disease" in 1962 near London, England, when approximately 100,000 turkey poults died.[1,2] This mysterious disease was later found to be caused by feeding peanut (groundnut) meal contaminated with the toxin of *Aspergillus flavus* named "aflatoxin." Chemically, aflatoxins are difuranocoumarin derivatives. Aflatoxins B_1, B_2, G_1, and G_2 (AFB_1, AFG_1, AFB_2, and AFG_2) (Fig. 11.1) are the four major aflatoxins named based on their colors of fluorescence emission under ultraviolet light (blue or green) after thin layer chromatographic separation. Aflatoxins are produced primarily by the filamentous fungi *Aspergillus flavus* and *A. parasiticus*[3,4] as well as some strains of *A. nomius*, two isolates of *A. pseudotamarii,*[5] nine of *A. bombycis,*[6] as well as one isolate of *A. ochraceoroseus*[7,8] and *Emericella venezuelensis* (Klich, unpublished data). *A. flavus* produces aflatoxins B_1 and B_2. Other toxic compounds produced by *A. flavus* are cyclopiazonic acid, kojic acid, β-nitropropionic acid, aspertoxin, aflatrem, and aspergillic acid.[9] *A. parasiticus* produces aflatoxin G_1 and G_2, in addition to B_1 and B_2, but not cyclopiazonic acid. *A. flavus* is more persistent in crop debris, produces higher levels of conidia early in the growing season, and is most commonly associated with preharvest aflatoxin contamination of food and feed crops. Although *A. flavus* is not an aggressive pathogen, under weather conditions favorable to its growth, the fungus can cause ear rot on maize, thus demonstrating characteristics of an "opportunistic" pathogen. Because of its ability to grow at low water activity, *A. flavus* is also well adapted to colonize seeds of grain and oil crops in storage, where exposure of seed to moisture is purposely limited. Control methods have been developed for post-harvest control of aflatoxin contamination, but there are no effective control strategies to prevent pre-harvest aflatoxin contamination.[10]

Health and Economic Impact of Aflatoxin Contamination

The diseases caused by fungal invasion into animal or human hosts are collectively called "mycoses," while the diseases or symptoms caused by exposures to toxic fungal metabolites are collectively called "mycotoxicoses." The aflatoxigenic strains of *Aspergillus flavus* can cause both mycoses and mycotoxicoses in animals and human beings. Aflatoxin is associated with both toxicity and carcinogenicity in human and animal populations.[11-14] The diseases caused by aflatoxin consumption are loosely called "aflatoxicoses." Aflatoxins are hepatotoxic, mutagenic, teratogenic, carcinogenic to animals and humans, and bisfuran-containing, polyketide-derived toxins that are produced by certain *Aspergillus* species.[14-19] Aflatoxin B_1 (Fig.11.1) is the most potent natural carcinogen known.[19] The short-term toxicity of aflatoxin has been recognized for 40 years,[20,21] and evidence depicting the chronic, low-level exposure leading to human hepatocarcinomas has been established in the last 10 years.[22-24] Acute aflatoxicosis results in death; chronic aflatoxicosis results in cancer, immune suppression, and other "slow" pathological conditions.[25] The liver is the primary target organ, with damage when animals are fed with aflatoxins. Studies in patients with liver cancer in Africa and China have shown a mutation in the p53 tumor suppressor at codon 249 associated with a G to T transversion.[22,23] Cytochrome P450 enzymes convert aflatoxins to the reactive 8,9-epoxide form (also referred to as aflatoxin-2,3 epoxide in older literature), which is capable of binding to both DNA and proteins.[14] Mechanistically, it is known that the reactive aflatoxin epoxide binds to the N^7 position of guanines.[14] Moreover, aflatoxin B_1-DNA adducts can result in the GC to TA transversions.[14] Inactivation of the p53 tumor suppressor gene is the culprit in the development of primary liver cancer.[22,23]

Aflatoxin contamination of agricultural commodities poses a potential risk to livestock and human health.[10,21,26-30] Contamination in feed for livestock and in food for human consumption has received significant attention since these compounds in food and feed are ubiquitous and occur in many parts of the world. Food safety is the paramount issue in developing countries where detection and decontamination policies are impractical. In countries where populations are facing starvation, or where regulations are either not enforced or nonexistent, routine ingestion of aflatoxin may occur.[31] This is reflected in the reported incidence (67%) of liver carcinomas in Senegal, as well as in China, Swaziland,[32] Mozambique,[22] and Mexico.[33] Worldwide, liver cancer incidence rates are 2 to 10 times higher in developing countries than in developed ones.[33] In developed countries, food safety and health of the general population is protected by regulations. The maximum allowable amount of aflatoxin in food and feed for human consumption and for livestock has been mandated by laws. Of the countries that attach a numerical value to their tolerance, the difference between the limits varies significantly. A guideline of 20 parts aflatoxin per billion parts of food or feed substrate (ppb) is the maximum allowable limit imposed by the U.S. Food and Drug Administration for interstate shipment of foods and feeds. European countries are expected to introduce more stringent guidelines that may restrict aflatoxin levels in imported foods to a much lower level (3-5 ppb). The crop is destroyed or decontaminated if the content exceeds the official regulatory levels, resulting yearly in billion dollar losses worldwide. Aflatoxin contamination is a chronic problem in some parts of the U.S.[18] e.g., in Arizona cotton growing areas and the Southeast peanut farming regions. However, sporadic severe outbreaks of aflatoxin contamination have occurred in the U.S. Midwest cornbelt in 1977, 1980, and 1988. In years conducive for aflatoxin

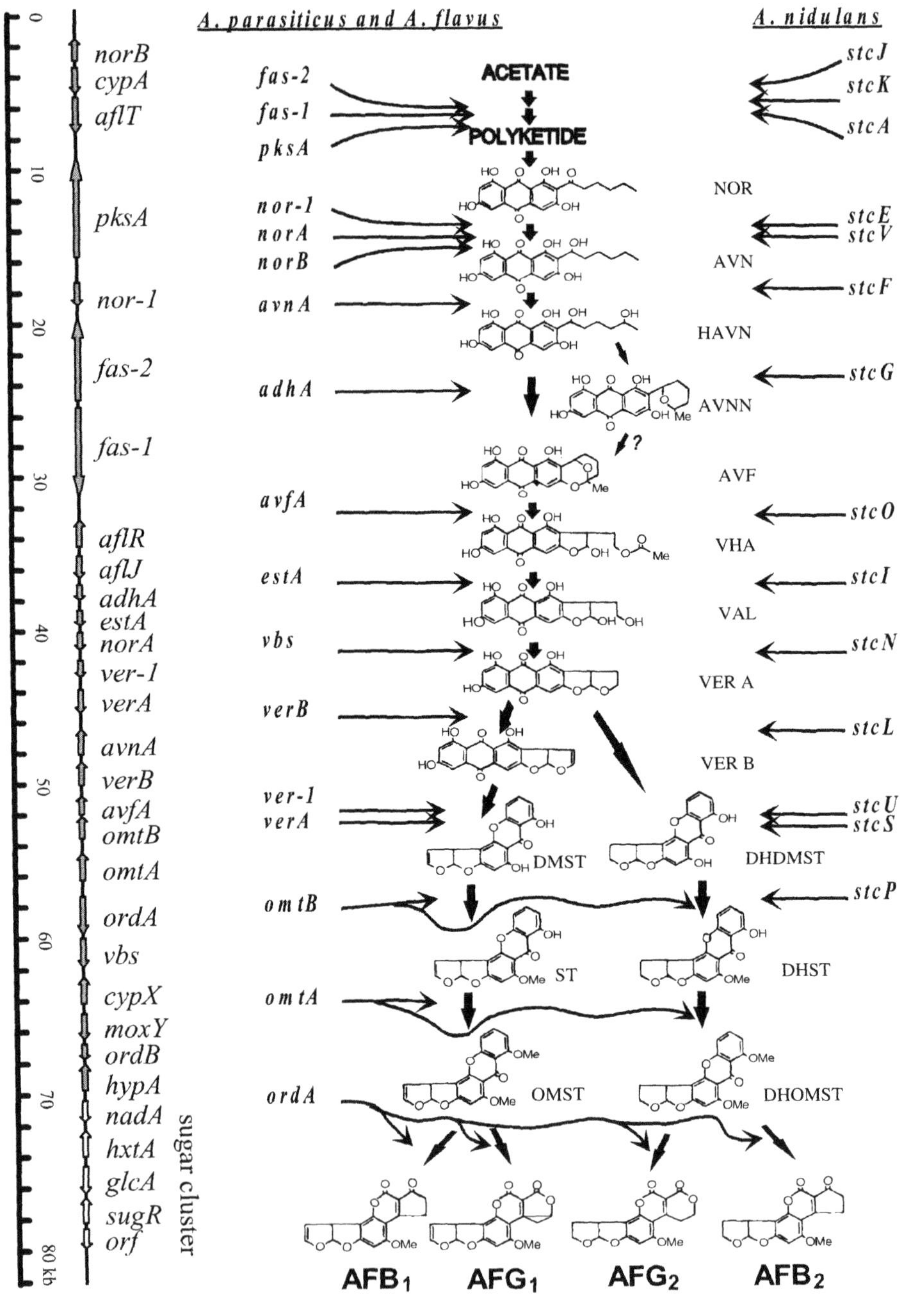
A. parasiticus and A. flavus
A. nidulans
ACETATE
POLYKETIDE
norB
cypA
aflT
pksA
nor-1
fas-2
fas-1
aflR
aflJ
adhA
estA
norA
ver-1
verA
avnA
verB
avfA
omtB
omtA
ordA
vbs
cypX
moxY
ordB
hypA
nadA
hxtA
glcA
sugR
orf
sugar cluster
80 kb
NOR
AVN
HAVN
AVNN
AVF
VHA
VAL
VER A
VER B
DMST
DHDMST
ST
DHST
OMST
DHOMST
stcJ
stcK
stcA
stcE
stcV
stcF
stcG
stcO
stcI
stcN
stcL
stcU
stcS
stcP
AFB1
AFG1
AFG2
AFB2

production, no control procedure is effective. Recent weather patterns in the U.S. have resulted in serious aflatoxin problems in a number of southern states, with enormous economic losses. Though, it is impossible to estimate the precise amount of losses of the value of food and feed, or the mitigation efforts, the potential economic costs of crop losses from mycotoxins (mainly aflatoxins, fumonisins, and deoxynivalenol) in the United States alone, are approximately $932 million per year.[34] The mean mitigation costs were estimated to be about $466 million, and the mean simulated livestock costs were about $6 million per year.[34] The human health costs are not included in the estimation. Thus, aflatoxin contamination is not only a serious food safety concern, but has significant economic implications for the agriculture industry worldwide.

Contamination of food with aflatoxin and its toxicity to humans and animals are not the only concerns. The fungus itself is an emerging health problem. Infections in humans due to *Aspergillus* species are occurring at a greater frequency in all developed countries. Aspergillosis is a term that encompasses a variety of diseases caused by members of the genus *Aspergillus*. These include invasive aspergillosis,

Figure 11.1: Proposed and generally accepted pathway for aflatoxin B_1, B_2, G_1 and G_2 biosynthesis and the corresponding genes are presented. The aflatoxin biosynthetic pathway genes for a specific conversion step in *A. parasiticus* and *A. flavus* are labeled on the left panel with their names listed. The homologous genes involved in sterigmatocystin biosynthesis in *A. nidulans* are labeled on the right. Note that no aflatoxins are produced in *A. nidulans* and the final conversion products are ST and DHST. The locations and relative order of the genes in the gene cluster are presented on the left. The arrows indicate the direction of gene transcription. The relative sizes of these genes are shown by the relative length of the arrows and the scale bar in kb. Abbreviations for the intermediates are: norsolorinic acid (NOR), averantin (AVN), 5'-hydroxyaverantin (HAVN), averufanin (AVNN), averufin (AVF), versiconal hemiacetal acetate (VHA), versiconal (VAL), versicolorin B (VER B), versicolorin A (VER A), demethylsterigmatocystin (DMST), sterigmatocystin (ST), *O*-methylsterigmatocystin (OMST), aflatoxin B_1 (AFB_1), aflatoxin G_1 (AFG_1), demethyldihydrosterigmatocystin (DMDHST), dihydrosterigmatocystin (DHST), dihydro-*O*-methylsterigmatocystin (DHOMST), aflatoxin B_2 (AFB_2), and aflatoxin G_2 (AFG_2).

pulmonary aspergilloma, allergic bronchopulmonary aspergillosis, and others.[35,36] *Aspergillus flavus* is the second leading cause of aspergillosis in humans next to *A. fumigatus*, and the leading causative agent of chronic indolent invasive sinonasal infection in immunocompetent patients.[36]

GENETICS AND MOLECULAR BIOLOGY OF *Aspergillus flavus*

Because of health concerns, *A. flavus* currently represents one of the most extensively studied plant pathogens in the U.S. Current areas of research include programs on evolutionary and population biology, epidemiology, biological control, regulation of secondary metabolism, host resistance, and fungal development. Research is also ongoing on the involvement of *A. flavus* in animal and human mycoses. Research on the natural occurrence, identification, characterization, chemistry, enzymology, biosynthesis, and genetic regulation of aflatoxins, as well as prevention and control of aflatoxin contamination of food and feed have been studied in great detail. Electrophoretic karyotyping studies by CHEF gel indicated that there are about 6-8 chromosomes in both *A. parasiticus* and *A.flavus* genomes.[37,38] Recent evidence through whole genome sequencing of *A. flavus* and related species supports the notion that there are 8 chromosomes in *A. flavus*. The estimated genome size is about 34-36 Mbp containing approximately 11,000 functional genes. Based on our research with *Aspergillus*, the *A. flavus* genome is compacted with fewer duplicated sequences or multiple copies of genes[39] (Yu, unpublished data). The non-coding sequences between genes are much shorter than in higher plants, and there are only a few small introns within each ORF, if any. Foutz *et al.*[37] assigned six linkage group (LG) markers to five chromosomes of *A.flavus*. The LG VII markers, arg7 and leu7, map to the 4.9-Mb chromosome containing the aflatoxin biosynthetic cluster. *Aspergillus flavus* can be transformed with DNA, and genes can be readily disrupted. Parasexuality has been described in the fungus,[40] and parasexual analysis has been used in *A. flavus* to map over 36 genes to 8 linkage groups.[3]

Significant progress has been made in the last decade in deciphering the aflatoxin biosynthetic pathway[15,16,41,42] (Fig.11.1). It has been estimated that at least 23 enzymatic reactions are involved in aflatoxin formation. As many as 15 structurally-defined aflatoxin intermediates have been identified in the aflatoxin/sterigmatocystin biosynthetic pathway starting with acetate and polyketide precursors.[3,15,16,42,43] A complete aflatoxin biosynthetic pathway gene cluster consisting of 25 genes has been cloned[39] (unpublished data). Details of these genes and their encoded enzymes as well as the regulation of gene expression have been reported.[39,42]

Sterigmatocystin (ST) or dihydrosterigmatocystin (DHST) (Fig.11.1), a related dihydrofuran toxin, is the penultimate precursor of aflatoxins, and is also produced as a final biosynthetic product by a number of species such as *Aspergillus versicolor* and *Aspergillus nidulans*.[44] Sterigmatocystin is hepatotoxic, carcinogenic, and mutagenic as well, but is less potent than aflatoxins.[45] The common biochemical pathway, homologous genes, and regulatory mechanism of ST synthesis in *A. nidulans* are compared and reviewed here.

BIOCHEMICAL PATHWAY OF AFLATOXIN FORMATION

The aflatoxin pathway is one of the best-studied pathways of fungal secondary metabolism. Attempts to decipher the pathway began with the discovery of the structure of these toxins. However, the major biochemical steps and the corresponding genetic components of AFB_1 biosynthesis have been elucidated at a molecular level only in the last decade. Significant progress has been made in deciphering the details of the aflatoxin/ST biosynthetic pathways.[3,16,39,41,42,46-49]

Various studies have determined that aflatoxins are synthesized in two stages from malonyl CoA, first with the formation of hexanoyl CoA, followed by formation of a decaketide anthraquinone.[46,48] Two fatty acid synthases (FAS) and a polyketide synthase (PKS) are involved in the synthesis of polyketide from the primary metabolite, acetate.[49-52] However, norsolorinic acid (NOR) is the first stable aflatoxin intermediate identified in the pathway.[53-57] With the use of NOR-accumulating mutants, it was demonstrated in *A. flavus*[55,56] and in *A. parasiticus*[53,58] that norsolorinic acid (NOR) was an intermediate in the aflatoxin biosynthetic pathway. A series of highly organized oxidation-reduction reactions then allows formation of aflatoxin.[46,47,59-62] The currently accepted scheme (Fig.11.1) is: hexanoyl CoA precursor --> norsolorinic acid, NOR --> averantin, AVN --> hydroxyaverantin, HAVN --> averufin, AVF --> hydroxyversicolorone, HVN--> versiconal hemiacetal acetate, VHA --> versiconal, VAL --> versicolorin B, VERB --> versicolorin A, VERA --> demethyl-sterigmatocystin, DMST --> sterigmatocystin, ST --> O-methylsterigmatocystin, OMST --> aflatoxin B_1, AFB_1 and aflatoxin G_1, AFG_1. A branch point in the pathway has been established, following VHA production, leading to different structural forms of aflatoxins B_2 and G_2, AFB_2 and AFG_2.[46,63-69] A number of metabolic grids may provide alternate pathways to aflatoxins.[46,54,65,70-73]

Several specific enzyme activities associated with precursor conversions in the aflatoxin pathway[27,46,47,67,71,74-77] have been partially purified[78-80] (Fig.11.1); whereas others such as methyltransferases[74,80,81] have been purified to homogeneity. Several other enzymes involved in aflatoxin biosynthesis, such as a reductase[82] and a cyclase,[83,84] have also been purified from *A. parasiticus*. A desaturase that converts VERA to VERB has been found in cell-free fungal extracts.[64,67] Matsushima *et al.*[85] have purified and characterized two versiconal hemiacetal acetate reductases involved in toxin synthesis, whereas Kusumoto and Hsieh purified to homogeneity an esterase that converts VHA to versiconal.[86]

MOLECULAR GENETICS OF AFLATOXIN BIOSYNTHESIS

Clustering of Aflatoxin Pathway Genes

The first experimental evidence showing the potential of clustering of aflatoxin pathway genes was demonstrated when the *nor-1* and *ver-1* genes were found to be linked in a cosmid clone with the regulatory gene *aflR*.[87,88] By mapping overlapping cosmid clones in *A. parasiticus* and *A. flavus*, it was established that at least nine aflatoxin pathway genes including *nor-1*, *aflR*, *ver-1*, and *omtA* were clustered.[89] The establishment of the aflatoxin biosynthetic pathway gene cluster accelerated the rate of gene discovery.[69,84,90-98] Recently, several additional aflatoxin biosynthetic pathway genes and open reading frames (ORFs) have been identified within the existing aflatoxin pathway gene cluster.[38] The complete 82 kb DNA sequence of the entire gene cluster harbors a total of 25 genes (or ORFs) characterized or proposed to be involved in aflatoxin biosynthesis and 4 genes proposed to be involved in sugar utilization[39,99] (Fig.11.1; Yu, unpublished data).

A primary advantage of gene clustering may be coordinated gene expression. Clustering of genes allows regulatory elements to be shared. Gene complementation experiments performed in this laboratory demonstrated that the aflatoxin pathway genes are expressed adequately only when they are targeted into the gene cluster.

In *A. parasiticus*, a partial duplicated aflatoxin gene cluster has been identified.[100, 101] This partial duplicated gene cluster consisting of seven duplicated genes has been cloned and characterized.[102] These duplicated genes were named with the addition of the number "2" indicating second copy, such as *aflR2*, *aflJ2*, *adhA2, estA2*, *norA2, ver1B, omtB2*. The genes within this partial duplicated cluster, due possibly to the chromosome location,[103] were found likely to be non-functional under normal conditions even though no apparent defects were identified in the sequences of at least some of them. Recent evidence (Linz, personal communication) measured the expression of *ver1B* gene; however, its translation pattern is yet to be investigated.

Genes Involved in Biosynthesis

The first aflatoxin pathway gene was identified through genetic complementation of a NOR-accumulating mutant of *A. parasiticus*.[104] It was named *nor-1* for the conversion of norsolorinic acid (NOR) to averantin (AVN). Further characterization of this gene demonstrated that it encodes an enzyme that functions as a ketoreductase for the conversion of NOR to AVN.[105] A *norA* gene, encoding an aryl-alcohol dehydrogenase in the aflatoxin pathway gene cluster, was demonstrated to have high amino acid homology to *nor-1*.[90] An additional gene, *norB*, was cloned and found to be homologous to the *norA* gene in the aflatoxin pathway gene cluster in *A. parasiticus* (Yu, unpublished data). The *nor-1* and *norA* gene homologs in *A.*

nidulans are *stcE* and *stcV*, respectively.[106] No *norB* gene homolog was identified in the ST gene cluster.[106]

The second important gene cloned was *ver-1*, involved in a key step in aflatoxin synthesis,[107] which is required for the conversion of versicolorin A (VERA) to demethylsterigmatocystin (DMST) and versicolorin B (VERB) to demethyldihydrosterigmatocystin (DMDHST) in *A. parasiticus*. The expression of this gene was also reported.[108] The *ver-1* homolog in the ST gene cluster is *stcU* in *A. nidulans,*[109] which encodes for a ketoreductase required for the conversion of VERA to DMST. A gene named *verA* was also identified in *A. parasiticus* SRRC 143 (Yu, unpublished), which is a homolog of *stcS*, encoding a cytochrome P-450 type monooxygenase, involved in VERA and VERB conversion in aflatoxin/ST synthesis.[110,111]

Another important gene involved in a later step of aflatoxin biosynthesis, named *omt-1*, encoding an *O*-methyltransferase for the conversion of sterigmatocystin (ST) to *O*-methylsterigmatocystin (OMST) and demethylsterigmatocystin (DMST) to dihydro-*O*-methylsterigmatocystin (DHOMST), was cloned by antibody screening of a cDNA expression library from *A. parasiticus.*[112] The enzyme was expressed in *E. coli*, and its activity for converting ST to OMST was demonstrated by substrate feeding studies.[112] The genomic DNA sequence (named *omtA*) was cloned from *A. parasiticus* and *A. flavus.*[113] The function of *omtA* was unambiguously demonstrated *in vivo* by disruption experiments.[114]

In the aflatoxin pathway gene cluster, two large genes (7.5-kb transcripts), *fas-1* (initially named *uvm8*, *fas-1A*) and *fas-2* (*fas-2A*), encoding beta (FASβ) and alpha-subunit (FASα) of fatty acid synthase, respectively, were identified.[49, 88, 94] The *fas-2* and *fas-1* genes were also named *hexA* and *hexB*, respectively, for hexanoate synthase alpha and beta subunits respectively (AF 391094, Hitchman, *et al.*, unpublished data). Watanabe *et al.* provided the biochemical evidence for the role of a fatty acid synthase and a polyketide synthase in the biosynthesis of aflatoxin.[115] Chang *et al.* confirmed that a PKS is required for aflatoxin biosynthesis by cloning the *pksA* gene encoding a PKS for the synthesis of polyketide from *A. parasiticus.*[91] Trail *et al.* demonstrated by knockout experiment that *pksA* was important for aflatoxin biosynthesis.[88] Feng and Leonard also isolated a *pksL1* gene for PKS, which is the equivalent of *pksA* gene.[116] Disruption of the *pksL1* gene produced neither aflatoxin nor any aflatoxin intermediates. The predicted amino acid sequences of these PKS contain typical four conserved domains common to other known PKS proteins: β-ketoacyl synthase (KS), acyltransferase (AT), acyl carrier protein (ACP), and thioesterase (TE).[91] The *fas-2*, *fas-1*, and *pksA* genes are directly involved in the backbone formation for the conversion from acetate to norsolorinic acid (NOR) in aflatoxin synthesis.

Since the establishment of the aflatoxin pathway gene cluster,[89] additional genes have been identified within the aflatoxin gene cluster, with their functions characterized or proposed based on homologies to genes in the ST gene cluster in *A.*

nidulans. These genes are briefly described bellow: the *avnA* gene encodes a cytochrome P450 monooxygenase for the conversion of averantin (AVN) to 5'hydroxyaverantin (HAVN);[89,96] the *adhA* gene encodes an alcohol dehydrogenase in *A. parasiticus* for the conversion of 5'hydroxyaverantin (HAVN) to averufin (AVF);[70] the *avfA* gene from *A. parasiticus*, *A. flavus* AVF-accumulating strains and the *A. sojae* strain encodes for the conversion of averufin (AVF) to versiconal hemiacetal acetate (VHA);[98] the *estA* gene encodes an esterase for the conversion of versiconal hemiacetal acetate (VHA) to versiconal (VAL)[117] (Chang *et al.*, unpublished data); the VERB synthase gene *vbs* encodes for the conversion of versiconal (VAL) to versicolorin B (VER B) in *A. parasiticus*.[73,84,95,118] It was demonstrated that the versicolorin B synthase catalyzes the side chain cyclodehydration of racemic VHA to VER B.[73,118] This is a key step in aflatoxin formation since it closes the bisfuran ring of aflatoxin for binding to DNA.

The conversion of versicolorin B (VERB) to versicolorin A (VERA) was proposed as requiring a desaturation of the bisfuran ring of VERB.[27,73] Disruption of the *stcL* gene encoding for a P-450 monooxygenase demonstrated its involvement in the conversion of VERB to VERA in *A. nidulans*.[119] The *stcL* homolog from *A. parasiticus* and *A. flavus* gene was cloned in the aflatoxin pathway gene cluster and is named *verB*[39] (Yu, unpublished data).

The *omtB* gene encoding an *O*-methyltransferase in *A. parasiticus* for the conversion of demethylsterigmatocystin (DMST) to sterigmatocystin (ST) and demethyldihydrosterigmatocystin (DMDHST) to dihydrosterigmatocystin (DHST)[68, 80, 120] was proposed based on disruption of its homolog in *A. nidulans*.[121] This gene was cloned and characterized in *A. parasiticus*[122] (named *dmtA* or *mt-I* for *O*-methyltransferase I) and concurrently cloned (named *omtB* for *O*-methyltransferase B) in *A. parasiticus*, *A. flavus*, and *A. sojae*.[97]

Enzymatic studies supported the hypothesis that there are separate pathways leading to B-Group (AFB_1 and AFB_2) and G-Group (AFG_1 and AFG_2) aflatoxin formation.[65] Prieto *et al.* reported in *A. flavus* that a cytochrome P-450 monooxygenase gene, *ord-1*, is required for this reaction.[123,124] This gene was cloned (named *ordA*) from *A. parasiticus* and an *A. flavus* mutant strain.[69] It was demonstrated by substrate feeding studies in a yeast system that this gene is responsible for the conversion of *O*-methylsterigmatocystin (OMST) to AFB_1 and AFG_1, and demethyldihydrosterigmatocystin (DMDHST) to AFB_2 and AFG_2.[69] The critical amino acids for the enzymatic activity and heme-binding motif were identified by site-directed mutagenesis.[69]

Several additional genes have also been identified in the gene cluster (Fig.11.1; Yu, unpublished data). Typical AflR binding motifs have been identified in the promoter regions of all of these newly identified genes (Yu, unpublished data), indicating potential evidence for the involvement in aflatoxin formation under *aflR* regulation.[125] These genes potentially encode an antibiotic efflux pump protein (*aflT*), cytochrome P450 type monooxygenases (*cypA*, *cypX*), monooxygenase (*moxY*),[98] oxidase (*ordB*), and a hypothetical protein (*hypA*), respectively (Yu, unpublished data). In the enzyme activity assay on oxidoreductase in a yeast system, it was demonstrated that at least one additional enzyme is required for G-group toxin synthesis in *A. parasiticus*.[69] However, the enzyme(s) and corresponding gene(s) for such reaction(s) so far have not been positively identified. There is a possibility that one or more of the above newly identified genes might be involved in the G-group toxin formation.

Genes Involved in Regulation

The fact that aflatoxin and sterigmatocystin biosynthetic pathway genes are tightly compacted on a single chromosome within a 82 kb DNA region in both *A. parasiticus* and *A. flavus* and in *A. nidulans*, respectively,[89,106] led to the presumption of gene expression in concert in the genome. In both the aflatoxin and sterigmatocystin gene clusters, there is a positive regulatory gene, *aflR* (originally named *afl-2* and *apa-2*), for activating pathway gene transcription. Disruption of *aflR* prevented the accumulation of structural gene transcripts for aflatoxin biosynthesis.[126] Introduction of an additional copy of the *aflR* caused the overproduction of aflatoxin biosynthetic intermediates.[126] The *aflR* gene, coding for a sequence specific zinc binuclear DNA-binding protein, has been shown to be required for transcriptional activation of most, if not all, of the structural genes.[126-134] The aflatoxin pathway gene transcription can be activated when the AflR protein binds to the palindromic sequence 5'-TCGN5CGA-3' (also called AflR binding motif) in the promoter region of the structural genes in *A. parasiticus, A. flavus*, and *A. nidulans*.[125,135,136] AflR binds in some cases to a deviated sequence rather than the typical motif, as shown in Table 11.1 (*avnA*, *aflR*). When there is more than one such motif in the promoter region of a gene, only one is a preferred binding site, such as the promoter of *pksA*.[125,135]

The protein encoded by *aflR* has major domains typical of fungal and yeast Gal4-type transcription factors.[126] One of these is an N-terminal cysteine-rich stretch, CTSCASSKVRCTKEKPACARCIERGLAC (Cys_6-Zn_2), which is required for DNA-binding.[126,133,137-141] Preceding the Cys_6-Zn_2 domain is the arginine-rich (RRARK) nuclear localization domain. In the C-terminus (residues 408-444) (HHPASPFSLLGFSGL**E**ANL**R**HRLRAVSS**D**II**D**YLHRE), several charged amino acid residues (Lysine, Histidine, Arginine, Aspartic acid, and Glutamic acid) constitute the transcription activation domain.[129] Studies by site-directed mutagenesis demonstrated that substitution of Arginine by Leucine, or of any of the three acidic amino acid residues (shown in bold face letters, E, R, D, D) by Lysine or Histidine destroys its protein function for transcription activation.[129] *A. sojae*, a non-toxigenic strain used in industrial fermentations, was found to contain a defective

Table 11.1: Aflatoxin biosynthetic pathway cluster genes

Gene name and other names used & accession #	ST gene homolog	Enzyme	Function in the pathway
fas-2 (*hexA*) (AF391094)	*stcJ*	Fatty acid synthase alpha subunit	Acetate → polyketide
fas-1 (*hexB*) (AF391094), *uvm8*, *fas1*, *fas-1A* (L48183)	*stcK*	Fatty acid synthase beta subunit	Acetate → polyketide
pksA (Z47198), *pksL1* (L42765, L42766)	*stcA* (*wA*)	Polyketide synthase	Acetate → Polyketide
nor-1 (L27801)	*stcE*	Reductase	NOR → AVN
norA (U24698, Q00049), *adh-2* (U32377), *aad* (U24698)	*stcV*	NOR reductase/dehydrogenase	NOR → AVN
norB (Yu et al., 2004a)	--	dehydrogenase	NOR → AVN
avnA (U62774), *ord-1* (L40839)	*stcF*	P450 monooxygenase	AVN → HAVN
adhA (U76621)	*stcG*	Alcohol dehydrogenase	HAVN → AVF or AVNN
avfA (AF154050, AF159789 in *A. flavus*)	*stcO*	Oxidase	AVF → VHA
estA (AF417002)	*stcI*	Esterase	VHA → VAL
vbs (U51327, AF169016)	*stcN*	VERB synthase	VAL → VERB
verB (AF106958, AF106959, AF106960, AAD50366, AAD50367)	*stcL*	Desaturase	VERB → VERA
ver-1 (M91369)	*stcU*	Dehydrogenase/Ketoreductase	VERA → DMST
verA (Yu et al., 2004a)	*stcS* (*verA*)	Monooxygenase	VERA → DMST
dmtA (*mt-I*) (AB022905, AB022906) *omtB* (AF154050, AF159789 in *A. flavus*),	*stcP*	*O*-methyltransferase I or *O*-methyltransferase B	DMST → ST & DHDMST → DHST
omtA (L25834, AF25836), *omt-1* (L22091 in *A. flavus*)	--	*O*-methytransferase A or *O*-methyltransferase II	ST → OMST & DHST → DHOMST
ordA (AF017151, AF169016), *ord-1* (U81806, U81807)	--	Oxidoreductase /P450 monooxygenase	OMST → AFB_1 & AFG_1, DHOMST → AFB_2 & AFG_2
aflR (L26222, AF427616, AF441429), *apa-2* (L22177), *afl-2*	*aflR*	Transcription activator	Pathway regulator
aflJ (AF002660, AF077975)	un-named	Transcription enhancer	Pathway regulator

Table 11.1 (cont.): Aflatoxin biosynthetic pathway cluster genes

Gene name and other names used & accession #	ST gene homolog	Enzyme	Function in the pathway
aflT (AF268071)	--	Transmembrane protein	Unassigned
cypA (Yu et al., 2003c, this issue)	--	P450 monooxygenase	Unassigned
cypX (AF169016)	*stcB*	P450 monooxygenase	Unassigned
moxY (AF169016)	*stcW*	Monooxygenase	Unassigned
ordB (Yu et al., 2003c, this issue)	*stcQ*	Monooxygenase/oxidase	Unassigned
hypA (Yu et al., 2003c, this issue)	--	Hypothetical protein	Unassigned
aflR2 (AF452809)	Second copy	Transcription activator	Pathway regulator
aflJ2 (AF452809, AF295204)	Second copy	Transcription enhancer	Pathway regulator
adhA2 (AF452809)	Second copy	Alcohol dehydrogenase	HAVN → AVF or AVNN
estA2 (AF452809)	Second copy	Esterase	VHA → VAL
norA2 (AF452809)	Second copy	Dehydrogenase (Early terminated)	NOR → AVN
ver-1B (AF452809)	Second copy	Dehydrogenase (Missing N-terminal)	VERA → DMST
omtB2 (AF452809)	Second copy	Methyltransferase B (Missing N-terminal)	DMST → ST & DHDMST → DHST

Notes:

a. The genes and their accession numbers are from *A. parasiticus* unless noted;

b. when 2 accession numbers appear next to a gene, the first one is for genomic DNA sequence followed by its cDNA sequence;

c. The *aflR2*, *aflJ2*, *adhA2*, *estA2*, *norA2*, *ver-1B*, *omtB2* genes are partial duplicated cluster genes (second copy) in *A. parasiticus*.

d. Abbreviations: NOR, norsolorinic acid; AVN, averantin; HAVN, 5'-hydroxy-averantin; AVNN, averufanin; AVF, averufin; VAL, versiconal; VHA, versiconal hemiacetal acetate; VERB, versicolorin B; VERA, versicolorin A; ST, sterigmatocystin; DHST, dihydrosterigmatocystin; OMST, *O*-methylsterigmatocystin; DHOMST, dihydro-*O*-methylsterigmatocystin; AFB_1, aflatoxin B_1; AFB_2, aflatoxin B_2; AFG_1, aflatoxin G_1; and AFG_2, aflatoxin G_2.

aflR transcription activation domain that is due to early termination of the codons for 62 amino acids from the C-terminal end of the AflR protein.[142-144] In addition, other defects may exist in aflatoxin pathway structural genes.[142-144] Thus, with the absence of the functional regulatory protein, no induction of aflatoxin can occur in this food grade *Aspergillus*.

Adjacent to the *aflR* gene in the aflatoxin gene cluster, a divergently transcribed gene, *aflJ*, was found that is also involved in the regulation of transcription.[102,131,145,146] This gene encodes a protein, AflJ, that binds to the carboxy terminal region of AflR and may affect AflR activity.[146] Disruption of *aflJ* in *A. flavus* resulted in a failure to produce any aflatoxin pathway metabolites.[145] It was also found that a transcription factor required for nitrate assimilations, AreA, binds to sites near the *aflJ* transcription start site in the *aflR-aflJ* intergenic region, suggesting that *aflJ* expression could be mediated by a nitrogen source via the action of AreA.[92] Therefore, AflJ may be an AflR coactivator.

Recent studies have discovered a gene that potentially controls the expressions of genes involved in not only ST but also penicillin in *A. nidulans*.[147] The new gene was named *laeA*, for lack of *aflR* expression (N. Keller, personal communication). Interruption of this *laeA* gene resulted in loss of not only *aflR* gene expression for ST synthesis, but also expression of the genes involved in penicillin biosynthesis in *A. nidulans* (Keller, personal communication). Disruption of the *laeA* homologous gene in *A. fumigatus* lost gliotoxin formation, and in *A. terreus* lost lovastatin production (Keller, personal communication). It is likely that the *laeA* gene is involved globally in the regulatory circuit of secondary metabolites, aflatoxins, ST, penicillin, gliotoxin, and lovastatin in several fungal species.

FACTORS AFFECTING AFLATOXIN FORMATION

Many nutritional and environmental factors, such as temperature, pH, carbon and nitrogen source, stress factors, lipids, and trace metal salts affect the production of aflatoxin by toxigenic *Aspergilli*.[131,148-156] The molecular mechanisms for these effects are still not clear despite numerous studies.[3,92,99,157,158] Some of these factors may affect expression of the aflatoxin regulatory gene, *aflR*, or structural genes, possibly by altering the expression of globally acting transcription factors that respond to nutritional and environmental signals.[159] Some of these nutritional and environmental factors may affect aflatoxin accumulation by altering the activity of one or more of the enzymes involved in aflatoxin biosynthesis.

Nutritional Factors

The relationship of carbon source and aflatoxin formation has been well established[3]. Simple sugars such as glucose, sucrose, maltose, but not peptone, sorbose, or lactose, support aflatoxin formation.[3,160] However, the role of carbon in the regulation of aflatoxin pathway gene expression is poorly defined. Expression is not expected to be subject to carbon catabolite repression by the transcription factor, CreA, due to the lack of the CreA sites in their promoters. However, an interesting possible role for CreA in *aflR* expression could be control of expression of the antisense *aflR* mRNA transcript (Ehrlich, unpublished observation), since at the start of this reported transcript are two tandem CreA-binding sites, GCGGGGaGTGGGG (Ehrlich, unpublished observation). Another transcription factor that responds to simple sugars is Rgt1, a positively acting factor that has been shown to be necessary for regulation of glucose transporter molecule expression.[161] A possible Rgt1 site is present in the promoter region of *A. parasiticus aflJ*, and may be involved in regulation of its expression. Such regulation may be necessary for production of aflatoxin pathway metabolites. Additionally, carbon source utilization could affect aflatoxin gene expression by inducing G-protein-dependent signaling in *Aspergillus* cells.[162] The G-protein signaling regulates fungal development and aflatoxin formation.[163, 164]

Another indirect role for an effect of glucose utilization on aflatoxin pathway gene expression could be related to the sugar cluster (Fig.11.1) of the four genes encoding NADH oxidase, hexose transporter, glucosidase, and Cys_6Zn_2-type regulator[99] adjacent to the aflatoxin gene cluster. Activation of genes in the sugar cluster by an external hexose signal could create a region of active chromatin that includes the neighboring aflatoxin gene cluster.[165] To support this observation, we and others have found that when individual aflatoxin biosynthetic genes insert at sites other than the aflatoxin gene cluster following fungal transformation, expression of these genes is much lower (>100-fold) than when these genes insert into the aflatoxin cluster[103] (Yu, unpublished data).

Whether or not nitrate suppresses aflatoxin production is unclear. Expression of genes involved in nitrate utilization is transcriptionally activated by the global positive-acting regulatory factor, AreA.[92,166] The nitrate effect on aflatoxin pathway gene expression may be caused directly by changes in the *aflR* or *aflJ* gene expression level since certain strains of aflatoxin-producing *Aspergilli* respond differently to nitrate than do other strains. The differences could be correlated with differences in the number of possible GATA sites (ranging from 5 to 9) in the *aflR*-*aflJ* intergenic region.[167] Nitrate could affect aflatoxin production by increasing the cytoplasmic NADPH/NADP ratio, which could favor biosynthetic reductive reactions and, thus, could promote utilization of malonyl coenzyme A and NADPH for fatty acid synthesis rather than for polyketide synthesis.[168]

Environmental Factors

Temperature, pH, water activity (drought stress) etc. are the environmental factors affecting aflatoxin production.[3,160,169] The biological and genetic mechanisms are not clear. Recent studies suggest that *aflR* transcription is responsive to a G-protein signaling cascade that is mediated by protein kinase A[163]. Such a signaling pathway may mediate some of the environmental effects on aflatoxin biosynthesis. The presence of a putative PacC-binding site in the region close to *aflR's* transcription start site may play some role in pH regulation on aflatoxin production. It was reported that the PacC-binding represses the transcription of acid-expressed genes under alkaline conditions[170] and aflatoxin biosynthesis in *A. flavus* occurs in acidic media, but is inhibited in alkaline media.[151] The PacC[170] and AreA[92] binding sites in the *aflR-aflJ* intergenic region are the potential evidence that gene expression is regulated by environmental signals (pH and nitrate).

Other genes in the aflatoxin biosynthetic cluster have also been found to contain AreA and PacC binding sites at key positions in their promoters that may affect their expression. For example, the 1.7 kb intergenic region separating the *nor-1* and *pksA* genes has two adjacent PacC sites nearly in the middle that, from site-directed mutagenesis studies, show that they affect expression of *pksA*, which encodes the pathway-specific polyketide synthase necessary for the first steps in formation of the polyketide backbone.[171]

Developmental Factors

Evidence exists that secondary metabolism is associated with fungal developmental processes such as sporulation and sclerotia formation.[164,172-176] It has been observed that the environmental conditions required for secondary metabolism and for sporulation are similar.[172,173] It has also been reported that spore formation and secondary metabolite formation occur at about the same time.[49,163] Mutants that are deficient in sporulation are unable to produce aflatoxins.[17] A *Fusarium verticillioides* mutation in the *FCC1* gene resulted in both reduced sporulation and reduced fumonisin B_1 production.[177] Certain compounds in *A. parasiticus* that exhibit the ability to inhibit sporulation have also been shown to inhibit aflatoxin formation.[178] Chemicals that inhibit polyamine biosynthesis in *A. parasiticus* and *A. nidulans* inhibit both sporulation and aflatoxin/ST biosynthesis.[174] A critical advance in this regard was the finding that the regulation of sporulation and ST production is by a shared G-protein mediated growth pathway in *A. nidulans.*[163] Mutations in *A. nidulans flbA* and *fadA* genes, early acting members of a G-protein signal transduction pathway, resulted in loss of ST production, ST gene expression, and sporulation.[163,179] It has been demonstrated that this regulation is partially mediated through protein kinase A.[180] This G-protein signaling pathway involving FadA in the regulation of aflatoxin production may also exist in other *Aspergilli* such as *A. parasiticus* (Keller, unpublished data).

GENOMICS APPROACHES TO PREVENT AFLATOXIN CONTAMINATION

So far we have a fairly good understanding of aflatoxin biosynthesis and its genetic regulation. However, our knowledge is limited only to the level of the aflatoxin biosynthetic pathway and the pathway genes within the gene cluster. Furthermore, the genes identified can not account for all the bioconversion steps of the aflatoxin pathway. Five aflatoxin cluster genes have no homologs identified in the ST gene cluster, and five genes in the ST gene cluster have no homologs identified in the aflatoxin pathway gene cluster. This indicates that some of the genes responsible for biosynthesis of aflatoxin and ST may reside outside of the gene cluster somewhere else in the genome. Many important questions remain: the functions of several genes in the cluster have not yet been fully characterized; no gene (s) residing outside the gene cluster that are potentially involved in afflatoxin formation have been identified; the exact mechanism by which *aflJ* modulates transcription of these pathway genes in concert with *aflR* is still unclear; the gene or genes controlling *aflR* and or *aflJ* expression have not been identified. To identify all of the genes responsible for aflatoxin formation is a daunting task and is difficult to accomplish by traditional cloning techniques. In order to identify all of the genes, identify the global regulatory elements (genes) beyond *aflR* and *aflJ* regulatory genes, understand the relationship of primary and secondary metabolisms, understand the relationship of development (sporulation) and aflatoxin formation, unravel the mechanism of the signal transduction pathway (as stimulated by nutritional shift, temperature, pH, volatile compounds from host plant), genomics, *i.e.*, whole genome sequencing, Expressed Sequence Tag (EST), and microarray technologies, provides a technological renovation for achieving such goals.[169,181-183] The *A. flavus* genomics is expected to provide valuable information on turning aflatoxin production on and off in fungal systems. This will provide vital clues for identifying anti-fungal gene(s) or aflatoxin-inhibitory gene(s).

Aspergillus flavus EST and Microarrays

Aspergillus flavus EST technology allows rapid identification of the majority, if not all, of the genes expressed in the fungal genome and helps better understand gene functions, regulation, coordination of gene expression in response to internal and external factors, the relationship between primary and secondary metabolism, plant-fungal interactions and fungal pathogenicity, as well as evolutionary biology. A microarray, made from the EST sequences, can be used to detect a whole set of genes expressed under specific environmental conditions. This technology allows us to study, simultaneously, a complete set of genes that is responsible for or related to toxin production.

A large-scale *A. flavus* EST/Microarray project is being carried out at the USDA/ARS, Southern Regional Research Center (SRRC), New Orleans, Louisiana, USA.[169,184-187] The strain of *A. flavus* used in this project is wild-type aflatoxin-producing strain NRRL 3357. BLAST results indicate that more than 7,000 expressed unique genes have been identified. Among those, many are rare copy genes, potentially involved in secondary metabolism and gene regulation. All known aflatoxin biosynthetic genes have been identified from among the sequenced clones in the library, an indication of the enrichment in genes of secondary metabolism in the library. Within the unique ESTs, we have identified many genes that may be involved directly or indirectly in aflatoxin formation.[165-187] Those of interest can be summarized in the following four categories: 1) aflatoxin biosynthetic pathway genes; 2) regulatory genes that have the potential to regulate aflatoxin production or signal transduction, *e.g.*, genes encoding DNA-binding proteins, RNA-binding proteins, zinc-finger proteins, transcription regulators, transducins, cAMP receptors, protein kinases etc.; 3) genes that have the potential to contribute to fungal virulence or pathogenicity; 4) genes involved in fungal development. The latter could be involved in processes such as sporulation, conidiation, and hyphal growth. Some unique genes in the EST library also show sequence homology to genes encoding hydrolytic enzymes, including amylase, cellulase, pectinases, proteases, chitinase, chitosanases, pectin methylesterases, endoglucanase C precursor, glucoamylase S1/S2 precursors, β–1,3-glucanase precursor, 1,4-β-D-glucan cellobiohydrolase A precursor, glycogen debranching enzyme, and xyloglucan-specific endo-β-1,4-glucanase precursor. Such hydrolytic enzymes could be highly expressed virulence factors during invasion of *A. flavus* into crops and, if so, have the potential to be useful targets for inhibiting aflatoxin production or for antifungal growth through genetic engineering.

Microarrays can be used to detect a whole set of genes transcribed under specific conditions and can be useful for studies, not only of biological functions of genes, and for studies of gene expression and regulation, but also for identifying factors involved in plant-microbe (crop-fungus) interactions.[188,189] By using microarray technologies, we can screen and identify the critical gene or genes involved in aflatoxin production and fungal invasion of host plants, as well as better understand the evolutionary biology of the aflatoxigenic and non-toxigenic strains and field isolates. A microarray containing all of the identified unique genes has been constructed at The Institute for Genomic Research (TIGR) for functional studies. *A. oryzae*, a close relative of *A. flavus*, is used in the fermentation industry for enzyme production. *A. oryzae* is highly homologous (98%-100% DNA identity) to *A. flavus* but has lost the ability to produce aflatoxin and survive in the field due to hundreds of years of "domestication." Genes cloned from *A. oryzae* can be used for studying aflatoxin synthesis using microarrays. Under a collaborative research agreement with the Japanese *A. oryzae* consortium, construction of a super array combining all of the unique genes identified from both *A. flavus* (7,214) and *A. oryzae* (6,710) is planned. The super array will contain over 11,000 gene elements from both species, which account for over 90% of the functional genes in the two *Aspergillus* species. This microarray will be used to screen genes that could be targeted in fungal systems for inhibiting aflatoxin formation or antifungal growth. Thus, microarray analysis could provide information that leads to the identification of potent antifungal genes or genes that inhibit aflatoxin formation. It may then be possible to engineer these genes into crop plants in an attempt to eliminate or reduce aflatoxin contamination in crops. The annotated sequence information will be available to the public once it is deposited into the GenBank Database.

Concurrent to the USDA/ARS/SRRC effort, Studies at North Carolina State University (NCSU) have been carried out in an *A. flavus* EST/Microarray project[190] (http://www.fungalgenomics.ncsu.edu/Projects/aspergillus.htm). Over 10,000 positive clones that were expressed under aflatoxin-producing conditions were identified. BLAST searches identified 753 unique ESTs within over 2,000 quality sequences.[190] Studies on gene expression profiling have identified six genes now being targeted for gene disruption to study the regulation of aflatoxin biosynthesis.[190]

Aspergillus flavus Whole Genome Sequencing

The definition of genome refers to a single, complete set of chromosomes in terms of "ploidy" in a single cell of an organism. However, with the technological advance of molecular genetics, the modern concept of genome now refers to all of the nucleotide sequence information in terms of mega base pairs (Mb) in a single cell. Sequencing and annotation of the entire genome is now called genomics. Genomics with the help of bioinformatics is a powerful tool to identify, and study all of the genes in an organism. The proposal to sequence the *A. flavus* genome has been funded by the U.S. Department of Agriculture (USDA), and the National Science Foundation (NSF), Microbial Genome Sequencing Program (MGSP). Sequencing of the *A. flavus* whole genome will begin by the end of 2003. The genome sequencing of several other *Aspergillus* species has been completed. These are: *A. nidulans* genome by the Genome Research Center of The Whitehead Institute (http://www.genome.wi.mit.edu); *A. oryzae* genome by the National Institute of Advanced Industrial Science and Technology (AIST), Japan; *A. fumigatus* genome by the Sanger Centre, the University of Salamaca, and the Pasteur Institute (http://www.tigr.org); and *A. niger* genome (DSM, The Netherlands). Additionally, the *A. oryzae* EST expressed under several conditions has been completed by Japanese consortium[191] (http://www.nrib.go.jp/ken/EST/db/blast.html). An *A. nidulans* gene index (AnGI) has been constructed at TIGR (http://www.tigr.org).

This large body of accumulating genomic information will allow comparative genomics studies and the development of comprehensive microarrays containing a complete set of genes that can be used in gene profiling studies to reveal regulatory networks, functional mechanisms, and evolutionary relationships. It will also facilitate comparison of toxigenic and non-toxigenic strains, and the screening and comparison of pathogeneicity and non-pathogeneicity factors.

SUMMARY

A. flavus is the most common cause of aflatoxin contamination in pre-harvest field crops and post-harvest grains. It is a characteristic "opportunistic" plant pathogen that infects corn, cotton, peanuts, and treenuts, and contaminates them with aflatoxins. There are no effective control strategies to prevent aflatoxin accumulation in the field when conditions are favorable for fungal growth. The mode of action and metabolism of aflatoxin biosynthesis have been extensively studied. Chemical binding of liver enzyme-activated aflatoxin molecules to animal DNA, causing mutations and possible carcinogenesis, has been demonstrated. The chemistry, biochemistry, molecular biology, and synthesis of aflatoxins B_1, and B_2, and the transcriptional activation of the genes involved are understood in significant detail. We have discovered an aflatoxin pathway gene cluster, a sugar utilization gene cluster, and a nitrogen pathway gene cluster. However, the mechanisms of aflatoxin formation in terms of global regulation, pathogenicity of the fungus, and crop-fungus interactions are poorly understood. A large research community has developed focused on understanding the biology of the fungus, and the biosynthesis of aflatoxins, with the goal of developing novel control strategies. *A. flavus* genomics will provide a powerful tool for identification of genes of interest. Microarrays containing *A. flavus* gene fragments can be used to study gene expression profiles under diverse conditions. *Aspergillus*, particularly, the species in the *flavus* group, has been recognized as a model system for studying fungal biology and pathogeneicity in solving food safety related issues. Identification and functional elucidation of those genes that are responsible for aflatoxin formation, regulation, signal transduction, pathogenicity, and the environmental effects on aflatoxin production by the fungus will provide vital information for devising new strategies to eliminate pre-harvest aflatoxin contamination, resulting in a safer, economically viable food and feed supply.

REFERENCES

1. BLOUT, W. P., Turkey "X" disease. *Turkeys,* 1961, **9,** 52-77.
2. FORGACS, J., CARLL, W.T., Mycotoxicoses, *Advan. Vet. Sci.*, 1962, **7**, 273-382.
3. PAYNE, G.A., BROWN, M.P., Genetics and physiology of aflatoxin biosynthesis, *Annu. Rev. Phytopathol.*, 1998, **36,** 329-362.
4. BENNETT, J.W., CHRISTENSEN, S.B., New perspectives on aflatoxin biosynthesis, *Adv. Appl. Microbiol.*, 1983, **29**, 53-92.

5. ITO,Y., PETERSON, S.W., WICKLOW, D.T., GOTO, T., *Aspergillus pseudotamarii*, a new aflatoxin producing species in *Aspergillus* section *Flavi*, *Mycol. Res.*, 2001, **105,** 233-239.
6. PETERSON, S.W., ITO, Y., HORN, B.W., GOTO, T., *Aspergillus bombycis*, a new aflatoxigenic species and genetic variation in its sibling species, *A. nomius*, *Mycologia*, 2001, **93,** 689-703.
7. FRISVAD, J.C., SAMSON, R.A., New producers of aflatoxin, *in*: Interaction of Molecular and Morphological Approaches to *Aspergillus* and *Penicillium* Taxonomy (R.A. Samson and J.I. Pitt, eds.), Harwood, Reading. 1999.
8. KLICH, M.A., MULLANEY, E.J., DALY, C.B., CARY, J.W., Molecular and physiological aspects of aflatoxin and sterigmatocystin biosynthesis by *Aspergillus tamarii* and *A. ochraceoroseus*, *Appl. Microbiol. Biotechnol.*, 2000, **53,** 605-609.
9. GOTO, T., WICKLOW, D.T., ITO, Y., Aflatoxin and cyclopiazonic acid production by sclerotium-producing *Aspergillus tamarii* strain, *Appl. Environ. Microbiol.,* 1996, **62**, 4036-4038.
10. PAYNE, G.A., Process of contamination by aflatoxin-producing fungi and their impacts on crops, *in:* Mycotoxins in Agriculture and Food Safety, Vol 9 (K.K. Sinha and D. Bhatnagar, eds.), Marcel Dekker, New York. 1998, pp. 279-306.
11. NEWBERNE, P.M., BUTLER, W.H., Acute and chronic effect of aflatoxin B_1 on the liver of domestic and laboratory animals: a review, *Cancer Res.*, 1969, **29**, 236-250.
12. SHANK, R.C., BHAMARAPRAVATI, N., GORDON, J.E., WOGAN, G.N., Dietary aflatoxins and human liver cancer. IV. Incidence of primary liver cancer in two municipal populations in Thailand, *Food Cosmet. Toxicol.*, 1972, **10**, 171-179.
13. PEERS, F.G., LINSELL, C.A., Dietary aflatoxins and human liver cancer – a population study based in Kenya, *British J. Cancer,* 1973, **27**, 473-484.
14. EATON, D.L., GROOPMAN, J.D. (eds.), The Toxicology of Aflatoxins: Human Health, Veterinary and Agricultural Significance. Academic Press, New York. 1994.
15. YU, J., Genetics and biochemistry of mycotoxin synthesis, *in:* Handbook of Fungal Biotechnology (2nd Edition) (D.K. Arora, ed.), Marcel Dekker, New York, in press.
16. BHATNAGAR, D., EHRLICH, K.C., CLEVELAND, T.E., Molecular genetic analysis and regulation of aflatoxin biosynthesis, *Appl. Miocrobiol. Biotechnol.*, 2003, **61,** 83-93.
17. BENNETT, J.W., PAPA, K.E., The aflatoxigenic *Aspergillus*, *in*: Genetics of Plant Pathology, vol 6 (D.S. Igram and P.A. Williams, eds.), Academic Press, London. 1988, pp. 264-280.
18. CAST, Mycotoxins: Economic and Health Risks, Council for Agricultural Science and Technology Task Force Report, No. 116, Ames, Iowa, 1989.
19. SQUIRE, R.A., Ranking animal carcinogens, a proposed regulatory approach, *Science*, 1981, **214**, 877-880.
20. DIENER, U.L., COLE, R.J., SANDERS, T.H., PAYNE, G.A., LEE, L.S., KLICH, M.A., Epidemiology of aflatoxin formation by *Aspergillus flavus*, *Annu. Rev. Phytopath.,* 1987, **25**, 249-270.
21. LANCASTER, M.D., JENKINS, F.P., PHILLIP, J.M., Toxicity associated with certain samples of groundnuts, *Nature*, 1961, **192,** 1095-1096.

22. BRESSAC, B., KEW, M., WANDS, J., OZTURK, M., Selective G to T mutations of p53 gene in hepatocellular carcinoma from southern Africa, *Nature*, 1991, **350**, 429-431.
23. HSU, I.C., METCALF, R.A., SUN, T., WELSH, J.A., WANG, N.J., HARRIS, C.C., Mutational hotspot in the p53 gene in human hepatocellular carcinomas, *Nature,* 1991, **350**, 377-378.
24. WOGAN, G.N., Aflatoxins as risk factors for hepatocellular carcinoma in humans, *Cancer Res*., 1992, **52**, 2114s-2118s.
25. HSIEH, D., Potential human health hazards of mycotoxins, *in:* Mycotoxins and Phytotoxins, Third Joint FAO/WHO/UNEP International Conference of Mycotoxins (S. Natori, K. Hashimoto, and Y. Ueno, eds.), Elsevier, Amsterdam. 1988, pp. 69-80.
26. BENNETT, J.W., Mycotoxins, mycotoxicoses, mycotoxicology and mycopathologia, *Mycopathlogia,* 1987, **100**, 3-5.
27. BHATNAGAR, D., EHRLICH, K.C., CLEVELAND, T.E., Biochemical characterization of an aflatoxin B_2 producing mutant of *Aspergillus flavus, FASEB J.,* 1993, **7,** A 1234.
28. CLEVELAND, T.E., BHATNAGAR, D., Molecular regulation of aflatoxin biosynthesis, *in*: Pennington Center Nutrition Series, Vol. 1, Mycotoxins, Cancer and Health (G.A. Bray, and D.H. Ryan, eds.), 1991, pp. 270-287.
29. CLEVELAND, T.E., BHATNAGAR, D., Molecular strategies for reducing aflatoxin levels in crops before harvest, *in*: Molecular Approaches to Improving Food Quality and Safety (D. Bhatnagar and T.E. Cleveland, eds.), Van Nostrand Reinhold, New York. 1992, pp. 205-228.
30. JELINEK, C.F., POHLAND, A.E., WOOD, G.E., Worldwide occurrence of mycotoxins in foods and feeds - an update, *J. Assoc. Off. Anal. Chem*., 1989, **72,** 223-230.
31. COTTY, P.J., BAYMAN, P., EGEL, D.S., ELIAS, K.S., Agriculture, aflatoxins and *Aspergillus*, *in:* The Genus *Aspergillus* (K.A. Powell, A. Renwick, and J.F. Peberdy, eds.), Plenum Press, New York and London. 1994, pp. 1-27.
32. PEERS, F.G., GILMAN, G.A., LINSELL, C.A., Dietary aflatoxins and human liver cancer. A study in Swaziland. *Int J. Cancer*, 1976, **17**, 167-176.
33. HENRY, S.H., BOSCH, F.X., TROXELL, T.C., BOLGER, P.M., Reducing liver cancer- global control of aflatoxin, *Science,* 1999, **286**, 2453-2454.
34. CAST, Mycotoxins: Risks in Plant, Animal, and Human Systems, Council for Agricultural Science and Technology Task Force Report, No. 139, Ames, Iowa, 2003.
35. STEVENS, D.A., Diagnosis of fungal infections, current status, *J. Antimicrob. Chemother.*, 2002, **49,** 11-19.
36. STEVENS, D.A., KAN, V.L., JUDSON, M.A., MORRISON, V.A., DUMMER, S., DENNING, D.W., BENNETT, J.E., WALSH, T.J., PATTERSON, T.F., PANKEY, G.A., Practice guidelines for diseases caused by *Aspergillus*, Infectious Diseases Society of America. *Clin. Infect. Dis*., 2000, **30**, 696-709.
37. FOUTZ, K.R., WOLOSHUK, C.P., PAYNE, G.A., Cloning and assignment of linkage group loci to a karyotypic map of the filamentous fungus *Aspergillus flavus*, *Mycologia,* 1995, **87,** 787-794.

38. KELLER, N.P., CLEVELAND, T.E., BHATNAGAR, D., A molecular approach towards understanding aflatoxin production, *in*: Mycotoxins in Ecological Systems Vol. 5 (D. Bhatnagar, E.B. Lillehoj, D.K. Arora, eds.), Marcel Dekker, Inc., New York. 1992, pp. 287-310.
39. YU, J., CHANG, P.-K., EHRLICH, K.C., CARY, J.W., BHATNAGAR, D., CLEVELAND, T.E., PAYNE, G.A., LINZ, J.E., WOLOSHUK, C.P., BENNETT, J.W., The clustered pathway genes in aflatoxin biosynthesis, *Appl. Environ. Microbiol.,* in press.
40. PAPA, K.E., The parasexual cycle in *Aspergillus flavus*, *Mycologia.*, 1973, **65**, 1201-1205.
41. BHATNAGAR, D., YU, J., EHRLICH, K.C., Toxins of filamentous fungi, *Chem. Immunol.,* 2002, **81,** 167-206.
42. YU, J., BHATNAGAR, D., EHRLICH, K.C., Aflatoxin biosynthesis, *Rev. Iberoam Microl. (RIAM*), 2002, **19,** 191-200.
43. BENNETT, J.W., KLICH, M., Mycotoxins, *Clin. Microbiol. Rev.,* 2003, **16,** 497-516.
44. COLE, R.J., COX, E.H., Handbook of Toxic Fungal Metabolites. Academic Press, New York. 1987.
45. BERRY., The pathology of mycotoxins, *J. Pathol.,* 1988, **154**, 201-311.
46. BHATNAGAR, D., EHRLICH, K.C., CLEVELAND, T.E., Oxidation-reduction reactions in biosynthesis of secondary metabolites, *in*: Handbook of Applied Mycology: Mycotoxins in Ecological Systems (D. Bhatnagar, E.B. Lillehoj, and D.K. Arora, eds.), Marcel Dekker, New York. 1992, pp. 255-286.
47. DUTTON, M.F., Enzymes and aflatoxin biosynthesis, *Microbiol. Rev*., 1988, **52,** 274-295.
48. MINTO, R. E., TOWNSEND, C.A., Enzymology and molecular biology of aflatoxin biosynthesis, *Chem. Rev*., 1997, **97,** 2537-2556.
49. TRAIL, F., MANHANTI, N., LINZ, J., Molecular biology of aflatoxin biosynthesis, *Microbiol.,* 1995, **141,** 755-765.
50. WATANABE, C.M., TOWNSEND, C.A., Initial characterization of a type I fatty acid synthase and polyketide synthase multienzyme complex NorS in the biosynthesis of aflatoxin $B_{(1)}$, *Chem Biol.,* 2002, **9,** 981-988.
51. TOWNSEND, C.A., CHRISTENSEN, S.B., TRAUTWEIN, K., Hexanoate as a starter unit in polyketide synthesis, *J. Am. Chem. Soc*., 1984, **106,** 3868-3869.
52. BROWN, D.W., ADAMS, T.H., KELLER, N.P., *Aspergillus* has distinct fatty acid synthases for primary and secondary metabolism, *Proc. Natl. Acad. Sci. USA*, 1996, **19,** 14873-14877.
53. BENNETT, J.W., Loss of norsolorinic acid and aflatoxin production by a mutant of *Aspergillus parasiticus, J. Gen. Microbiol.,* 1981, **124**, 429-432.
54. BENNETT, J.W., CHANG, P.-K., BHATNAGAR, D., One gene to whole pathway: The role of norsolorinic acid in aflatoxin research, *Adv. Appl. Microbiol.,* 1997, **45**, 1-15.
55. PAPA, K.E., Genetics of *Aspergillus flavus*: Complementation and mapping of aflatoxin mutants, *Genet. Res*., 1979, **34,** 1-9.
56. PAPA, K.E., Norsolorinic acid mutant *of Aspergillus flavus*, *J. Gen. Microbiol*., 1982, **128,** 1345-1348.

57. YABE, K., NAKAMURA, Y., NAKAJIMA, H., ANDO, Y., HAMASAKI, T., Enzymatic conversion of norsolorinic acid to averufin in aflatoxin biosynthesis, *Appl. Environ. Microbiol.,* 1991, **57,** 1340-1345.
58. DETROY, R.W., FREER, S., CIEGLER, A., Aflatoxin and anthraquinone biosynthesis by nitrosoquanidine-derived mutants of *Aspergillus parasiticus, Can. J. Microbiol.,* 1973, **19,** 1373-1378.
59. TOWNSEND, C.A., PLAVCAN, K.A., PAL, K., BROBST, S.W., IRISH, M.S., ELY, E.W., BENNETT, J.W., Hydroxyversicolorone: Isolation and characterization of a potential intermediate in aflatoxin biosynthesis, *J. Org. Chem.,* 1988, **53,** 2472-2477.
60. TOWNSEND, C.A., Progress towards a biosynthetic rationale of the aflatoxin pathway, *Pure Appl. Chem.*, 1997, **58,** 227-238.
61. YABE, K., Pathway and genes of aflatoxin biosynthesis, *in:* Microbial Secondary Metabolites: Biosynthesis, Genetics and Regulation, (F. Fierro and J. Francisco, eds.), Research Signpost, India. 2003, pp. 227-251
62. YABE, K., CHIHAYA, N., HAMAMATSU, S., SAKUNO, E., HAMASAKI, T., NAKAJIMA, H., BENNETT, J.W., Enzymatic conversion of averufin to hydroxyversicolorone and elucidation of a novel metabolic grid involved in aflatoxin biosynthesis, *Appl. Environ. Microbiol.*, 2003, **69,** 66-73.
63. CLEVELAND, T.E., Conversion of dihydro-*O*-methylsterigmatocystin to aflatoxin B_2 by *Aspergillus parasiticus, Arch. Environ. Contam. Toxicol.*, 1989, **18**, 429-433.
64. MCGUIRE, S.M., BROBST, S.W., GRAYBILL, T.L., PAL, K., TOWNSEND, C.A., Partitioning of tetrahydro- and dihydrobisfuran formation in aflatoxin biosynthesis defined by cell-free and direct incorporation experiments, *J. Am. Chem. Soc.*, 1989, **111**, 8308-8309.
65. YABE, K., ANDO, Y., HAMASAKI, T., Biosynthetic relationship among aflatoxins B_1, B_2, G_1, and G_2, *Appl. Environ. Microbiol.*, 1988, **54,** 2101-2106.
66. YABE, K., NAKAMURA, H., ANDO, Y., TERAKADO, N., NAKAJIRNI, H., HAMASAKI, T., Isolation and characterization of *Aspergillus parasiticus* mutants with impaired aflatoxin production by a novel tip culture method, *Appl. Environ. Microbiol.,* 1988, **54**, 2096-2100.
67. YABE, K., ANDO, Y., HAMASAKI, T., Desaturase activity in the branching step between aflatoxins B_1 and G_1 and aflatoxins B_2 and G_2, Agric. Biol. Chem., 1991, **55,** 1907-1911.
68. YABE, K., NAKAMURA, M., HAMASAKI, T., Enzymatic formation of G-group aflatoxins and biosynthetic relationship between G- and B-group aflatoxins, *Appl. Environ. Microbiol.,* 1999, **65,** 3867-3872.
69. YU, J., CHANG, P.-K., CARY, J.W., EHRLICH, K.C., MONTALBANO, B., DYER, J.M., BHATNAGAR, D., CLEVELAND, T.E., Characterization of the critical amino acids of an *Aspergillus parasiticus* cytochrome P450 monooxygenase encoded by *ordA* involved in aflatoxin B_1, G_1, B_2, and G_2 biosynthesis, *Appl. Environ. Microbiol.*, 1998, **64,** 4834-4841.
70. CHANG, P.-K., YU, J., EHRLICH, K.C., BOUE, S.M., MONTALBANO, B.G., BHATNAGAR, D., CLEVELAND, T.E., The aflatoxin biosynthesis gene *adhA* in *Aspergillus parasiticus* is involved in conversion of 5'-hydroxyaverantin to averufin, *Appl.Environ. Microbiol.,* 2000, **66,** 4715- 4719.

71. YABE, K., ANDO, Y., HAMASAKI, T., A metabolic grid among versiconal hemiacetal acetate, versiconol acetate, versiconol and versiconal during aflatoxin biosynthesis, *J. Gen. Microbiol.*, 1991, **137,** 2469-2475.
72. YABE, K., MATSUYAMA, Y., ANDO, Y., NAKAJIMA, H., HAMASAKI, T., Stereochemistry during aflatoxin biosynthesis: Conversion of norsolorinic acid to averufin, *Appl. Environ. Microbiol.*, 1993, **59,** 2486-2492.
73. YABE, K., HAMASAKI, T., Stereochemistry during aflatoxin biosynthesis: Cyclase reaction in the conversion of versiconal to versicolorin B and racemization of versiconal hemiacetal acetate, *Appl. Environ. Microbiol.*, 1993, **59,** 2493-2500.
74. BHATNAGAR, D., ULLAH, A.H.J., CLEVELAND, T.E., Purification and characterization of a methyltransferase from *Aspergillus parasiticus* SRRC 163 involved in aflatoxin biosynthetic pathway, *Prep. Biochem.*, 1988, **18,** 321-349.
75. HSIEH, D.P., WAN, C.C., BILLINGTON, J.A., A versiconal hemiacetal acetate converting enzyme in aflatoxin biosynthesis, *Mycopathologia,* 1989, **107,** 121-126.
76. ANDERSON, J.A., CHUNG, C.H., CHO, S.-H., Versicolorin A hemiacetal, hydroxydihydro-sterigmatocystin and aflatoxin G_2 a reductase activity in extracts from *Aspergrillus parasiticus*, *Mycopathologia,* 1990, **111**, 39-45.
77. MCGUIRE, S.M., TOWNSEND, C.A., Demonstration of a Baeyer-Villiger oxidation and the time course of cyclization in bisfuran ring formation during aflatoxin B_1 biosynthesis, *Bioorgan. Medicin. Chem. Lett.,* 1993, **3,** 653-656.
78. BHATNAGAR, D., CLEVELAND, T.E., KINGSTON, D.G.I., Enzymological evidence for separate pathways for aflatoxin B_1 and B_2 biosynthesis, *Biochemistry*, 1991, **30,** 4343-4350.
79. CHUTURGOON, A.A., DUTTON, M.F., BERRY, R.K., The preparation of an enzyme associated with aflatoxin biosynthesis by affinity chromatography, *Biochem. Biophys. Res. Comm.,* 1990, **166,** 38-42.
80. YABE, K., ANDO, Y., HASHIMOTO, J., HAMASAKI, T., Two distinct *O*-methyltransferases in aflatoxin biosynthesis, *Appl. Environ. Microbiol.*, 1989, **55,** 2172-2177.
81. KELLER, N.P., DISCHINGER, J.H.C., BHATNAGAR, D., CLEVELAND, T.E., ULLAH, A.H.J., Purification of a 40-kilodalton methyltransferase active in the aflatoxin biosynthetic pathway, *Appl. Environ. Microbiol.*, 1993, **59,** 479-484.
82. BHATNAGAR, D., LAX, A.R., PRIMA, B., CARY, J.W., CLEVELAND, T.E., Purification of a 43 Kda enzyme that catalyzes the reduction of norsolorinic acid to averantin in aflatoxin biosynthesis, *FASEB J.,* 1996, **10,** A1522.
83. LIN, B.K., ANDERSON, J.A., Purification and properties of versiconal cyclase from *Aspergillus parasiticus, Arch. Biochem. Biophys.*, 1992, **293,** 67-70.
84. SILVA, J.C., MINTO, R.E., BARRY, C.E., HOLLAND, K.A., TOWNSEND, C.A., Isolation and characterization of the versicolorin B synthase gene from *Aspergillus parasiticus:* Expansion of the aflatoxin B_1 biosynthetic cluster, *J. Biol. Chem.*, 1996, **271,** 13600-13608.
85. MATSUSHIMA, K., ANDO, Y., HAMASAKI, T., YABE, K., Purification and characterization of two versiconal hemiacetal acetate reductases involved in aflatoxin biosynthesis, *Appl. Environ. Microbiol.*, 1994, **60,** 2561-2567.

86. KUSUMOTO, K., HSIEH, D.P., Purification and characterization of the esterases involved in aflatoxin biosynthesis in *Aspergillus parasiticus*, *Can. J. Microbiol.*, 1996, **42**, 804-810.
87. SKORY, C.D., CHANG, P.-K., LINZ, J.E., Regulated expression of the *nor-1* and *ver-1* genes associated with aflatoxin biosynthesis, *Appl. Environ. Microbiol.*, 1993, **59**, 1642-1646.
88. TRAIL, F., MAHANTI, N., RARICK, M., MEHIGH, R., LIANG, S.H., ZHOU, R., LINZ, J.E., Physical and transcriptional map of an aflatoxin gene cluster in *Aspergillus parasiticus* and functional disruption of a gene involved early in the aflatoxin pathway, *Appl. Environ. Microbiol.*, 1995, **61**, 2665-2673.
89. YU, J., CHANG, P.-K., CARY, J.W., WRIGHT, M., BHATNAGAR, D., CLEVELAND, T.E., PAYNE, G.A., LINZ, J.E., Comparative mapping of aflatoxin pathway gene clusters in *Aspergillus parasiticus* and *Aspergillus flavus*, *Appl. Environ. Microbiol.*, 1995, **61**, 2365-2371.
90. CARY, J.W., WRIGHT, M., BHATNAGAR, D., LEE, R., CHU, F.S., Molecular characterization of an *Aspergillus parasiticus* dehydrogenase gene, *norA*, located on the aflatoxin biosynthesis gene cluster, *Appl. Environ. Microbiol.*, 1996, **62**, 360-366.
91. CHANG, P.-K., CARY, J.W., YU, J., BHATNAGAR, D., CLEVELAND, T.E., *Aspergillus parasiticus* polyketide synthase gene, *pksA*, a homolog of *Aspergillus nidulans wA*, is required for aflatoxin B_1, *Mol. Gen. Genet.*, 1995, **248**, 270-277.
92. CHANG, P.-K., YU, J., BHATNAGAR, D., CLEVELAND, T.E., Characterization of the *Aspergillus parasiticus* major nitrogen regulatory gene, *areA*, *Biochem. Biophys. Acta*, 2000, **1491**, 263-266.
93. CLEVELAND, T.E., CARY, J.W., BROWN, R.L., BHATNAGAR, D., YU, J., CHANG, P.-K., CHLAN, C.A., RAJASEKARAN, K., Use of biotechnology to eliminate aflatoxin in preharvest crops, *Bull. Inst. Compr. Agr. Sci. Kinki Univ.*, 1997, **5**, 75-90.
94. MAHANTI, N., BHATNAGAR, D., CARY, J.W., JOUBRAN, J., LINZ, J.E., Structure and function of *fas-1A*, a gene encoding a putative fatty acid synthetase directly involved in aflatoxin biosynthesis in *Aspergillus parasiticus*, *Appl. Environ. Microbiol.*, 1996, **62**, 191-195.
95. SILVA, J.C., TOWNSEND, C.A., Heterologous expression, isolation, and characterization of versicolorin B synthase from *Aspergillus parasiticus*, *J. Biol. Chem.*, 1996, **272**, 804-813.
96. YU, J., CHANG, P.-K., CARY, J.W., BHATNAGAR, D., CLEVELAND, T.E., *avnA*, a gene encoding a cytochrome P-450 monooxygenase is involved in the conversion of averantin to averufin in aflatoxin biosynthesis in *Aspergillus parasiticus*, *Appl. Environ. Microbiol.*, 1997, **63**, 1349-1356.
97. YU, J., WOLOSHUK, C.P., BHATNAGAR, D., CLEVELAND, T.E., Cloning and characterization of *avfA* and *omtB* genes involved in aflatoxin biosynthesis in three *Aspergillus* species, *Gene*, 2000, **248**, 157-167.
98. YU, J., CHANG, P.-K., BHATNAGAR, D., CLEVELAND, T.E., Genes encoding cytochrome P450 and monooxygenase enzymes define one end of the aflatoxin pathway gene cluster in *Aspergillus parasiticus*, *Appl. Microbiol. Biotechnol.*, 2000, **53**, 583-590.

99. YU, J., CHANG, P.-K., BHATNAGAR, D., CLEVELAND, T.E., Cloning of sugar utilization gene cluster in *Aspergillus parasiticus*, *Biochim. Biophys. Acta.*, 2000, **1493,** 211-214.
100. LIANG, S.-H., SKORY, C.D., LINZ, J.E., Characterization of the function of the *ver-1A and ver-1B* genes, involved in aflatoxin biosynthesis in *Aspergillus parasiticus, Appl. Environ. Microbiol.,* 1996, **62,** 4568-4575.
101. CARY, J.W., DYER, J.M., EHRLICH, K.C., WRIGHT, M.S., LIANG, S.H., LINZ, J.E., Molecular and functional characterization of a second copy of the aflatoxin regulatory gene, *aflR-2*, from *Aspergillus parasiticus*, *Biochim. Biophys. Acta.*, 2002, **1576,** 316-323.
102. CHANG, P.-K., YU, J., Characterization of a partial duplication of the aflatoxin gene cluster in *Aspergillus parasiticus* ATCC 56775, *Appl. Microbiol. Biotechnol.*, 2002, **58,** 632-636.
103. CHIOU, C.H., MILLER, M., WILSON, D.L., TRAIL, F., LINZ, J.E., Chromosomal location plays a role in regulation of aflatoxin gene expression in *Aspergillus parasiticus*, *Appl. Environ. Microbiol.,* 2002, **68,** 306-315.
104. CHANG, P.-K., SKORY, C.D., LINZ, J.E., Cloning of a gene associated with aflatoxin B1 biosynthesis in *Aspergillus parasiticus, Curr. Genet.*, 1992, **21,** 231-233.
105. TRAIL, F., CHANG, P.-K., CARY, J., LINZ, J.E., Structural and functional analysis of the *nor-1* gene involved in the biosynthesis of aflatoxins by *Aspergillus parasiticus, Appl. Environ. Microbiol.,* 1994, **60,** 4078-4085.
106. BROWN, D.W., YU, J.-H., KELKAR, H.S., FERNANDES, M., NESBITT, T.C., KELLER, N.P., ADAMS, T.H., LEONARD, T.J., Twenty-five coregulated transcripts define a sterigmatocystin gene cluster in *Aspergillus nidulans, Proc. Natl. Acad. Sci. USA,* 1996, **93,** 1418-1422.
107. SKORY, C.D., CHANG, P.-K., CARY, J., LINZ, J.E., Isolation and characterization of a gene from *Aspergillus parasiticus asso*ciated with the conversion of versicolorin A to sterigmatocystin in aflatoxin biosynthesis, *Appl. Environ. Microbiol.*, 1992, **58,** 3527-3537.
108. LIANG, S.-H., WU, T.S., LEE, R., CHU, F.S., LINZ, J.E., Analysis of mechanisms regulating expression of the *ver-1* gene, involved in aflatoxin biosynthesis, *Appl. Environ. Microbiol.*, 1997, **63,** 1058-1065.
109. KELLER, N.P., KANTZ, N.J., ADAMS, T.H., *Aspergillus nidulans verA* is required for production of the mycotoxin sterigmatocystin, *Appl. Environ. Microbiol.*, 1994, **60,** 1444-1450.
110. KELLER, N.P., SEGNER, S., BHATNAGAR, D., ADAMS, T.H., *stcS,* a putative P-450 monooxygenase, is required for the conversion of versicolorin A to sterigmatocystin in *Aspergillus nidulans, Appl. Environ. Microbiol.,* 1995, **61,** 3628-3632.
111. KELLER, N.P., BROWN, D., BUTCHKO, R.A.E., FERNANDES, M., KELKAR, H., NESBITT, C., SEGNER, S., BHATNAGAR, D., CLEVELAND, T.E., ADAMS, T.H., A conserved polyketide mycotoxin gene cluster in *Aspergillus nidulans*, *in*: Molecular Approaches to Food Safety Issues Involving Toxic Microorganisms (J.L. Richard, ed.), Alaken, Fort Collins, Colorado. 1995, pp. 263-277.

112. YU, J., CARY, J.W., BHATNAGAR, D., CLEVELAND, T.E., KELLER, N.P., CHU, F.S., Cloning and characterization of a cDNA from *Aspergillus parasiticus* encoding an *O*-methyltransferase involved in aflatoxin biosynthesis, *Appl. Environ. Microbiol.*, 1993, **59,** 3564-3571.
113. YU, J., CHANG, P.-K., PAYNE, G.A., CARY, J.W., BHATNAGAR, D., CLEVELAND, T.E., Comparison of the *omtA* genes encoding *O*-methyltransferases involved in aflatoxin biosynthesis from *Aspergillus parasiticus* and *A. flavus*, *Gene.*, 1995, **163,** 121-125.
114. LEE, L.W., CHIOU, C.H., LINZ, J.E., Function of native OmtA in vivo and expression and distribution of this protein in colonies of *Aspergillus parasiticus*, *Appl. Environ. Microbiol.,* 2002, **68,** 5718-5727.
115. WATANABE, C.M.H., WILSON, D., LINZ, J.E., TOWNSEND, C.A., Demonstration of the catalytic roles and evidence for the physical association of type I fatty acid syntheses and a polyketide synthase in the biosynthesis of aflatoxin B_1, *Chem. Biol.*, 1996, **3,** 463-469.
116. FENG, G.H., LEONARD, T.J., Characterization of the polektide synthase gene (*pksLl*) required for aflatoxin biosynthesis in *Aspergillus parasiticus, J. Bacteriol.*, 1995, **177,** 6246-6254.
117. YU, J., CHANG, P-K, BHATNAGAR, D., CLEVELAND, T.E., Cloning and functional expression of an esterase gene in *Aspergillus parasiticus*, *Mycopathologia*, 2003, **156,** 227-234.
118. MCGUIRE, S.M., SILVA, J.C., CASILLAS, E.G., TOWNSEND, C.A., Purification and characterization of versicolorin B synthase from *Aspergillus parasiticus.* Catalysis of the stereodifferentiating cyclization in aflatoxin biosynthesis essential to DNA interaction, *Biochemistry*, 1996, **35,** 11470-11486.
119. KELKAR, H.S., HERNANT, S., SKLOSS, T.W., HAW, J.F., KELLER, N.P., ADAMS, T.H., *Aspergillus nidulans stcL* encodes a putative cytochrome P-450 monooxygenase required for bisfuran desaturation during aflatoxin/sterigmatocystin biosynthesis, *J. Biol. Chem.*, 1997, **272,** 1589-1594.
120. YABE, K., MATSUSHIMA, K., KOYAMA, T., HAMASAKI, T., Purification and characterization of *O*-methyltransferase I involved in conversion of demethylsterigmatocystin to sterigmatocystin and of dihydrodemethylsterigmatocystin to dihydrosterigmatocystin during aflatoxin biosynthesis, *Appl. Environ. Microbiol.,* 1998, **64,** 166-171.
121. KELKAR, H.S., KELLER, N.P., ADAMS, T.H., *Aspergillus nidulans stcP* encodes an *O*-methyltransferase that is required for sterigmatocystin biosynthesis, *Appl. Environ. Microbiol.*, 1996, **62,** 4296-4298.
122. MOTOMURA, M., CHIHAYA, N., SHINOZAWA, T., HAMASAKI, T., YABE, K., Cloning and characterization of the *O*-methyltransferase I gene (*dmtA*) from *Aspergillus parasiticus* associated with the conversions of demethylsterigmatocystin to sterigmatocystin and dihydrodemethylsterigmatocystin to dihydrosterigmatocystin in aflatoxin biosynthesis, *Appl. Environ. Microbiol.*, 1999, **65,** 4987-4994.
123. PRIETO, R., YOUSIBOVA, G.L., WOLOSHUK, C.P., Identification of aflatoxin biosynthesis genes by genetic complementation in an *Aspergillus flavus* mutant lacking the aflatoxin gene cluster, *Appl. Environ. Microbiol.*, 1996, **62,** 3567-3571.

124. PRIETO, R., WOLOSHUK, C.P., *ordl*, an oxidoreductase gene responsible for conversion of 0-methylsterigmatocystin to aflatoxin in *Aspergillus flavus, Appl. Environ. Microbiol.*, 1997, **63,** 1661-1666.
125. EHRLICH, K.C., MONTALBANO, B.G., CARY, J.W., Binding of the C6-zinc cluster protein, AFLR, to the promoters of aflatoxin pathway biosynthesis genes in *Aspergillus parasiticus*, *Gene,* 1999, **230,** 249-257.
126. CHANG, P.-K., EHRLICH, K.C., YU, J., BHATNAGAR, D., CLEVELAND, T.E., Increased expression of *Aspergillus parasiticus aflR*, encoding a sequence-specific DNAbinding protein, relieves nitrate inhibition of aflatoxin biosynthesis, *Appl. Environ. Microbiol.*, 1995, **61,** 2372-2377.
127. CHANG, P.-K., CARY, J.W., BHATNAGAR, D., CLEVELAND, T.E., BENNETT, J.W., LINZ, J.E., WOLOSHUK, C.P., PAYNE, G.A., Cloning of the *Aspergillus parasiticus apa-2* gene associated with the regulation of aflatoxin biosynthesis, *Appl. Environ. Microbiol.*, 1993, **59,** 3273-3279.
128. CHANG, P.-K., YU, J., BHATNAGAR, D., CLEVELAND, T.E., Repressor-AFLR interaction modulates aflatoxin biosynthesis in *Aspergillus parasiticus*, *Mycopathologia,* 1999, **147,** 105-112.
129. CHANG P.-K., YU, J., BHATNAGAR, D., CLEVELAND, T.E., The carboxy-terminal portion of the aflatoxin pathway regulatory protein AFLR of *Aspergillus parasiticus* activates *GAL1:lacZ* gene expression in *Saccharomyces cerevisiae*, *Appl. Environ. Microbiol.,* 1999, **65,** 2058-2512.
130. EHRLICH, K.C., MONTALBANO, B.G., BHATNAGAR, D., CLEVELAND, T.E., Alteration of different domains in AFLR affects aflatoxin pathway metabolism in *Aspergillus parasiticus* transformants, *Fungal Genet. Biol.*, 1998, **23,** 279-287.
131. FLAHERTY, J.E., PAYNE, G.A., Overexpression of *aflR* leads to upregulation of pathway gene expression and increased aflatoxin production in *Aspergillus flavus, Appl. Environ. Microbiol.,* 1997, **63,** 3995-4000.
132. PAYNE, G.A., NYSTROM, G.J., BHATNAGAR, D., CLEVELAND, T.E., WOLOSHUK, C.P., Cloning of the *afl-2* gene involved in aflatoxin biosynthesis from *Aspergillus flavus*, *Appl. Environ. Microbiol.,* 1993, **59,** 156-162.
133. WOLOSHUK, C.P., FOUTZ, K.R., BREWER, J.F., BHATNAGAR, D., CLEVELAND, T.E., PAYNE, G.A., Molecular characterization of *aflR,* a regulatory locus for aflatoxin biosynthesis, *Appl. Environ. Microbiol.,* 1994, **60,** 2408-2414.
134. YU, J.-H., BUTCHKO, R.A., FERNANDES, M., KELLER, N.P., LEONARD, T.J., ADAMS, T.H., Conservation of structure and function of the aflatoxin regulatory gene *aflR* from *Aspergillus nidulans* and *A. flavus, Curr. Genet.*, 1996, **29,** 549-555.
135. EHRLICH, K.C., CARY, J.W., MONTALBANO, B.G., Characterization of the promoter for the gene encoding the aflatoxin biosynthetic pathway regulatory protein AFLR, *Biochim. Biophys. Acta.,* 1999, **1444,** 412-417.
136. FERNANDES, M., KELLER, N.P., ADAMS, T.H., Sequence-specific binding by *Aspergillus nidulans* AflR, a C6 zinc cluster protein regulating mycotoxin biosynthesis, *Mol. Microbiol.*, 1998, **28,** 1355-1365.
137. LAMB, H.K., NEWTON, G.H., LEVETT, L.J., CAIRNS, E., ROBERTS, C.F., HAWKINS, A.R., The QUTA activator and QUTR repressor proteins of *Aspergillus*

nidulans interact to regulate transcription of quinate utilization pathway genes, *Microbiol.,* 1996, **142**, 1477-1490.

138. BURGER, G., STRAUSS, J., SCAZZOCCHIO, C., LANG, B.F., *nirA*, the pathway-specific regulatory gene of nitrate assimilation in *Aspergillus nidulans*, encodes a putative GAL4-type zinc finger protein and contains four introns in highly conserved regions, *Mol. Cell. Biol.,* 1991, **11,** 5746-5755.
139. TODD, R.B., ANDRIANOPOUL, O.S.A., DAVIS, M.A., HYNES, M.J. FacB, the *Aspergillus nidulans* activator of acetate utilization genes, binds dissimilar DNA sequences, *EMBO J.,* 1998, **17**, 2042-2054.
140. SUAREZ, T., OESTREICHER, N., PENALVA, M.A., SCAZZOCCHIO, C., Molecular cloning of the *uaY* regulatory gene of *Aspergillus nidulans* reveals a favoured region for DNA insertions, *Mol. Gen. Genet*., 1991, **230**, 369-375.
141. KULMBURG, P., SEQUEVAL, D., LENOUVEL, F., MATHIEU, M., FELENBOK, B., Identification of the promoter region involved in autoregulation of the transcriptional activator ALCR in *Aspergillus nidulans*, *Mol. Cell. Biol.,* 1992, **12,** 1932-1939.
142. MATSUSHIMA, K., CHANG, P.-K., YU, J., ABE, K., BHATNAGAR, D., CLEVELAND, T.E., Pre-termination in *aflR* of *Aspergillus sojae* inhibits aflatoxin biosynthesis, *Appl. Microbiol. Biotechnol.,* 2001, **55,** 585-589.
143. MATSUSHIMA, K., YASHIRO, K., HANYA Y., ABE, K., YABE, K., HAMASAKI, T., Absence of aflatoxin biosynthesis in koji mold (*Aspergillus sojae*), *Appl. Microbiol. Biotechnol*., 2001, **55,** 771-776.
144. TAKAHASHI, T., CHANG, P.-K., MATSUSHIMA, K., YU, J., KOYAMA, Y., ABE, K., BHATNAGAR, D., CLEVELAND, T.E., The non-functionality of *Aspergillus sojae aflR* in *Aspergillus parasiticus aflR* disrupted strain, *Appl. Environ. Microbiol*., 2002, **68,** 3737-3743.
145. MEYERS, D.M., O'BRIAN, G., DU, W.L., BHATNAGAR, D., PAYNE, G.A., Characterization of *aflJ*, a gene required for conversion of pathway intermediates to aflatoxin, *Appl. Environ. Microbiol.,* 1998, **64,** 3713-3717.
146. CHANG, P.-K., The *Aspergillus parasiticus* protein AFLJ interacts with the aflatoxin pathway-specific regulator AFLR, *Mol. Genet. Genomics,* 2003, **268,** 711-719.
147. BUTCHKO, R.A., ADAMS, T.H., KELLER, N.P., *Aspergillus nidulans* mutants defective in stc gene cluster regulation, *Genetics,* 1999, **153**, 715-720.
148. DEMAIN, A.L., Cellular and environmental factors affecting the synthesis and excretion of metabolites, *J. App. Chem. Biotechnol.,* 1972, **22,** 345-372.
149. FENG, G.H., LEONARD, T.J., Culture conditions control expression of the genes for aflatoxin and sterigmatocystin biosynthesis in *Aspergillus parasiticus* and *A. nidulans*, *Appl. Environ. Microbiol*., 1998, **64,** 2275-2277.
150. KELLER, N.P., HOHN, T.M., Metabolic pathway gene clusters in filamentous fungi, *Fungal Genet. Biol.,* 1997, **21,** 17-29.
151. COTTY, P. J., Aflatoxin and sclerotial production by *Aspergillus flavus:* Influence of pH, *Phytopathology,* 1988, **78,** 1250-1253.
152. KACHHOLZ, T., DEMAIN, A.L., Nitrate repression of averufin and aflatoxin biosynthesis, *J. Nat. Prod.,* 1983, **46**, 499-506.

153. SHIM, W.B., WOLOSHUK, C.P., Nitrogen repression of fumonisin B_1 biosynthesis in *Gibberella fujikuroi*, *FEMS Microbiol. Lett.*, 1999, **177**, 109-116.
154. CUERO, R., OUELLET, T., YU, J., MOGONGWA, N., Metal ion enhancement of fungal growth, gene expression and aflatoxin synthesis in *Aspergillus flavus*: RT-PCR characterization, *J. Appl. Microbiol.*, 2003, **94**, 953-61.
155. PAYNE, G.A., HAGLER, Jr., W.M., Effect of specific amino acids on growth and aflatoxin production by *Aspergillus parasiticus and Aspergillus flavus* in defined media, *Appl. Environ. Microbiol.*, 1983, **46**, 805-812.
156. YU, J., MOHAWED, S.M., BHATNAGAR, D., CLEVELAND, T.E., Substrate-induced lipase gene expression and aflatoxin production in *Aspergillus parasiticus* and *Aspergillus flavus*, *J. Appl. Microbiol.*, in press.
157. BENNETT, J.W., RUBIN, P.L., LEE, L.S., CHEN, P.N., Influence of trace elements and nitrogen sources on versicolorin production by a mutant strain of *Aspergillus parasiticus*, *Mycopathologia*, 1979, **69**, 161-166.
158. CHANG, P.-K., EHRLICH, K.C., LINZ, J.E., BHATNAGAR, D., CLEVELAND, T.E., BENNETT, J.W., Characterization of the *Aspergillus niaD* and *niiA* gene cluster, *Curr. Genet.*, 1996, **30**, 68-75.
159. TAG, A., HICKS, J., GARIFULLINA, G., AKE Jr., C., PHILLIPS, T.D., BEREMAND, M., KELLER, N., G-protein signalling mediates differential production of toxic secondary metabolites, *Mol. Microbiol.*, 2000, **38**, 658-665.
160. YU, J., CHANG, P.-K., BHATNAGAR, D., CLEVELAND, T.E., Genetic, nutritional and environmental factors affecting aflatoxin biosynthesis, *Mycopathologia*, 2002, **155**, 70.
161. OZCAN, S., LEONG, T., JOHNSTON, M., Rgt1p of *Saccharomyces cerevisiae*, a key regulator of glucose-induced genes, is both an activator and a repressor of transcription, *Mol. Cell. Biol.*, 1996, **16**, 6419-6426.
162. DANIEL, P.B., WALKER, W.H., HABENER, J.F., Cyclic AMP signaling and gene regulation, *Annu. Rev. Nutr.*, 1998, **18**, 353-383.
163. HICKS, J.K., YU, J.H., KELLER, N.P., ADAMS, T.H., *Aspergillus* sporulation and mycotoxin production both require inactivation of the FadA G alpha protein-dependent signaling pathway, *EMBO. J.*, 1997, **16**, 4916-4923.
164. CALVO, A.M., WILSON, R.A., BOK, J.W., KELLER, N.P., Relationship between secondary metabolism and fungal development, *Microbiol. Mol. Biol. Rev.*, 2002, **66**, 447-459.
165. MURO-PASTEUR, M.I., GONZALEZ, R., STRAUSS, J., NARENDJA, F., SCAZZOCCHIO, C., The GATA factor AreA is essential for chromatin remodelling in a eukaryotic bidirectional promoter, *EMBO J.*, 1999, **18**, 1584-1597.
166. KUDLA, B., CADDICK, M., LANGDON, T., MARTINEZ-ROSSI, N.M., BENNETT, C.F., BIBLEY, S., DAVIS, R.W., ARST, JR., H.N., The regulatory gene *areA* mediating nitrogen metabolite repression in *Aspergillus nidulans*. Mutations affecting specificity of gene activation alter a loop residue of a putative zinc finger, *EMBO J.*, 1990, **9**, 1355-1364.
167. EHRLICH, K.C., COTTY, P.J., Variability in nitrogen regulation of aflatoxin production by *Aspergillus flavus* strains, *Appl. Microbiol. Biotechnol.*, 2002, **60**, 174-178.

168. NIEHAUS, W.G. Jr., JIANG, W., Nitrate induces enzymes of the marmitol cycle and suppresses versicolorin synthesis in *Aspergillus parasiticus, Mycopathologia,* 1989, **107**, 131-137.
169. GUO, B.Z., YU, J., HOLBROOK, C.C., LEE, R.D., LYNCH, R.E., Application of differential display RT- PCR and EST/Microarray technology to the analysis of gene expression in response to drought stress and aflatoxin contamination. *J. Toxicology, Toxin Reviews*. 2003, pp. 287-312.
170. TILBURN, J., SARKAR, S., WIDDICK, D.A., ESPESO, E.A., OREJAS, M., MUNGROO, J., PENALVA, M.A., ARST, Jr. H.N., The *Aspergillus* PacC zinc finger transcription factor mediates regulation of both acid-and alkaline expressed genes by ambient pH, *EMBO J.,* 1995, **14**, 779-790.
171. EHRLICH, K.C., MONTALBANO, B.G., CARY, J.W., COTTY, P.J., Promoter elements in the aflatoxin pathway polyketide synthase gene, *Biochim. Biophys. Acta.* 2002, **1576**, 171-175.
172. BU'LOCK, J.D., Intermediary metabolism and antibiotic synthesis, *Adv. Appl. Microbiol.,* 1961, **3**, 293-342.
173. SEKIGUCHI, J., GAUCHER, G.M., Conidiogenesis and secondary metabolism in *Penicillium urticae*, *Appl. Environ. Microbiol.,* 1977, **33**, 147-158.
174. GUZMAN-DE-PENA, D., AGUIRRE, J., RUIZ-HERRERA, J., Correlation between the regulation of sterigmatocystin biosynthesis and asexual and sexual sporulation in *Emencella nidulans*, *Antonie Leeuwenhock,* 1998, **73,** 199-205.
175. KALE, S.P., BHATNAGAR, D., BENNETT, J.W., Isolation and characterization of morphological variants of *Aspergillus parasiticus* deficient in secondary metabolite production, *Mycol. Res.* 1994, **98**, 645-652.
176. KALE, S.P., CARY, J.W., BHATNAGAR, D., BENNETT, J.W., Characterization of experimentally induced, nonaflatoxingenic variant strains of *Aspergillus parasiticus*, *Appl. Environ. Microbiol.* 1996, **62**, 3399-3404.
177. SHIM, W.B., WOLOSHUK, C.P., Regulation of fumonisin B1 biosynthesis and conidiation in *Fusarium verticillioides* by acyclin-like (C-type) gene, *FCC1*, *Appl. Environ. Microbiol.*, 2001, **67,** 1607-1612.
178. REIB, J., Development of *Aspergillus parasiticus* and formation of aflatoxin B1 under the influence of conidiogenesis affecting compounds, *Arch. Microbiol.*, 1982, **133**, 236-238.
179. YU, J.H., WEISER, J., ADAMS, T.H., The *Aspergillus* FlbA RGS domain protein antagonizes G-protein signaling to block proliferation and allow development, *EMBO J.*, 1996, **15,** 5184-5190.
180. SHIMIZU, K., KELLER, N.P., Genetic involvement of camp-dependent protein kinase in a G-protein signaling pathway regulating morphological and chemical transitions in *Aspergillus nidulans*, *Genetics,* 2001, **157**, 591-600.
181. CLEVELAND, T.E., YU, J., BHATNAGAR, D., CHEN, Z., BROWN, R., CHANG, P.-K., CARY, J.W., Deciphering the *Aspergillus* genome for controlling preharvest aflatoxin contamination of crops. *J. Toxicology, Toxin Reviews*, in press.
182. YU, J., PROCTOR, R.H., BROWN D.W., ABE, K., GOMI, K., MACHIDA, M., HASEGAWA, F., NIERMAN, W.C., BHATNAGAR, D., CLEVELAND, T.E., Genomics of economically significant *Aspergillus* and *Fusarium* species, *in*: Applied

Mycology & Biotechnology, An International Series. Vol. 4, (D.K. Arora, ed.), Fungal Genomics, Elsevier Science, in press.

183. BENNETT, J.W., ARNOLD, J., Genomics for fungi, *in*: The Mycota VIII Biology of the Fungal Cell (R. Howard and N. Gow, eds.), Springer-Verlag, Berlin, Heidelberg. 2001, pp. 267-297.
184. BHATNAGAR, D., YU, J., CLEVELAND, T.E., Applying the genomic wrench - New tool for an old problem, *Mycopathologia*, 2002, **155**, 9.
185. YU, J., BHATNAGAR, D., CLEVELAND, T.E., *Aspergillus flavus* genomics for elimination of aflatoxin contamination, *Mycopathologia*, 2002, **155,** 10.
186. YU, J., BHATNAGAR, D., CLEVELAND, T.E., NIERMAN, W.C., *Aspergillus flavus* EST technology and its applications for eliminating aflatoxin contamination., *Mycopathologia*, 2002, **155,** 6.
187. YU, J., WHITELAW, C.A., BHATNAGAR, D., CLEVELAND, T.E., NIERMAN, W.C., Report on *Aspergillus flavus* EST project — a tool for eliminating aflatoxin contamination. Proceedings of the 2nd Fungal Genomics, 3rd Fumonisin Elimination and 15th Aflatoxin Elimination Workshops, San Antonio, Texas, October 23-25, 2002.
188. GOFFEAU, A., BARRELL, B.G., BUSSEY, H., DAVIS, R.W., Life with 6000 genes, *Science,* 1996, **274,** 546-567.
189. GROSS, C., KELLEHER, M., VYER, V.R., BROWN, P.O., WINGE, D.R., Identification of copper regulation of *Saccharomyces cerevisiae* by DNA microarrays, *J. Biol. Chem.*, 2000, **275**, 32310-32316.
190. O'BRIAN G.R., FAKHOURY, A.M., PAYNE, G.A., Identification of genes differentially expressed during aflatoxin biosynthesis in *Aspergillus flavus* and *Aspergillus parasiticus*, *Fungal Genet. Biol.*, 2003, **39,** 118-127.
191. MACHIDA, M., AKITA, O., KASHIWAGI, Y., YAMAGUCHI, S., Analysis of ESTs and the promoters of useful expression patherns from *Aspergillus oryzae*, *Int. Symp. Mol. Biol. Filamentous Fungi Aspergilli,* 2000, p. 3.

INDEX

www.ingramcontent.com/pod-product-compliance
Ingram Content Group UK Ltd.
Pitfield, Milton Keynes, MK11 3LW, UK
UKHW020248290926
15470UKWH00002B/812

9 780080 445014